Planck's Constant and Pi

A Unified Wave Theory for Particles and Bioforms

By

Irwin Wunderman

1stBooks – rev. 12/03/03

KEYWORDS: Planck's Constant, Curved Space, Waves, Natural Numbers, Irrational numbers, Prime Numbers, π, e, √-1, ∞, Radians, Cycles, Frequency, Wavelength, Wave Polarization, Dimensional Units, Non-locality, Time, Time's Arrow, Root Mean Square Sums, Population Dynamics, Uncertainty Principle, Cosmological Constant, Orthogonality, Expanding Universe, Red Shift, Electromagnetic Radiation, Action, Photon, Wave/Particle Dualism, Wave Propagation, Harmonics, Unit Circle, Exponential Function, Motional Conservation Laws, Hidden Variables, String Theory, Multiple Extra Dimensions, Quadrature Components, Enumeration, Counting, Monads, Symmetry, Supersymmetry, Invariance, Double Helix

The author is grateful to Allen Amaro and Jim Farned for help with computer graphics and editorial comments.

TABLE OF CONTENTS

Two and a half millennia ago followers of Pythagoras believed numbers and mathematics could explain mysteries of the universe. These pages pay tribute to their insight. Scientific generalizations sometimes necessitate concepts not previously conceived or utilized. They could not know the dominion of their mathematics was inadequate owing to the physical reality of Planck's constant.

1) ABSTRACT

This work introduces a new interpretation for the space between integers of the natural number counting system. The material is not easily distilled into a simple, cogent Abstract and Introduction. A mathematical theory is developed built upon the thesis that variables representing a collection of n smallest non-divisible entities called monads or quanta need to be treated as a population of objects. Such variables only occupy integer values and in the variable's progression occupancy of the space between integers is disallowed. For something comprised of non-divisible entities, variables can make headway only by orthogonal transition between integers, which transitions are intangible. Such right-angle advancements between integers avoid increase through fractional values relative to the origin. The consequence of this "**orthogonality axiom**" will be shown to yield vector-transitions between Integers n of magnitude $s_n = \sqrt{(2n+1)}$ and a spiral progression that embodies harmonic waves. The periodic harmonics emerge because a generic synchronicity exists within the sequence of ordinal counting numbers whenever two consecutive integers $n + (n+1) = (2n+1)$ sum to a prefect square. This occurs whenever n = 0, 4, 12, 24, 40, 60, 84, etc., having respective paired sums, 1, 9, 25, 49, 81, 121, 169, etc. whose square roots are the consecutive odd integers, 1, 3, 5, 7, 9, 11, 13, etc. That circumstance conveys a harmonic periodicity that exists throughout all numbers. It makes accrual of non-divisible monads (as a population variable that increases discretely by integers), behave as a generalized harmonic wave. This **unified wave** based on summing smallest monads under the orthogonality axiom, is conjectured as the phenomena affiliated with how forces act and particles interact. Exactly what the smallest entities being summed are need not be specified so long as those entities are not divisible. The harmonics occur as a corollary of the natural numbers when orthogonal transitions between integers are accounted for. Smallest units of energy-time for example, can only take on integer values along traditional scales signified by Cartesian-Coordinate Euclidean-type axes. Resultant orthogonal transitions between axis integers are shown to create new cyclic processes surrounding each traditional axis. When counting small objects that are partially divisible, inter-integer orthogonal transitions occur at the fixed fraction $0 \leq \xi < 1$ of the interval between consecutive integers and transition magnitudes become $s_n = \sqrt{(2n\xi+1)}$. Harmonic properties sustain for all ξ and encompass conventional exponential harmonics (sinusoids).

Virtually all processes in nature can interpret through the periodicity of waves. This treatise attempts to show that such periodicities derive from enumerating non-divisible entities, "***monads***", and entities that are partially divisible where $0 \leq \xi < 1$. Using only natural numbers to illustrate these smallest repeated constituents that accrue to a population, four-quadrant-symmetrical waves emerge that exhibit 4-radian rectilinearity, (in contrast to quadrants associated with 2π radians). The key postulate that brings this periodic symmetry about is called the ***orthogonality axiom***. It stipulates that for counting monads, transitions between integer numbers of them must be orthogonal to (independent of) the prevailing population. The inquiry purportedly provides a foundation for, and a bridge between,

2

classical mechanics, quantum mechanics, and possibly string theory. Many dilemmas of physics may elucidate and simplify through the newly-derived waves. The theory requires no arbitrary constants. Many conventional mathematical constants like e, π, i, ∞, become unnecessary to represent physical reality using this development for waves. Derivation of the traditional harmonic exponential for sinusoids [exp (iωt)], is shown to **violate causality**. **It evokes a solution (a frequency) that begins at minus infinity in time, which response precedes the precipitating event that caused the solution**. By contrast the newly developed sinusoidal harmonic waves begin at the precipitating event. Exponential type harmonics can however be analytically generated at one non-physical limit of the new unified wave. This grants a correspondence between the more general new waves formed by accrued monads, and conventional exponentials. Sample waveforms of harmonic modes are presented. They require three dimensional representations and portray feasible standing wave solution modes amongst an enormous number of possible "random modes". The basis of natural-numerical-periodicity stems from a mathematical property easily checked through Table 4 and the discussion of equation 21*8. Applications of the analysis are briefly discussed with references provided for further studies. To visualize the composite picture it is suggested that this work be initially scanned in its entirety without justifying the concepts or mathematics. It may then be re-read with any level of detailed critique.

Irwin Wunderman

4

2) INTRODUCTION AND BASIC POSTULATE

This work aims to furnish a new basis for how waves manifest in physical reality. It analyzes variables comprised of monads, rather than being continuous. The mathematics is offered as self-consistent independent of the interpreted physics, which is comparatively speculative. Understanding of the mathematics suggests altered meaning for enumerating spacetime variables subsequent to an origin event. Scalar quantities describing a population of monads become vectors. Using an integer adaptation inherent to the ordinal numbers, a logical basis is provided for diverse phenomena such as matter waves and wave-origins in general. The theory bases on one axiom for monadic population growth. **For non-sub-divisible entities, transitions where population n increases to (n+1) must be orthogonal to the direction between origin and n**. Such transitions are tangential to whatever scale depicts population n. By not originating at (referencing) the zero origin these transitions are *"intangible"*, akin to imaginary values. **Here orthogonal means mutually independent of; irrelevant to; in a right-angle direction from**. It will be seen that the postulate itself could be considered a direct outgrowth of energy-time and momentum-position being quantized into smallest non-divisible units.

The orthogonal method of variable advance arises since fractional values of unit things are prohibited by definition. Fragmented magnitudes along an axis become forbidden and continuity through fractional values between integers

5

disallowed. The resultant missing gap incurs a loss of outright predictability and lack of an objective state of affairs when the system is unobserved. Presumption of **orthogonal progression** between consecutive integers along an axis constitutes the essence of this theory[1]. As smallest possible non-divisible entities, quanta may only assume whole numbers of minimal energy-time or momentum-position units and as such exemplify changing populations. The need to maintain dimensional consistency for cycles emerges in the photon energy equation $h = E/v =$ (**energy**)(**time**)/**cycle**. A mathematical origin for absolute radians is shown as a basic unit-of-rotation. **Traditional rectilinearity of Cartesian coordinates becomes replaced through cycles comprised from <u>four symmetrical radians</u> that partition around a conic surface**. Populations of energy-time units are shown to interpret as characterizing the compounding of an excess (indeterminate or noise-level) angle within each complete cycle of a wave. These can construe as arising from the curvature of spacetime. Such curvature may only be a metaphor for the missing information associated with each monad. Planck's constant h satisfies that excess-angle per 360° and the prevailing indeterminism (loss of objective certainty) that results within the wave. It infers dimensional units for energy-time are not arbitrary, but dependent upon the absolute numerical value of h.

The waves characterize population dynamics for the accumulation of n monadic entities of minimum divisibility. **For n = 0 to ∞, peak and zero crossing locations in the new sinusoidal-type harmonic waves are shown to <u>never associate with a population value n that is a prime number</u>. That provides an**

alternative possible definition for pure periodicity in terms of natural numbers. In descriptions of the wave as straight-chord-lengths traversing around a cone (or circle), harmonic periodicity minutely increases inversely with smaller n and the number of prior cycles squared. The source of periodicity is shown to be **perfect and absolute for all n** when differential chord-lengths between adjacent transitions fit a circular perimeter. Consecutive transition magnitudes of $\sqrt{(2n+1)}$ subtracted from each other produce the relationship outlined in Table 4. **I.e, the sum of terms $\Delta t = 2*[\sqrt{(2n+3)} - \sqrt{(2n+1)}]$ taken between consecutive odd integers of $\sqrt{(2n+1)}$, will each equal exactly 4 for all integers n between 0 and** ∞. Each such Δt represents the incremental time to traverse Δt radians on a circle comprising 4 radians inscribed on a unit-cone or unit-sphere. Every 4 radians constitutes one cycle resulting in perfect harmonic periodicity.

Under these analyses all waves (called *unified waves*) are suggested as the inextricable consequence of properties embedded within natural numbers. Those integer numbers signify monads, each number being the possible quantity of non-divisible things forming a population or ensemble. The unified wave is shown to encompass conventional exponential harmonics at one limit where the excess angle of space goes to zero (or n→∞). The mathematics also suggests a natural number origin for the format of the DNA helix ladder. Many additional consequences are derived and tabulated in the conclusions. Further details of the theory's development can be found in the above reference[1].

3) MODIFYING EUCLIDEAN AXES

The natural numbers assert that one describes a number of n things as 1, 2, 3, 4, 5, - - - n → ∞. The numbers declare nothing directly about the space between those integers. However, suppose each "thing" was divisible into say two half-things. The natural numbers **do not imply** that one could have ½, 1, 1 ½, 2, 2 ½, 3, 3 ½ - - - etc. things. They state one could have 1, 2, 3, 4, 5, - - - half-things. If the half-things are divided in half again they state one may have 1, 2, 3, 4, 5, - - - etc. quarter-things. No matter how divisible each original thing might have been, the natural numbers only deal with the number of smallest entities one might have, wherein those entities are no longer sub-dividable. In short, the sequence of numbers really describes the integer amount of non–divisible entities (monads) in any population or assembly, with that population potentially progressing to very large numbers. Monads by definition are the smallest repeated constituents that larger things can be comprised of. They are the most fundamental building blocks of Nature without sub-attributes, so that further description of them is difficult. Articulating the number of monads in an ensemble might typically yield integers so large they would require 20 or 30 digits. To be practical, exposition in scientific notation would occur, wherein the integer quality then tends to get "lost". However inconvenient it is to express such large integers, **that is what natural numbers or populations describe; integer numbers of non-divisible entities**. To make such descriptions pragmatic, people have historically reduced the number of digits handled by dividing the many-digit number by a billion, a trillion, or whatever.

Conceptually that then seems to allow a billion or a trillion partitions between each integer. That maneuver for operational pragmatism is not what the natural numbers describe. They model the number of smallest monads and remain unconcerned with how large the integer digits must get to treat normally encountered empirical objects. More emphatically, it must be stipulated that **the space between integer numbers of smallest monads making up larger things cannot be occupied in modeling**. Natural numbers assert that only integer values are allowed, the consecutive integers of n, requiring however many digits as necessary to portray that population of n monads. They demand that space between integer numbers of represented monads be physically "disallowed", rather than characterized by a continuum. The integers themselves provide no yardstick which describes how progression occurs from one integer to the next. The natural numbers only imply that the applicable criterion not allow a continuum as traditionally envisioned to include all contiguous points along a Euclidean axis line with integers marked upon it.

So how can progression occur from one integer n to the next higher (n+1)? One way is for that progression to be orthogonal to whatever n signifies. Integer n represents an extent from zero to n. It designates a vector from which, progression to (n+1) might be in any right-angle direction. Then that "tangential-going" transition between radial vector n and radial vector (n+1) can delineate a different category of "allowability" than integers along a radial-going scale. The radial integers represent the "statically allowable" integer magnitudes, with dynamical transitions between them being tangential vectors, not emanating from the origin. Each tangential transition then only references that specific advance from vector population n it began from, terminating at population (n+1). The quantity of monads n and (n+1) must reference the same starting zero-origin, so they must both emanate radially from that origin, as must all integers of the number system. By contrast, transitions between the integers must all lie in normal directions to the respective n from which they originate. This bears some analogue to imaginary values being tangential to their real counterparts. Transitions must at least possess an orthogonal component to depict the intangibility of each transition process, in comparison to unidirectional radial-outward advance away from the origin for population increases. However, **<u>no constraint can be made as to the direction in which orthogonal transitions might occur within the respective plane normal to vector n</u>**. It must be at right angles to radial-going vector n, without prescription of which bearing in that normal-going plane it heads.

Orthogonality allows each transition from radial vector n the freedom to get to radial extent (n+1) by heading in any direction normal to the origin-to-n vector until (n+1) units from the origin is reached. To extend from radial (n+1) to radial (n+2) they need progress in any tangential bearing within the plane normal to radial direction origin-to-(n+1). No directional restraint is imposed requiring that transition from (n+1) to (n+2) **retain any relationship** to the transition direction from n to (n+1), or anything else. Thus the population may potentially occupy any integer radial distance from the origin as n, (n+1), (n+2), (n+3), - - - etc. to n → ∞. **Those**

potentialities represent "possible concentric-spherical-surfaces" at each integer removed from the origin and that is what the natural numbers decree. The population can be any of the concentric spherical surfaces from the origin that have radius a unit integer apart. <u>**Natural numbers describe a population as possibly occupying locations that are nesting consecutive spheres of integer magnitude separation**</u>. Where on the sphere of radius n any population is, defies description, unless the direction of (n-1) was specifically known. Even then the location on sphere n would only be knowable as being somewhere on a circular locus defined by where a plane at (n-1) normal to that zero-to-(n-1) vector intersects the concentric sphere of radius n. Thus, specific "location" on any possible spherical surface rapidly becomes totally unknown. <u>**However unusual this description may be of what the natural numbers mean, it asserts what they in fact designate for monadic entities**</u>. **The smallest building blocks of Nature exhibit properties that differ from ultra-large ensembles of those same non-divisible entities**. That is what natural numbers express, integer separations from an origin with transitions between those values having nebulous delineation. **Each natural number retains an affiliated bundle of angular possibilities that associate it with only immediate-neighbor natural numbers**. Over any range of numbers the potential "scattering" of those angles becomes so diverse as to prevent description of any absolute direction. The result is that numeric integer values assume concentric spherical locations about an origin. All is not lost however since the numbers unfold beautiful elegance in this circumstance. To aid picturing this condition it is worthwhile to initially consider the simpler special-case setting where all radial and tangential vectors are portrayed within a single plane, rather than anywhere spherical. Angles between the integer radial vectors and respective tangential transitions are then easily analyzed graphically. Each transition between arbitrary n and (n+1) always form a triangle of vectors. In this setting all concatenated triangles joining consecutive numbers can inter-compare within the same plane. That is a more restrictive picture then the more-general **actual case** where the plane of each concatenated triangle would only be constrained to intersect the origin. Planes of separate triangles could actually pass through the origin concatenated in whatever random or however directions. Keeping transition vectors in a single plane for analysis purposes results in a spiral progression of integer-magnitude radial-vectors interconnected by tangential transition vectors.

The harmonic exponential constitutes an ad hoc definition for wave generation through forced cyclic advance of time t (or x) about a unit circle. In any normal {non-relativistic} context, time, (and distance) interpret as a linear independent variable that progresses along a straight-line abscissa or ordinate axis, not around a circuit. The exponential function exp $(i\omega t)$ = lim n $\to\infty$ of $(1+i\omega t/n)^n$ furnishes no logical justification for coercing the time dimension to run in circles, rather than linearly along its coordinate-axis-type course. **The function is "reverse engineered" to artificially form sinusoids by steering the temporal (or spatial) variable along a rotary path**. No phenomenological or mathematical principle establishes why this should occur, other than the desired end-result of an expression for harmonic waves. Analysis to follow hopes to demonstrate conceptual and mathematical methods for a more appropriate wave-generation mechanism. It includes a logical transformation from variable progression along a linear axis to circular advance and provides waves having more degrees of freedom than conventional sinusoids.

As will be demonstrated, a monadic population is appropriately portrayed as integer-magnitude concentric spherical surfaces of possibility, rather than along an axis line. The detailed mechanism by which this occurs can emerge by starting with a familiar axial representation, subsequently revealing that it transforms into the nested-concentric-sphere picture. A typical Euclidean energy-time-axis exemplifies analysis to follow, as would pertain for describing any single quanta, (monad) which can only have one discrete-energy and specific frequency ν. However, the relevant Euclidean axis always represents

(energy)x(time) $\equiv$ action [or (momentum)x(position) for spatial axes], so that appropriate timescale diminishes for quanta of increased energy or wherever energy might change integer-to-integer. (Energy)x(time) = unity = the interval between axis integers that signify the smallest possible units. This forms a population of action units represented by the axis integers. Having disallowed diminution of the smallest constituents to nothingness (to Newtonian continuity), each population member would veil "unavailable missing information" associated with its "inaccessible interior". **For a monad, unit size is the only attribute that can associate with that and every population member, further sub-divisional properties being missing**. Such missing information between axis integers at the onset cannot be recovered or re-inserted. This guarantees that conjugate product variables comprise a population of equal units, (rather than a continuous variable) each member of which constitutes a single non-divisible entity without sub-attributes.

Energy-time yields a particularly interesting conjugate pair in a dominion where energy $E = h\nu$. It will be shown that the smallest-unit product depicted as each integer along a Euclidean axis is commensurate with a fixed minimum-angular-traversal of a cyclic-temporal-wave. One interpretation of a smallest energy-time unit then involves an angular segment of a "sinusoidal wave", a tiniest angular fragment, (or angular "noise level") of a cycle. A smallest unit of energy-time becomes the <u>minimal discernable fraction of a complete cycle</u>. It will be demonstrated here as an "excess angle" in each otherwise conventional complete cycle of Euclidean space. It will associate with a dimensionless Planck's constant h to be defined herein as (smallest angle)/2π. It might then be anticipated that in subsequent analysis a smallest excess segment-of-a-cycle would arise and associate with the minimal energy-time unit utilized, justifying its selection. It should prove worthwhile when attempting to comprehend this theory to envision energy-time as a min "excess angle" associated with the non-Euclidean curvature of space, with energy-time/cycle $\equiv h \equiv E/\nu$ thus being dimensionless {radians/radian} in the employed description. Energy E and frequency ν are then dimensionally synonymous. While the approach below presents in terms of an energy-time axis, analogous treatment for distance directions X, Y, Z, would have integers respectively signify (momentum)x(position) in relation to spatial frequency $1/\lambda$. For objects without rest mass, modes for the space axes can match their integers n with time axis integers when momentum-position = energy-time = action, or energy/momentum = position/time = velocity of light = c.

Under the presumption requiring orthogonal transition between axis integers, progression from integer n to (n+1) must always occur at a right angle to the origin-to-n direction. Each such advance forms a triangle structure of three vectors having lengths n, (n+1) plus an intermediary right-angle transition vector from n equaling

$$\sqrt{[(n+1)^2 - n^2]} = \sqrt{(2n+1)}. \qquad (3*1)$$

The next transition from vector (n+1) to (n+2) will have right angle transition length $\qquad \sqrt{[(n+2)^2 - (n+1)^2]} = \sqrt{(2n+3)}. \qquad (3*2)$

Continuing the sequence from n = 0 upwards results in the sequence of transition lengths $\sqrt{(\text{odd \#})}$. Geometric portrayal of radial plus tangential vectors within a plane results in a spiral whose construction is depicted in the plane of figure 1. For now angles like θ and ϕ in the figures can interpret as being in radians or degrees for visualization simplicity. Certain Greek-symbol nomenclature used can later be mathematically treated as normalized through division by 2π or 360° to become dimensionless, (evaluated per cycle). Each tangential-going vector of

The initial triangle of sides n = 0, (n+1) = 1 and $s_n = \sqrt{(2n+1)} = 1$ is a unity-height vertical line representing overlapping s_n and (n+1)

$(n+1) = 1$ $s_n = 1 = \sqrt{(2n+1)}$

Origin→

n=0

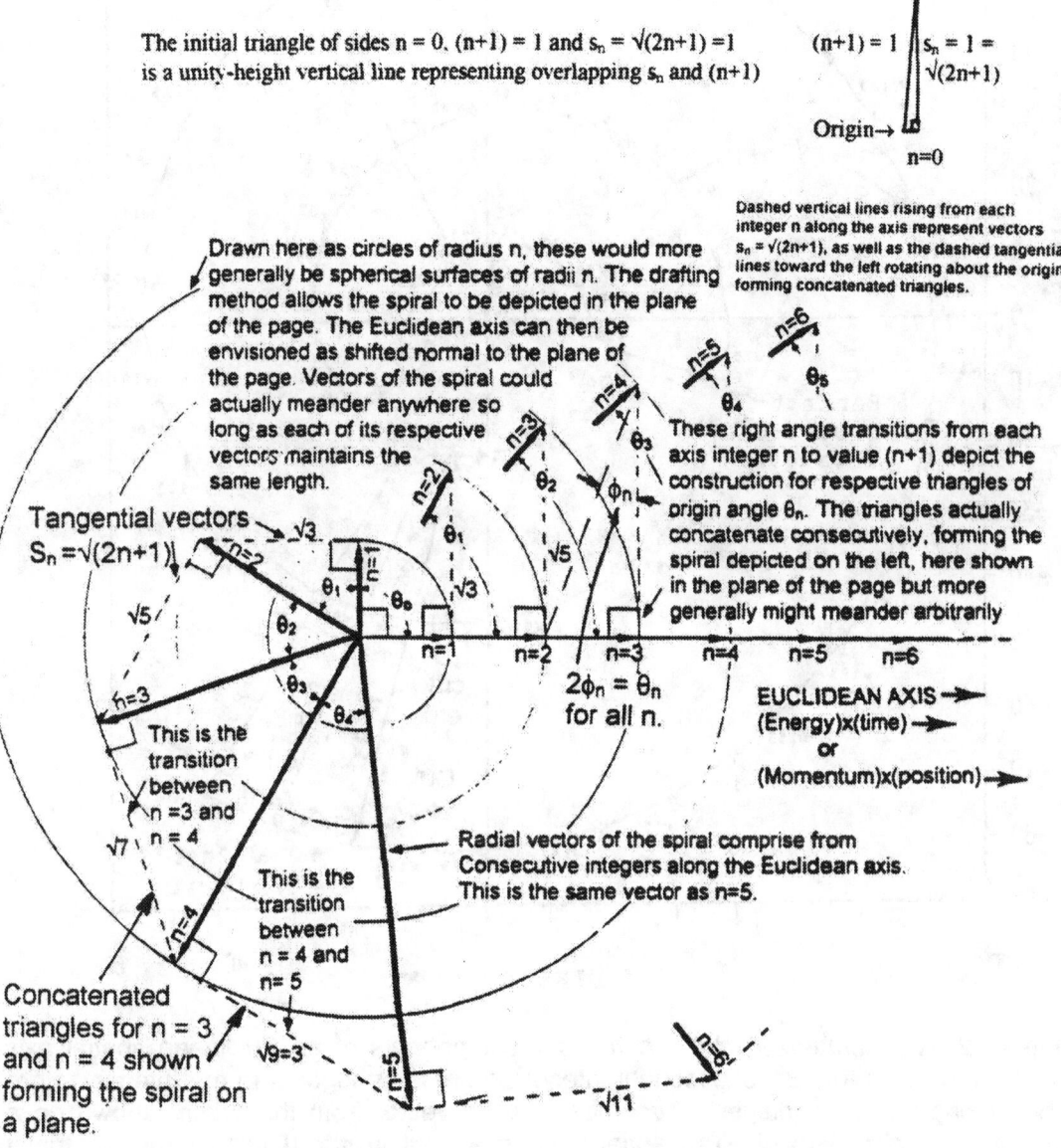

Dashed vertical lines rising from each integer n along the axis represent vectors $s_n = \sqrt{(2n+1)}$, as well as the dashed tangential lines toward the left rotating about the origin forming concatenated triangles.

Drawn here as circles of radius n, these would more generally be spherical surfaces of radii n. The drafting method allows the spiral to be depicted in the plane of the page. The Euclidean axis can then be envisioned as shifted normal to the plane of the page. Vectors of the spiral could actually meander anywhere so long as each of its respective vectors maintains the same length.

These right angle transitions from each axis integer n to value (n+1) depict the construction for respective triangles of origin angle θ_n. The triangles actually concatenate consecutively, forming the spiral depicted on the left, here shown in the plane of the page but more generally might meander arbitrarily

Tangential vectors $S_n = \sqrt{(2n+1)}$

$\sqrt{3}$

$\sqrt{5}$

This is the transition between n = 3 and n = 4

$\sqrt{7}$

Concatenated triangles for n = 3 and n = 4 shown forming the spiral on a plane.

This is the transition between n = 4 and n= 5

$\sqrt{9}=3$

$\sqrt{11}$

$2\phi_n = \theta_n$ for all n.

Radial vectors of the spiral comprise from Consecutive integers along the Euclidean axis. This is the same vector as n=5.

EUCLIDEAN AXIS →
(Energy)x(time) →
or
(Momentum)x(position)→

n=1   n=2   n=3   n=4   n=5   n=6

Figure 1. This drawing demonstrates a construction method to establish the length for vectors of a spiral about the origin. Those vectors need not confine to the plane of the page as rendered but can meander anywhere, so long as the size of consecutive triangles comprising the spiral remain intact. Then each triangle would preserve its right angle and fulfill the axiom for orthogonal transition to the next higher integer. The Euclidean axis could also shift to being normal to the page, rather than within it as shown.

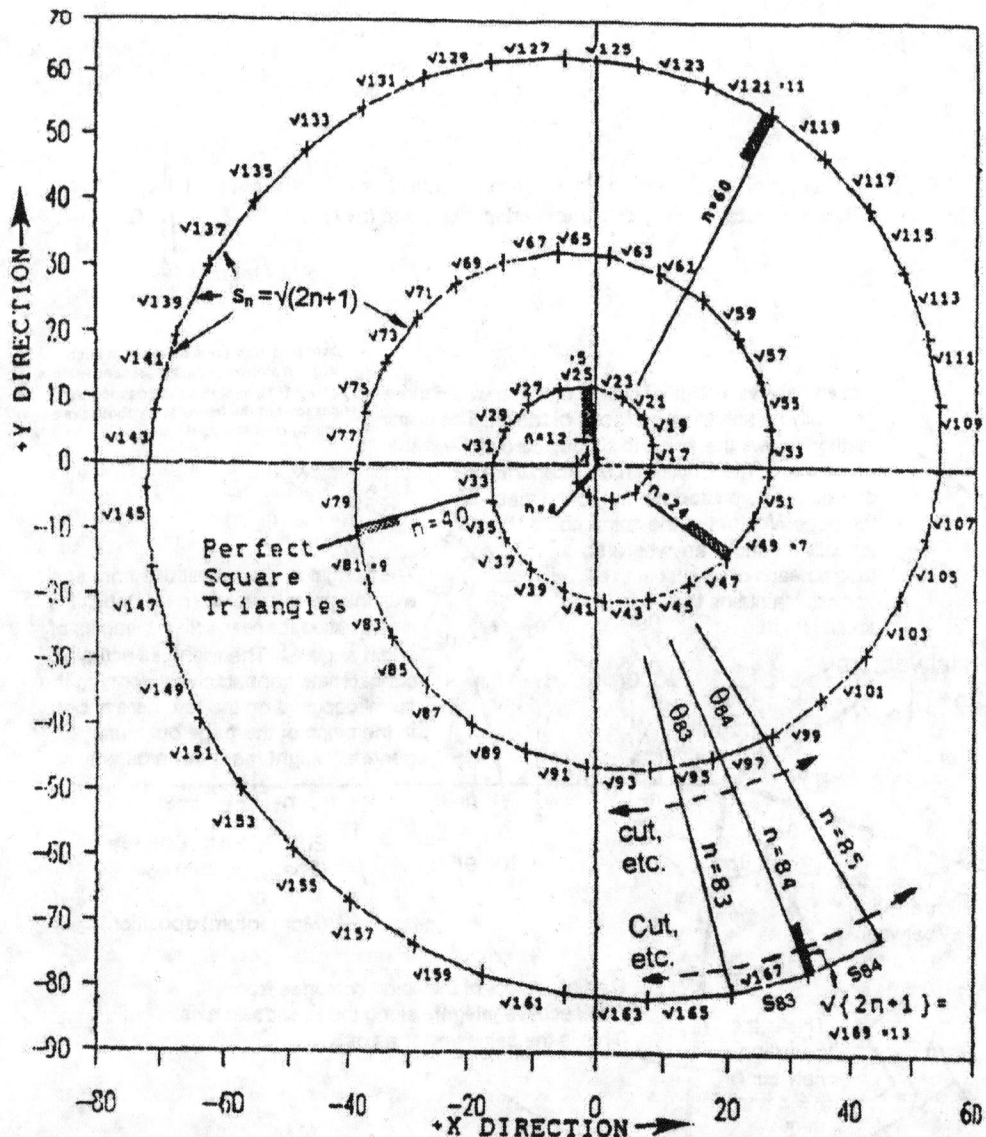

Figure 2. A computerized plot showing XY components of an Euclidean spatial axis when transitions to each subsequent integer progress orthogonal to the integer, rather than along the axis. Integer n becomes a radial vector from the origin (shown on a plane) from n = 0 to 85. Tangential vectors are all irrational except for the triplet triangles which asymptote to being precisely four radians apart toward large n. In this plot n represents (momentum) x (position) but could equally well depict (energy) x (time). The originating Euclidean axis can progress into the page at the origin since each node n actually represents a possible spherical surface about the origin, rather than being on a plane as drawn. Radial vectors for the shaded perfect square triangles [where $s_n = \sqrt{(2n+1)}$ = (odd number)] appear $2/\pi$'ths revolution (or 4 radians) apart in this figure. Plotted on the appropriate conic surface those triplet vectors could all emerge aligned in a given direction. That would occur for example if the spiral were rendered on the surface of a cone of apex angle $\beta = \sin(2/\pi)$, or $\beta/2$, $\beta/3$, $\beta/4$, etc. It will later be shown that each magnitude s_n can represent time from the origin in a harmonic wave.

the spiral, [called **sector** $s_n \equiv \sqrt{(2n+1)}$], progresses at right angles to its associated

radial vector n from the origin to reach the next radial vector (n+1).

Only about 5 triangles of the spiral appear in figure 1 while figure 2 continues

the process to n = 85. In that rendering most radial-vector-lines and their respective

magnitudes n from origin-to-tangential-vector-nodes are deleted for reasons of

clarity. **Values are articulated for each sequential inter-integer tangential**

vector s_n, which always has magnitude equaling the square root of a

consecutive odd number. This spiral portrayed within a single plane delineates

one possible mode of orthogonal progression under the orthogonality axiom. It

interconnects consecutive positive integers n along an otherwise Euclidean energy-

time axis by alternatively utilizing orthogonal transitions between the integers. The

spiral exhibits a unique periodicity whenever its constituent triangles possess three

integer sides. With only slight "deviation" at very small values of energy-time n (later

shown as irrelevant), <u>**periodic perfect-square triplet-triangles occur every**</u>

<u>**$(2/\pi)$'ths of a revolution, or every four radians**</u>. Each such benchmark defines a

"cycle" of the spiral and P cycles transpire whenever

$$n+(n+1) = 2n+1 = (\text{odd integer})^2 = 1, 3^2, 5^2, 7^2, 9^2, 11^2, ---(2P+1)^2 = s_n^2. \qquad (3*3)$$

Odd-integers for the magnitude of s_n in the sequence of numbers have particular

significance toward forming P cycles of the most basic harmonics. <u>**Moreover, the**</u>

<u>**orthogonality axiom for monadic population growth mandates that population**</u>

n will always equal the root mean square (RMS) summation of all prior transition magnitudes. That verifies directly from figure 1 since,

$$2^2 = 1^2 + (\sqrt{3})^2; \quad 3^2 = 2^2 + (\sqrt{5})^2; \quad 4^2 = 3^2 + (\sqrt{7})^2; \quad 5^2 = 4^2 + (\sqrt{9})^2, \text{ etc.} \qquad (3*4)$$

Every n is also the RMS sum of any smaller population value and all intervening transition magnitudes to reach n. These interrelating properties between transition magnitudes and n, calculated here for the restricted case where all concatenated triangles remain within a single plane also apply for the more general case where consecutive planes of each concatenated triangle occur at random angles. To simplify comprehension, initial discussions calculate each angle $\theta_n = \cos^{-1}[n/(n+1)]$ in figure 1 employing straight-line chords for $s_n = \sqrt{(2n+1)}$. This leads to slight periodicity inaccuracies in the resultant wave at small n values. Later arguments associated with equation 21*8 and Table 4 show the actual radians of each θ_n would instead comprise exactly $2[\sqrt{(2n+3)} - \sqrt{(2n+1)}]$ radians of a 4-radian cycle. This results in **perfect harmonic periodicity** from n = 0 to n = ∞.

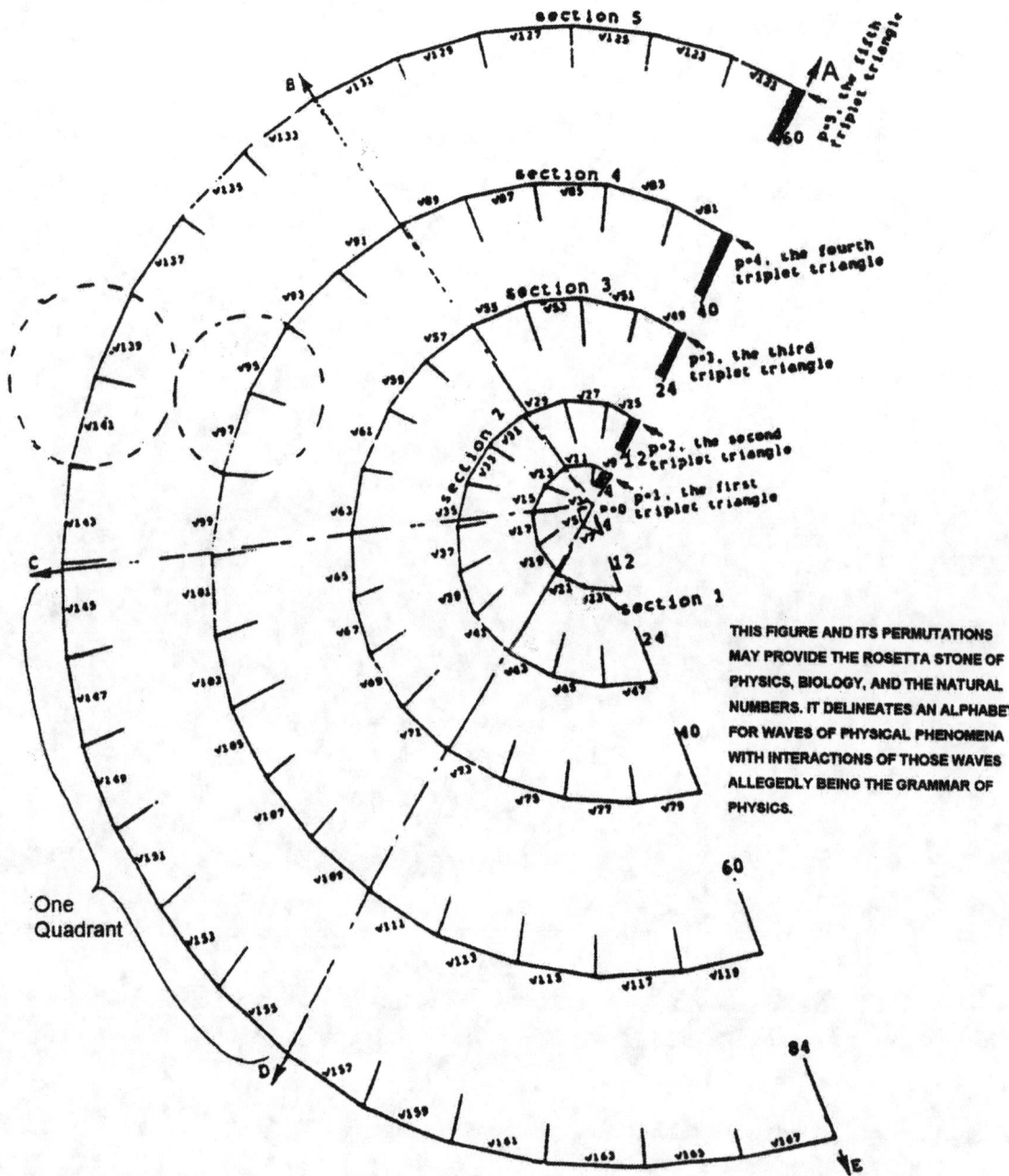

THIS FIGURE AND ITS PERMUTATIONS MAY PROVIDE THE ROSETTA STONE OF PHYSICS, BIOLOGY, AND THE NATURAL NUMBERS. IT DELINEATES AN ALPHABET FOR WAVES OF PHYSICAL PHENOMENA WITH INTERACTIONS OF THOSE WAVES ALLEGEDLY BEING THE GRAMMAR OF PHYSICS.

Figure 3. This plot depicts the identical spiral to figure 2 with sections between triplet triangles cut out and those radial vectors replaced in-line at direction A (and E). Radial extents are values of n. Sections comprise from a string of tangential vectors-s_n, each being the square root of a consecutive odd integer, i.e., of $\sqrt{(2n+1)}$. At direction A that square root itself becomes an odd integer, $s_{n@A} = \sqrt{(2n+1)} = 1, 3, 5, 7, 9, ---$. Tangential vectors s_n that straddle direction C approximate even integers out to infinity and at B and D they portray (even+1/2) and (odd+1/2) respectively. Each section occupies $2/\pi$'ths of a complete cycle, or four radians. Were the missing angle of the "fan" between directions A and E cut out, and those cuts joined, each section would delineate a cycle on a cone and there are six such cycles shown to n = 84. The cyclic process continues to indefinitely large n.

22

4) SPIRAL MODES SURROUND TRADITIONAL AXES

Figure 3 renders the identical spiral of figure 2 with each individual $(2/\pi)$'th revolution (called a section) separated out by itself, and consecutive triplet triangles aligned at direction A. Figure 4 delineates an axial end-view of this

$2/\pi$'th of $2\pi = 4$ radian "fan" occupying the surface of a cone (4*1)

of apex angle $\beta = \sin^{-1}(2/\pi)$. (4*2)

Radial directions A and E of figure 3 join to form this cone, and this axial view simulates the "conical helix description" given in the caption of figure 2. (The cone can be slightly "bullet shaped" to correct for small deviation at small n, not highly relevant and neglected for the moment.) Only radial vectors of figure 3 are considered to fall on the conic surface, with tangential vectors behaving as straight-line chords interior to the conic surface. Each triangle of origin angle θ_n is treated to lie along "chords" inside the conic surface, except for later analyses and when transparencies of figure 3 fold so that directions A and E overlap to form a conic surface. In mapping figure 3 within a conic surface, each triangle described by θ_n remains unchanged in the mapping with s_n being a straight-line vector. On this conic surface nutating 4 radians per repeated triplet triangle, spiral values divided by n would portray a cyclic wave (called a **unified wave**) analogous to sine waves of frequency ν. **This wave derives exclusively from properties of the natural numbers in accrual of monadic entities under the orthogonality axiom**. As n $\rightarrow$ ∞, the function will be shown to approach the harmonic exponential, $\exp(i\omega t)$.

Otherwise parameter n is retained in the expression, at the expense of "discontinuous gaps" (the transitions) associated with the discreteness of the wave. It will subsequently be shown that as n → ∞, each quadrant in figure 4 defined by the angle between directions A,B,C,D will comprise **<u>exactly one radian-of-angle</u>** around the cone making this topological adaptation of natural numbers an **absolute mathematical definition for the unit radian**. A complete cycle comprises exactly 4-radians for all n.

The X and Y coordinates of the figure 4 spiral's vector nodes mapped on to the conic surface exhibit periodicity, readily observable when those coordinates are respectively divided by n. Such division forms the equivalent of unit-length-vectors falling on the circle (or sphere) formed within the cone's surface everywhere unity from the origin. Justification for that operation will be developed. It suggests how progression of integer n creates a harmonic wave via repetitive circular traversal through four quadrants possessing rectilinear symmetry. Figure 5 compares quadrature components for the node positions of that wave. Data points depict consecutive X and Y coordinates of figure 2, each respectively divided by the n value for that coordinate. Conventional "amplitude" of the waveform would be indeterminate without displaying it normalized through division by n, which simulates unit-circle traversals. Such division by n also takes place in the orthogonal-defined progression-limit of exponential harmonics; i.e.,

$$\exp(i\omega t) \equiv \lim n \to \infty \text{ of } (1+i\omega t/n)^n. \tag{4*3}$$

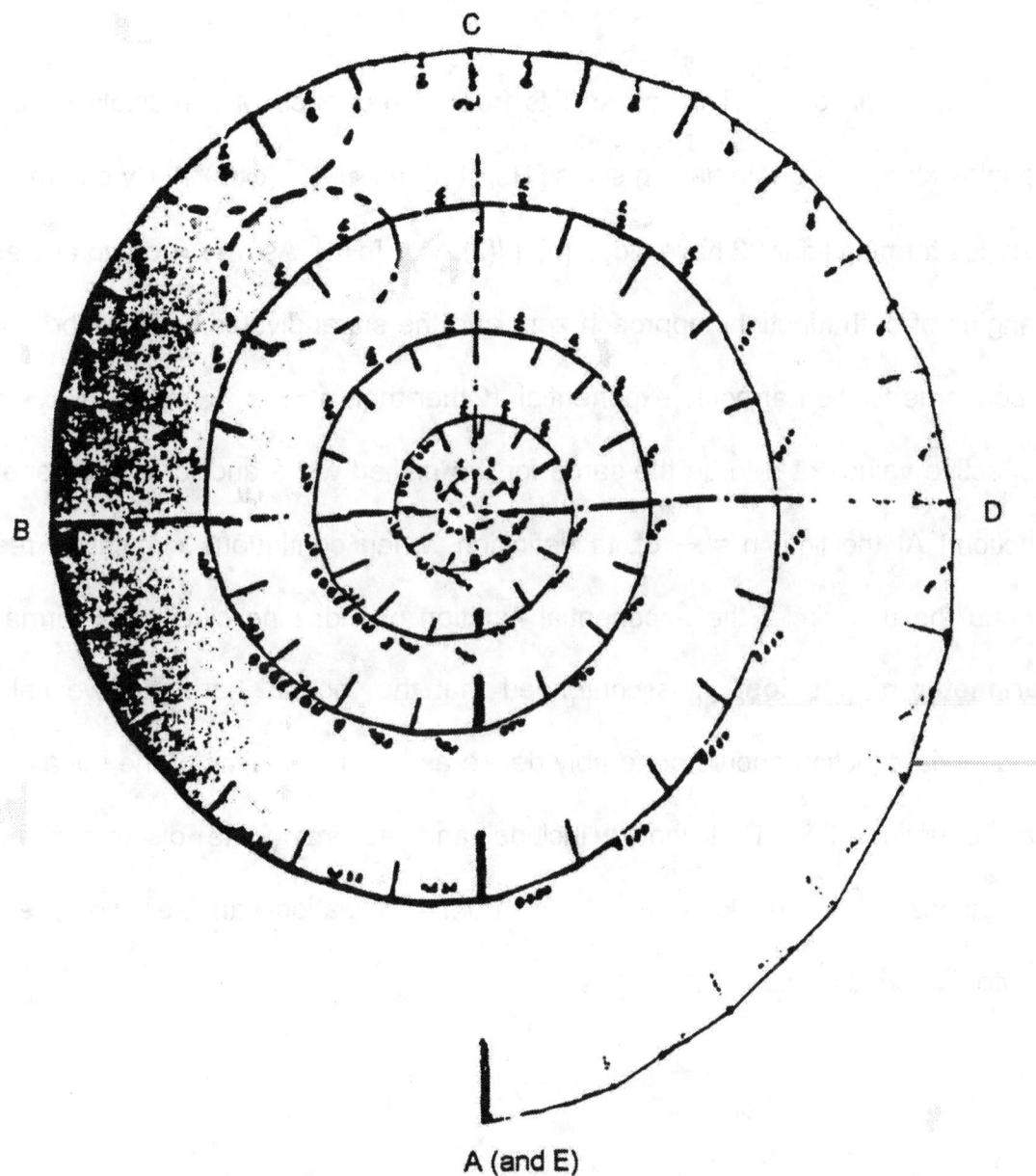

A (and E)

Figure 4. This shows an axial photograph of a figure 3 transparency with (2π-4) radians removed between directions A and E and those directions joined to form a cone. The Euclidean axis could for example progress into the page at the origin with this cone surrounding it at apex angle β = sin $^{-1}$(2/π). Resulting cycles on the cone would exhibit four-quadrant symmetry attaining exactly four symmetrical radians per cycle as n → ∞, with low-n deviating therefrom in proportion to $1/P^2$. Deviation from precisely four radians is less than 0.000416849- - - radians at P = 20 and less than 0.000104178- - - radians at P = 40. For a one-micron photon, 40 cycles would constitute a distance of 40 microns from an origin emission source. Such deviation is not significant however since the pictured spiral is not confined to the conic surface. It can meander into and around the conic surface via allowed "folds" along each radial vector, so long as each transition to the next integer remains at right angles. That is, so long as each triangle of origin angle θ_n with side n and (n+1) defines a plane.

This function similarly constructs from a sequence of n multiply cascaded triangles about the origin having sides [1], [$\omega t/n$], and [$\sqrt{(1+(\omega t/n)^2)}$]. By contrast, the triangles forming figure 2 have sides [n], [$\sqrt{(2n+1)}$], [n+1]. As $n \rightarrow \infty$ origin angles for triangles of both functions approach zero and the spiral divided by n will be shown to converge to the harmonic exponential. [Other than at $n \rightarrow \infty$, no need exists for respective values of n to be the same for the unified wave and for the exponential function.] At the limit $n = \infty$ of its definition, when continuous circulation results around the unit circle, the exponential function provides no clue to its formation. **Parameter n gets lost**. It is contended that the "actual" harmonic we call the exponential function should preferably define as the $n \rightarrow \infty$ limit of the spiral rather than equation 4*3. That would include and accommodate discrete member populations leading up to $n \rightarrow \infty$, at which limit population variables would exhibit Newtonian continuity.

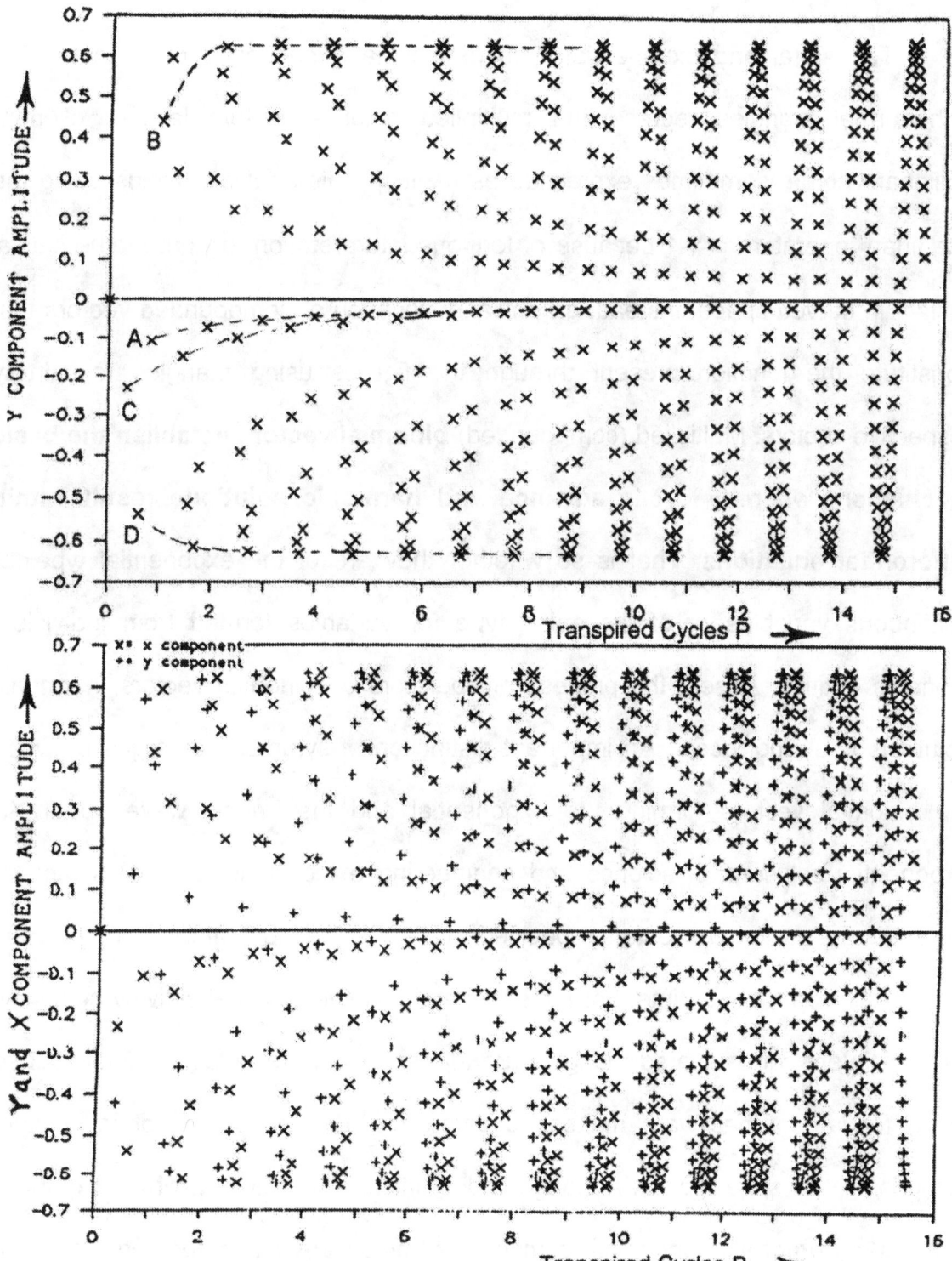

Figure 5. This is a comparison of the Y and X components of the conic spiral. These data points are generated by multiplying the sum angle in figure 2 by $\pi/2$ and every n by $2/\pi n$. Data points depicting the wave track transpired half-cycles or periods with each tangential vector-length $s_n = \sqrt{(2n+1)} = 2P_n+1$.

The spiral and exponential both define via a sequence of concatenated orthonormal triangles representing multiplied binomial-vectors [a pair of non-collinear vectors sometimes expressed as (A+iB)]. This analysis avoids using the imaginary operator $i \equiv \sqrt{-1}$ because of tenuous interpretation for that mathematical artifact in curved space. Instead, graphical description of compounded vectors that constitute the function present through the figures using triangles to portray respective vectors. Multiplied (compounded) **binomial vectors establish the basic mechanism whereby cyclic advance and harmonic solutions result within differential equations**. That is so whether they are of the exponential type for continuous variables or of the spiral type for variables formed from indivisible monadic entities. Absent the process of operators on binomial vectors, harmonic solutions could not occur. An important distinction of symmetry emerges between the binomial vectors forming the exponential and the unified wave spiral. All exponential solutions of science and engineering exhibit time reversal symmetry because the binomial vectors specified under the exponential's limit process emerge as uniform over time. For the exponential function, n is effectively at infinity. Its constituent concatenated triangles are all essentially isosceles and identical so going forward or backward makes no difference. By contrast, binomial vectors comprising the spiral solution change monotonically with time **prohibiting time-reversal-symmetry**[2]. Whenever compounded-binomial-vectors forming harmonic solutions are non-uniform, those solutions should display time irreversibility. The uneven spacing in figure 5 indicates such an arrow of time.

Time's arrow provides one basic distinction between the two types of harmonic waves, although at least ten other possible inadequacies of the exponential function are identifiable[1]. Absence of time's arrow in traditional classical, quantum and relativistic solutions is deemed due to the exponential function's time-symmetry, which function ideally should be replaced by a unified wave as described herein. All processes in Nature base on some smallest repeated constituent forming a population, thereby requiring the attendant representational relationship. These most elementary monads may be as small as entities of the sea of quantum that comprise the virtual vacuum. The analysis method is generic however and should prove useable when extended to any constituent membership. Toward $n \rightarrow \infty$, the exponential function articulates an infinite number of right-angles within the same plane as referenced from the fixed real axis. Those assigned directions requiring endlessly perfect rectilinearity on an idealized plane embody mathematical whimsy. No reason prevails to a priori assume space is perfectly rectilinear or a hypothetical unit-circle-surface perfectly flat. By contrast, concatenated radial vectors of the spiral **reference from each other** and the choice to portray them within a single plane merely provided analytical simplification to visualize the process. The vectors are not confined to reference from one absolute real-axis direction and thus can better accommodate the circumstance of non-Euclidean spacetime. Tangential vectors of the unified wave described herein remain unconstrained from referencing one axis or the plane of the paper. Each consecutive tangential vector only **bears orthogonal regard to the local radial**

vector n from which it emanates, requiring no continuity of direction with

planes defined by any prior radial vectors. Besides the time-irreversibility

feature, the new unified wave can accommodate significantly greater cyclical

phenomena than exponentials.

Figures 2 and 3 show that from n = 0 to n = ∞, each sequential outward

section of the spiral has four more nodes than its prior section. Each successive

cycle on the cone executes **exactly one added node per quadrant**. In conjunction

therewith, as measured on the plane of figures 2 and 3, each consecutive origin

angle $\theta_n = [\cos^{-1}\{n/(n+1)\}]$ (4*4)

gets smaller with increased n. As consequence, origin angles θ_n of the conic spiral

occur in an "affiliated time" δt_n such that $\theta_n/\delta t_n = 4\nu$ radians/second,   4*5)

contributing 4 symmetrical radians per cycle at large n. Time per section emerges

as constant yielding a periodic frequency analogous to but contrasting radian

frequency $\omega = 2\pi\nu$ radians/second applicable for traditional sinusoids of exponential

harmonics. Radial vectors of the conic spiral can appear like an axial view of a

uniformly-rotating clock-hand circling the central axis of the cone, strobed at

consecutive intervals proportional to θ_n. Each strobed radial vector will be unity

greater than its predecessor, which, after division by n as in figure 5 creates

uniformly-appearing sine waves similar to progression around the unit circle.

For monads, modes described by the conic spiral of figures 3 to 5 introduce

eight unfamiliar degrees of freedom that are called "dimensions" for lack of alternative. For each Euclidean axis X, Y, Z, t, "surrounding harmonic waves" could emerge as consequence of the orthogonality axiom. It may well be that the word dimension is inappropriate, but it provides a simple narrative for properties otherwise difficult to describe. Non-divisible quanta may only occupy integers. These bring-forth orthogonal and circumambient dimensions, which alter progression along each Euclidean axis into what manifests as four-quadrant centrally arranged "four-radian-cycles" of nutation from the origin. It is plausible that those tangential-going vectors symbolize a "specific category of time", rather than represent new dimensions. Parsing them as "time to reorient members of the ensemble affiliated with each new population member" provides one interpretation in chapter 19. Whether that constitutes a new unfamiliar degree-of-freedom or a dimension orthogonal to conventional axes may be semantic. In either case the phrase new dimension will sometimes be used and explored, whatever the additive new variables actually signify. A purpose for presenting the fan of figures 3 and the cone of figure 4 was to portray the unique 4-quadrant and 4-radian mathematical symmetry of natural numbers when interpreted for monadic accrual under the orthogonality axiom. It permits examination of the periodicity intrinsically related to all concatenated planar triangles specified by n and (n+1). This periodicity and symmetries vested within the sequence of triangles are a fundamental consequence of the succession of numbers n. Such harmony applies whether each tangential vector orthogonal to its radial vector remains in the same plane (as treated in the above analysis), or exists in any unrelated arbitrary plane as would be allowed under physical reality. Each triangle θ_n is identical in either case and so their sum to

4 radians between each odd integer of s_n is also the same. **However they might be graphically derived, plotted, or displayed, the harmonics described herein arise exclusively from the sequence of natural numbers**.

By double-folding, triple-folding or ½-folding the fan of figures 3 into different cones, various modes of symmetry are possible for conic spirals of different apex angles. From equation 4*2, "modes" envisaged at conic apex angles $\beta/N = (1/N)\arcsin(2/\pi)$ with N integer for example, can also yield periodicity with synchronous overlap of directions A, B, C, and D. Overlap of the four quadrant directions bring-forth harmonic conditions that grant wave-like interference attributes. Whenever combinations of vectors in directions A, B, C, D "coincide", orbits can be "stationary periodic" and may constructively or destructively interfere as with conventional periodic standing waves. Thus a host of different harmonic modes are feasible amongst the totality of all possible modes, including those where directions A,B,C,D, end up randomized and non-overlapping. As will be elaborated, various stationary standing wave solutions can exist amongst the entirety of ways that a population of monads can accrue. Those standing wave solutions can represent stationary states of possibility. It will subsequently be indicated (within the limits of prevailing indeterminism), that successful interaction (interrogation) of a unified wave by other entities might easily alter its harmonic cadence destroying its symmetrical-mode and interference properties.

32

Figures 2 through 5 depict a specific periodicity that can associate with a single particle of frequency ν. As will be elaborated, each radial vector n may portray units of (energy)x(time), (rather than time alone) such that the greater the quanta's energy, the less time signified by each integer n. From equation 4*5, for higher energy quanta, the ν of $\theta_n/4\delta t_n = \nu$ must correspondingly increase with decreased time t_n. Energy and time are mutually exclusive, the product of their smallest increment equaling each unity interval between integers of n, so energy proportions to reciprocal-time (or frequency) rather than amplitude squared. No fixed "amplitude" actually exists in this unified wave as n continually increases with time. Only through mathematical division by n does the wave appear of uniform amplitude as in figure 5. <u>Cycles established by cumulating all θ_n angles between perfect squares of (2n+1) can inherently be periodic however, **irrespective of direction of those angles in 3-dimensions**</u>. That is a mathematical property of natural numbers and inextricable from the feasible formation of stationary-state harmonic modes amongst the totality of all possible modes in the progression of monadic variables.

Figure 4 having a spin of one rotation per consecutive perfect-square triplet-triangle at direction A, purports to classically depict a generic waveform that might give rise to quantum mechanical standing waves in a potential well, or possibly the "waveform" of a particle. Rectilinear symmetry is apparent, with directions A, B, C, D, respectively affiliated with those in figure 3. Out to infinity, triplet triangles at direction A all possess consecutive odd-integer magnitudes for tangential vector s_n.

They are the only rational tangential-vector-extents of the spiral. Such distinction, allows that associated-vector direction A to signify the wave's polarization. **All other tangential vector magnitudes are irrational**. Tangential-vectors are thus called **intangible**, as they are irrational, remote from, not directly referencing the origin, and within a "new dimension". The dimension will depict time but is difficult for now to reconcile as conventional time and thus called a separate dimension for convenience. Treating them separately always allows the possibility of later fusing both time descriptions. That unfamiliar circulatory dimension denotes the only possible route for monadic populations to reach consecutive integer magnitudes. Non-divisible quanta cannot occupy fractional positions along the Euclidean axis and are thereby "forced to progress around it" creating these unified waves. Radial vectors emanating from the origin are integer and termed **tangible**. Whenever directions ABCD synchronously overlap as displayed in later photographs, cadence and interference properties of the wave can manifest. Those overlapping conditions allow "standing wave solutions" to quantum mechanical type problems, for example as a wave function characterized within potential boundaries.

Always equaling an <u>odd integer at direction A</u>, tangential vector-lengths of the spiral algebraically increase by <u>exactly two per revolution</u>. Figure 3 reveals that tangential vectors straddling direction C everywhere register <u>one integer greater than at A, with directions B and D respectively $^1/_2$ and $^3/_2$ greater</u>. Partition directions ABCD therefore everywhere associate respectively with (odd), (odd+1/2), (even), (even+1/2) proximal magnitudes of

$s_n = \sqrt{(2n+1)}$. Those odd, even, and mid-way between values of s_n depict quadrants of a cycle. From origin to infinity, the magnitude of each tangential-vector proportionally designates <u>cumulative half-cycles</u> traversed from n = 0 to that respective vector n. Within the unfamiliar dimensions surrounding the Euclidean axis, **tangential vector lengths automatically register the accumulated number of periods, cycles, or quadrants that occurred to that point**. Those half-cycles of variable s_n relate to n and transpired cycles P_n as:

$s_n = \sqrt{(2n+1)} = 2P_n+1$.   (4*6)

Parameters that delineate the wave, namely its radial vectors n, tangential vectors s_n, and transpired cycles P_n all interrelate in the simplest monotonic fashion. Tangential magnitude s_n equals the number of transpired half-cycles and (2n+1) registers all perfect square values encountered as the odd integers of s_n. Through n, essential properties of the harmonic wave everywhere reference the   n = 0 origin. Within its structure, each cycle (and fractions thereof) that emerges through the concept of orthogonality, embodies only integer values n. <u>All reference to the origin is through rational integer values, while all tangential magnitudes are irrational, except at direction A where they follow the sequence of odd numbers</u>. The irrational tangential magnitudes however are all $\sqrt{(\text{odd integer})} = s_n$. This wave synthesis avoids the mathematical operator $i = \sqrt{-1}$. Respective integers of n register discrete units of energy-time. The new off-axis dimensions portrayed by inter-integer transitions s_n analogously affiliate with those same smallest energy-time dimensional units. This tracking feature between the magnitude of s_n, unit-energy-time increments, and transpired half-periods creates a unique property and trigonometry wherein **total extractable information contained within the wave exists at every value of n, or s_n, or P_n.** Via a harmonic mode, the sequence of natural numbers n specify cycles and half-cycles from an origin event at n = 0. Given any of these values would grant maximum available knowledge that can be gleaned about the state of the situation. Further properties are unknowable due to inherent indeterminism obscuring in which direction consecutive tangential vectors s_n progress. That uncertainty effectively stems from inability to determine sub-attributes about each smallest population member. The uncertainty manifests as the "void interval" between integers of the constituent population. There cannot be fractional-values of smallest things whose "interior" cannot be further examined. Though these parameters are absent from the harmonic exponential function and thus unfamiliar, they represent an intrinsic part of how elementary waves advance. Harmonic exponentials derive from a limit corresponding to n = ∞ in this unified wave thereby deleting parameters n, P_n, s_n, and inter-integer indeterminism from expression. Parameters n, P_n, and s_n, would then go to infinity and vanish, while indeterminism reaches zero. That mathematical condition of them all vanishing is thus also available to the unified wave when n → ∞.

Parameters n, s_n, and P_n emerge in the unified wave at the expense of included indeterminism, created by concealed information about the innards of each indivisible population member. That missing information can never be recovered and its absence manifests in different forms throughout the analysis. It is noteworthy that the phase of a traditional sine wave applies within the comparative extent of a single cycle. Total propagation delay (or cycles past) remains unspecified by relative "phase shift". In s_n, n, and P_n this **<u>unified wave establishes absolute phase</u> <u>with reference to its origin as well as within each cycle</u>**. Besides traditional phase delineation available at those fractions of a cycle designated by radial vectors, through n the wave bears a message of the time, distance and cycles P_n from inception, a message sine waves do not convey. It mathematically upholds this unequivocal phase data at each n throughout the scope of its existence since the origin event.

The original Euclidean-type axis that was modified to allow orthogonal transitions between integers no longer confines to one specific line, or even a single plane as displayed in figures 2 and 3. The fan of figure 3 could in modes other than shown, alternatively fold itself anywhere within, outside or near the interior volume of the conic surface. **Each right-angle triangle of the spiral need exist within a plane, but orientation of adjacent-triangle planes need not relate. Nothing constrains the relative orientation of adjacent planes**. A right angle is necessary from radial vector n to (n+1), but the next right angle progressing from (n+1) to (n+2) **<u>can orient in any arbitrary plane through the origin</u>**. This grants a non-locality environment surrounding the original axis, allowing advance anywhere and everywhere within or around the cone, or wherever concatenated vectors might meander in their cyclical progression. The vectors could possibly "flood" the region surrounding the origin and energy-time axis in a spherical manner. Time itself may "occupy the seemingly extra two dimensions around the axis", or around the origin, permitting numerous modes of the wave within. As will be shown, prior to a terminal outcome from an event at n = 0, **all achievable directions-through-time remain feasible**, though possibly in random "location", rather than via the traditional straight-line course through time. This means the wave retains capacity to take all possible temporal trajectories within or around the origin, but only exhibits synchronicity when displayed on a conic surface. **The only requirement is that all concatenated right triangles remain contiguous**.

Modes of the spiral within the new dimensions can thus be either totally random (the general or background case) or stationary periodic, permitting interference and stationary state "standing-wave" type solutions when directions ABCD align. A mode constitutes any possible assemblage of concatenated triangles from 0 to n everywhere maintaining the appropriate respective angles $\theta_n = \cos^{-1}[n/(n+1)]$. Like a flexible toy snake hinged at the joints, for n joints there would be an enormous number of possible modes. Only a relative few of those modes exhibit synchronous overlap of directions ABCD and they tend to form a spiral. Examples are displayed in figures herein. In discussions here, the word time often refers to intervals incurred by traversal of consecutive angles θ_n or progression along vector s_n. It is also loosely utilized for the variable energy-time since energy is not a volatile component of their product and time is easier to visualize. **Regardless of the individual triangle-plane-orientations depicted as co-planar in the spirals of figures 1 to 3, the sum of the θ_n angles, (or s_n**

magnitudes) would remain unchanged. Whatever angles exist between the planes defined by consecutive tangential vectors s_n, the sum of all θ_n angles to any n is explicitly defined by n. Thus, harmonic features within the wave are "**always potentially present**" based on recurrence of ABCD overlap, even when such overlying cadence does not emerge in presentation of the wave on a plane or cone. **The periodicity exists within the natural number system itself, not the graphics of the spiral being used to elucidate the phenomena!**

At this point, the region surrounding the original Euclidean axis would be subject to several interpretations. It is reasonable to assume that θ_n and s_n would relate to time, but their being able to progress in different or arbitrary directions makes it difficult to quantify precise meaning. An artifact will thus be employed to distinguish the Euclidean axis (which is more or less conventional describable territory) from the region surrounding it, or more generally from the region surrounding the origin. The artifact chosen may knowingly have negligible physical validity, but it allows categorization of the different regions for purposes of discussion. These interpretations can always later be altered when greater clarity or meaning emerges. The Euclidean axis will tend to signify the domain of **macroscopic** processes and the surroundings to **microscopic** phenomena that derive from degrees of non-divisibility of the analyzed entity.

5) PROPERTIES OF WAVES IN FOUR PLUS EIGHT NEW

"DIMENSIONS"

Regions along the original Euclidean axis-line distinguish from surrounding orbital regions. Macroscopic objects are the large physical things we see and handle every day in familiar X, Y, Z, t, spacetime. Every cubic centimeter of them may subdivide into at least 10^{23} partitions representing comparable divisions in energy-time. Such ensemble "parallel numbers" grant occupancy for fractional integer positions along dimensional axis when representing macroscopic entities, which positions would be unavailable to smallest microscopic quanta. As conventionally interpreted, macroscopic objects might subsist at any rational point along X, Y, Z, and t dimensional axes. Conversely, non-divisible objects would be confined to regions "around" the axes, which is not the typical domain of divisible entities on their energy-time scale. Quantum objects would thus remain off the Euclidean axis between integers. The potential habitation of available energy-time sites along a scale thus **bifurcates in relation to "divisibility"** of the entity under consideration. This creates different dimensional hierarchies based on particle divisibility, with smallest quanta experiencing new "unfamiliar off-the-axis" regions.

To accommodate this partitioning of occupyable sites, we label traditional Cartesian Coordinate X, Y, Z, t, type Euclidean-axis sites **macrozones**, contrasting the **microzones** of unfamiliar off-axis dimensions. The dualism of Bohr's complementarity[4] may allegedly identify with the presence of two separate "dimensional zones", regions of macroscopic classical physics signified through conventional energy-time axes, versus domains "surrounding" each Euclidean axis allocated exclusively to smallest quanta. This association may only be an epistemological viewpoint subject to later interpretation, but will suffice for now. Part of the justification for this partitioning stems from the separate rational and irrational values that analysis encounters within the two regions. For smallest quanta, all transitions occur only in the intangible territory surrounding (or remote from) the axis while highly visible objects may occupy rational-number tangible-locations from the origin. An axis as in figure 1 effectively intersects the concentric spherical surfaces of integer radius only at the integer values 1, 2, 3, --- etc. along that axis. Each such axis thus vacates its space between integers while acquiring two surrounding degrees of freedom or "dimensions". Conventional X, Y, Z, directions, (in addition to the previously discussed axis accommodating "time"), would produce three unit-integer momentum-position axes. Thus each of the four axes X, Y, Z, t should grant two unfamiliar action-type dimensions for smallest quanta surrounding it. The result would be 2x4 = 8 unfamiliar degrees of freedom or "dimensions" that remain obscure from classical macroscopic observers, systems, or instruments. The act of observation or measurement of those quantum arenas can only transpire when our familiar macro world penetrates those abstract quantum dimensions. The word "dimensions" is again loosely used in discussion as "time" for energy-time and distance XYZ for respective momentum-position axes. The 8 new dimensions might actually be capable of inclusion within conventional dimensional morphology or however, but discussing them in separation should minimize confusion. Further justifications are available for these assumptions. By no means does the validity or invalidity of conjecture for extra dimensions or bifurcated macro and micro regions alter the mathematical rigor upon which these harmonic waves are based. Such nomenclature and presentation merely provides a tool enabling one means for discussion of the processes.

In free space, electromagnetic radiation and components of un-interrogated free particles can then be interpreted to exist only in classically "inaccessible" microzones of the unfamiliar dimensions. Unobserved-quantum-states in atoms would similarly be inaccessible. To observe or measure those entities requires that they interact with the macro world, making them susceptible to that act of observation. Abstraction would even exist concerning classical descriptions within those unfamiliar dimensions. Distinction between the two different dimensional zones is conjectured to provide the essence of quantum measurement dilemmas. In the macroscopic world of humans and instruments, knowledge will always be devoid of quantum zone status to the extent of each population member's shroud, to a least quantum of action. Understanding the origin and effects of this shroud, and the un-seeable dimensional zone of activity it creates, should give quantum mechanics "enhanced classical understanding" in the same sense that Newton's action-at-a distance was also a conceptually-surmounted non-visual impediment. That understanding might require overcoming traditions and routine concepts by acknowledged existence of those new dimensions via the orthogonality axiom. The difficulty thereby introduced may be violation of the fundamental scientific credo requiring measurable experimental evidence to validate a physical theory. One might then counter argue the flip side, namely that irrational numbers should therefore never be allowed as part of any physical theory.

Considering only the energy-time Euclidean axis (or that "dimensional direction"), the unified wave has at least one additional degree of freedom compared to sinusoids that traverse the unit circle. Cyclic occurrences are not confined to a single plane as in a circle, but can also prevail in at least one more direction elucidated through a three-dimensional cone. In addition, random, plus synchronously overlapping, plus circular polarization-type modes all become feasible. Though a great range of possibilities conceptually exist for the location of consecutive radial vectors, uncertainty hides detailed knowledge of their whereabouts. The wave comprises of outward vectors from the origin, any ray of which provides a possible "particle-like" description to that endpoint. Again these are unsubstantiated possible interpretations and proof of the pudding will rest in the mathematics. When the wave annihilates at a given n, it could conceivably appear like a particle between origin and that terminus. Unless there is an interaction, radial vectors from a given origin are not inclined to react with radial vectors from another origin. A wave "exists unto itself" and only when synchronous overlap conditions along directions ABCD occur would standing-wave solutions or a self-interference condition likely result. Through its degrees of freedom the wave describes all

44

possibilities that might occur within say a constraining field boundary, with standing-wave solutions possible within those constraints only when directions ABCD overlap. **For such overlap conditions and only those conditions, "numbers of the number system" "synchronize" for the wave as in figure 3**. Similarly, interference within the reflectors of an interferometer cavity could only occur for conditions of ABCD overlap. To examine features of the orthogonality axiom, discussion in the next several chapters consider only indivisible monadic quanta like zero rest mass particles characterized by waves that propagate at velocity c.

Large ranges of different modes appear feasible within the new dimensions though uncertainty shrouds further knowledge of exactly where radial vectors end up. Some modes exhibit maximum disorder while others have great symmetry and order. These modes characterize possible "routes" of a population's progression through energy-time [or momentum-position]. To the extent s_n is intangible and inaccessible, the wave is not continuous but comprised of discrete "strobed vectors" n. However it will be seen that s_n can signify time as a "continuous spinning vector" and may provide "perpetuation" in addition to discreteness. The function proceeds from one n to the next and only one n "applies at one time", [or in one "place"]. This suggests a speculative interpretation. Conventional continuity does not seem to prevail between related points along space and time as inheres in Maxwell's-equation type descriptions.

It will turn out that two descriptions of energy-time might be identifiable in this process which can lead to confusion. One-description associates with conventionally interpreted smallest units of energy-time, each integer n constituting one such unit affiliated with a least quantum of action. The other derives from the "**dimensionless**" constant of proportionality between energy and frequency as $h = E/\nu$, **equaling the energy-time per cycle or, as interpreted herein, a minute excess-angle per cycle**. It will be shown that a quarter of a cycle or quadrant associates with each natural division of a cycle and each quadrant characterizes unity extent (one radian) in the new unified wave. Thus, conventionally interpreted energy-time as a unit-of-action can perhaps be represented as energy-time/quadrant under this theory and that description would have unity depicting the quadrant in the denominator. Energy-time [action] and energy-time/quadrant would be the same. Issues in relation to figures 1 to 3 could in concept be discussed in conventional terms of integer n units of energy-time within the spiral without bringing in the notion of excess angle. However excess-angle is deemed the **de facto** **"cause"** of the phenomenon and therefore warrants treatment in those terms. **The** **importance of preserving dimensional consistency for cycles in the** **denominator of $h = E/\nu$, (and for all applicable relationships) will be addressed** **throughout this work**. Various distinctions concerning dimensional consistency for cycles will hopefully be clarified by later analysis.

Each intangible vector orthogonal to radial vector n has mathematical extent $\sqrt{(2n+1)} \equiv s_n$. **Figure 3 indicates its magnitude mathematically equals the**

number of traversed half-cycles at n. Frequency of the wave would be the reciprocal time to traverse a cycle on the cone. Because they delineate the dimensional-unit intervals between consecutive integers n, each tangential vector s_n also associates with those (energy)x(time) units within the new microscopic dimensions. The discontinuous energy-time gap between two adjacent tangential vectors is

$s_2 - s_1 \equiv \sqrt{(2n+3)} - \sqrt{(2n+1)}$. For large n this equates to (5*1)

$s_2 - s_1 = \sqrt{(2n+3)} - \sqrt{(2n+1)} \approx 1/\sqrt{(2n+2)}$, which approximates $1/\sqrt{(2n+1)} = 1/s_1$.

Also, $s_{n+1}^2 - s_n^2 = 2$, **indicating that throughout all integers n, there exists a constant increment in the difference between sectors squared**.

The gap between tangential vectors, $s_2 - s_1 = 1/s_1$ also expresses in terms of triangle angle θ_1 at s_1. Wherever $[n_1 \gg s_1]$,

$\theta_1 s_1 \approx (s_1/n_1)s_1 = s_1^2/n_1 = (2n_1+1)/n_1 \approx 2$. (5*2)

Within the inequality limits imposed primarily by the latter brackets $[n_1 \gg s_1]$, this characterizes energy-time space by relating triangle angle θ_1 to tangential vectors as $s_2 - s_1 = 1/s_1 = \theta_1/2$. (5*3)

For large n, this "spacing" between integers applies for smallest quanta, and allegedly to some degree for all entities.

An argument exists that the dimension of cycles can be neglected, specifically in the denominator of h = energy-time/cycle within the photon Energy equation   E = hν. The allegation is that cycles are supposedly dimensionally similar to radians (constituting the same dimensions), with both purportedly

47

dimensionless because radians are a ratio of distance/distance as perimeter/diameter. However, this only applies for a perfect circle on a perfect plane. In a higher order dimensional hierarchy, distance-along-a-curve and distance along a straight-line-direction need not distinguish as "the same dimensional units". For example, distance along a straight-line-direction/second is velocity, which is quite disparate in effect from distance along a helical trajectory whose central axis is a straight-line or any direction. Comparing a curved route to a straight route may make the applicable dimensional units dissimilar, though comparable. Perhaps the simplest proof that radians (or cycles, or degrees) are not dimensionless derives from **mathematically redefining radians directly from the ordinal numbers independent of perimeter, diameter, π, or the arbitrarily chosen circle**. A **standard unit-radian will emerge as *angular rotation*** incurred by ¼ of a section in figure 3 as $n \rightarrow \infty$, or as ¼'th the **angular rotation between perfect squares in figure 2 as $n \rightarrow \infty$**.

From equation 5*3, $s_2 - s_1 = 1/s_1 = \theta_1/2$, or $s_n = 2/\theta_n$ (5*4)

    Toward large n half the radial angle of each triangle in radians [$\theta_1/2$], equals the change in sector length from the adjacent larger triangle [s_2-s_1], and equals the reciprocal of the smaller sector length [$1/s_1$]. This equivalence of a differential sector length, a reciprocal sector length, and an angle, will later give rise to a multiplicity of different dimensional units for the same variables. Equation 5*4 provides a key relationship concerning the mathematical foundation of cycles, radians, and waves. The included angle $\Sigma\theta_{ak}$ of many triangles between θ_a and θ_k

$$n=k \qquad\qquad n=k$$

would be $\Sigma\theta_{ak} = \sum \theta_n = \sum[\theta_a + \theta_{a+1} + \theta_{a+2} + \text{------}+\theta_k] = \sum \arccos[n/(n+1)]$ (5*5)

$$n=a \qquad\qquad n=a$$

For large n this becomes:

$$n=k$$

$2(s_{a+1}-s_a) + 2(s_{a+2}-s_{a+1}) + 2(s_{a+3}-s_{a+2}) + \text{---}2(s_k-s_{k-1}) = 2(s_k-s_a) = 2(s_k-1) = \sum \theta_n \equiv \Sigma k,$

$$n=0$$

since s = 1 for n= 0. Recognizing an offset angle from the abscissa exists at small n, the cumulative angle from 0 to any n expresses as, $\Sigma\theta_n = 2(s_n-1)$. (5*6)

From equations 5*4 and 5*6, $\Sigma\theta_n + 2 = 2s_n = 4/\theta_n$. (5*7)

<u>The length of each sector minus unity equals half the cumulative spiral angle from the start of the spiral to the right angle end of the sector</u>, (for $s_n >> 1$, $s_n = \Sigma\theta_n/2$ in radians). Proceeding outward one cycle in figure 2 along a radial line from any n having associated sector s_n, angle $(\Sigma\theta_n + 2\pi)$ must be 2π greater, so the sector length there must be $2\pi/2 \cong \pi$ greater, relative to its closer in predecessor one cycle before. <u>In any given radial direction on figure 2, sector length s increases by approximately π per revolution</u>. (5*8)

Because approximations neglected the offset angle from the abscissa toward small n, prediction accuracy of this relationship becomes increasingly imperfect near the origin, accumulating an offset relative to large n of about 0.3 radians at n = 1. That offset is negligible toward large n and an artifact of straight-line chords toward small n. It will effect the relative polarization of the wave and not its symmetry or periodicity. The relationship becomes highly precise at large n and $\Sigma\theta_n$ can serve as conceptual reference by "offsetting" the start of the spiral ≈ 0.3 radians clockwise from the X-axis, (while the spiral spins CCW). Each consecutive tangential sector $s_n = \sqrt{(2n+1)}$ depicts the square root of a consecutive odd integer $\sqrt{1}$, $\sqrt{3}$, $\sqrt{5}$, $\sqrt{7}$, $\sqrt{9}$, $\sqrt{\sigma}$, etc. Perfect Square triangles occur whenever sector

$$s_P = \sqrt{(2n_P+1)} \equiv \sqrt{\sigma_P} \equiv \text{odd integer} = 1, 3, 5, 7, \text{- - -} \equiv 2P+1, \qquad (5*9)$$

Where, P = 0, 1, 2, 3, 4, - - - etc.

Parameter P and subscript P reference that triangle where P becomes integer and $\qquad \sigma_P = (2P+1)^2 = 2n_P+1 = (\text{odd integer})^2$. $\qquad (5*10)$

[For now, P characterizes a constant integer for the entire interval between perfect square triangles. However, it also applies for a defined fractional value at each integer n.] For simplicity now, unless otherwise noted all contiguous sectors until the next perfect square triangle will characterize by an integer value of P. To reiterate, P = 0 when n = 0, where s = 1, and extends to n = 3 where s = $\sqrt{7}$; P = 1 when n = 4, where s = 3, and extends to n = 11 where s = $\sqrt{23}$; etc. **Each consecutive integer of P thus registers the number of prior perfect squares encountered in s_n,** above s = 1.Thus, $s_n^2 = 2n+1 = (2P+1)^2 = 4P^2+4P+1$, or

$$n_P = 2P^2 + 2P = 2P(P+1). \text{ Consecutive} \qquad (5*11)$$

perfect square triangles occur where $s_P = 2P+1$ and $s_{P-1} = 2P-1$. $\qquad (5*12)$

The respective values of n at those consecutive perfect square triangles are:

$$n_P = 2P^2 + 2P \text{ and } n_{P-1} = 2P^2 - 2P \qquad (5*13)$$

The difference δ_n between n_P and n_{P-1} equals the number of sectors included between those two consecutive perfect square triangles as well as their difference in radial extent. $\delta_n = n_P - n_{P-1} = 2P + 2P = 4P$ (5*14)

From equations 5*6 and 5*8, in figure 1 s_n increases by π per revolution and $\Sigma\theta_n = 2(s_n - 1)$; $\Sigma\theta_n/2\pi \equiv R_n \equiv (s_n-1)/\pi$ (5*15)

must indicate the number of revolutions the spiral has made from its start out to position s_n or n. From 5*11, R would then also proportion to P as:

$R_n = [2P+1-1]/\pi = 2P_n/\pi$ (5*16)

The change in R between two consecutive perfect square triangles becomes,

$2(P+1)/\pi - 2P/\pi = 2/\pi$'ths of a revolution = 0.6366198 - - - revolution.   (5*17)

This interesting result shows how throughout all ordinal numbers n on figure 2, perfect square triangles are uniformly spaced from each other in angular rotation, converging with increased n toward $2/\pi$'ths of a revolution, or exactly $(2/\pi)(2\pi) = 4$ radians. It provides an absolute definition for radians as a degree of rotation. It will be shown to not be an artifact of the graphic depictions of figures 1 to 4, but a fundamental property intrinsic to natural numbers. For all waves of physical reality the four radians allegedly delineate the four quadrants of every cycle, contrasting exponential harmonics based on Cartesian Coordinates and transcendental values. **In conic format the unit radian takes on implicit significance as a quadrant of every physical harmonic. Those quadrant extents constitute the energy-interchange-intervals of every**

Irwin Wunderman

wave. **Energy transfers from one form to its alternate and then back during each consecutive quadrant of every wave. The conic spiral will demonstrate as the basis for all wave harmonics, with its n → ∞ condition explicitly defining an absolute 4-radian angle for the 4 quadrants of each wave**. Any other degree of rotation can be comparative to this **standard unit of rotation**. All **arbitrary angular-rotation-amounts thus bear dimensional units of radians relative to this standard-unit-radian derived here mathematically from sequential integers, just as arbitrary meter amounts bear their dimensional-unit comparative to the standard meter**. Comparing relative significance, the unit-radian so defined is an ***absolute***-standard reference **related to an intrinsic property of natural numbers**, while the unit-meter as utilized is a ***totally arbitrary*** standard!

Radians or cycles are plausibly the most absolute-standard unit among dimensional units currently in use. They signify the point of closure of repetitive events with 4 mathematically defined radians on the cone equaling one complete cycle. No justification exists to call radians dimensionless simply because, for the one unique circumstance of a perfect circle on a perfect plane they can also equal perimeter/diameter. That circumstance entails a singularity inherent within the circle as (radius)/(deviation in radius). Even if radian units for a "perfect circle" were considered dimensionless, the dimensions of cycles are not dimensionless. A cycle is not based on a ratio of perimeter to diameter but **to the complete traversal of a rotation**. It may typically be expressed as 2π radian units, but it is **cyclic closure** (characterizing a unit) that depicts the dimensional entity. Units of radians defined through perimeter/diameter would seemingly be a subordinate case to the higher order definition of cycles since the cycle comprises the closure denomination [or repeat condition specifying the unit]. In exponential harmonic descriptions values of radians are transcendental at the closure point for the unit depicting each complete cycle. **Such radians are not a physical unit involving closure** under exponential harmonic definition. I.e., where a full cycle identifies closure with 2π radians/cycle; [in spacetime without curvature and only for an ideal circle on an ideal plane]. In the unified wave as $n \to \infty$, both mathematically and physically, radians and cycles entail **integer-unit-description** at the closure points noted by directions ABCD. It is salient that the four-absolute-radian-definition derived from the spiral at $n \to \infty$ **is not constrained to a plane or to any surface**. It represents the accumulation of an enormous number of θ_n angles between consecutive perfect squares of $(2n+1)$, **which separate angles are not specified, or obliged to follow any direction**. **The four radians are a property of the perfect square number sequence, not of the graphics employed here for illustration**. The perimeter/diameter definition for radians is subordinate to the mathematical circle. Because of that planar requirement, its definition on a flat plane however familiar encompasses an idealistic surface utilizing more functional space than real-world physics allows. Though mathematics can functionally represent a region defined through an infinitum limit process, does not guarantee that region depicts part of the physical world. **Particularly about limit processes, mathematics incorporates a greater dominion than physics!**

It may seem unusual for a periodic wave described on a conic surface (or anywhere) to continually diverge from the axis it seemingly emerged from or surrounds. By contrast, electric and magnetic field sinusoids comprehend as traversing the same unit circle all along their course. Helical waves also interpret to sustain constant breath parallel to a central axis. The unified wave portrayed here appears to assert an ever-widening extent in new dimensions surrounding that axis. That extent occurs in energy-time, for which time continually increases. Occurring exclusively off the Euclidean axis, exactly how that manifests in macroscopic X, Y, Z, t, space may involve subtle complexities. These waves describe as a single "three-dimensional" entity, in contrast to a pair of two-dimensional components like electric and magnetic. Breakdown into Cartesian coordinate components is not a viable option in spacetime with an excess angle (in curved space). **The only possible "zero location" for a wave progressing in 3 dimensions is its origin**. Using Cartesian coordinates, only hypothetical **components of the wave** pass through zero in X, Y, Z dimensions for example. No zero origin is referenced within actual three-dimensional space except in terms of breakdown rectilinear components, a hypothetical man-made construct of questionable validity in curved space. Since the wave must start from zero at its point of origin, that represents **the only true "zero reference"** for continuity of the wave. That constraint manifests here in the form of the n vectors emanating from the origin throughout the entire graphical composition of the unified wave. These energy-time waves in unfamiliar dimensions exhibit "spreading from the origin event". That is totally consistent with the **divergent behavior of waves** and conjectured as more appropriate than sinusoids cycling the unit circle to **provide only rectilinear components of a wave**. As will be elaborated, these waves represent possible stationary state harmonic conditions within concentric spherical surfaces that emanate from the origin. The new waves will also be shown as describable in terms of traversing Planck-time (or Planck-length) chords of the unit circle when put into that conventional context.

It is worthwhile to consider the solution to a basic second order equation that generates harmonic [1]waves, $d^2V/dt^2 = -kV$ (5*18)

Contrasting sign waves that traverse the unit circle, the spiral can serve as solution to a simple harmonic oscillator equation because, within constraints of the prevailing uncertainty (or angle $\Delta\theta$), the spiral's second derivative proportions to the negative of itself. The quadrature phase relationship inferring those derivatives is implicit in figure 5. The + data points in the lower figure simulate the first derivative

---

[1] It has been demonstrated[1] that circumstances where curved space possesses an indeterminate angle $\Delta\theta$ = Planck's constant $\equiv$ h, is equivalent to "widening" the equal sign into a window of indeterminism of width h in classical equations like $d^2V/dt^2 = -kV$. Equations written with conventional perception for the equal sign, namely both sides being precisely equal at only a zero-width point, inadvertently imply that infinite information is known about the described system. Granting an indeterminism-gap of fixed aperture within the equal sign's meaning (a band of uncertainty about perfect balance) negates such implied infinite-information assumption and circumvents the need for infinity or infinitely iterative operations. The required width of the information-gap absent within the equal sign will in general represent the information void attributable to the missing shroud from each non-divisible iterated monad of the system; it will represent the space between integers of the smallest-element population being described. Equations interpreted with conventional zero-width equal signs have a priori assumed the smallest constituents of the represented system are of zero size. Interpreting the equal sign with a band of indeterminism is not the same as adding an uncertainty term to either side of the equation because the equal sign's band extends to the equation's exponentiation processes limiting the necessity for infinite exponentiation, as occurs in exponential functions. Also, the increased multiplicity of solutions that can satisfy the new equations defined within the constraints of the allowed indeterminism-band all constitute "simultaneously feasible" solutions, which generally include both harmonic and non-harmonic modes as described for unified waves herein. Analysis of equations defined with uncertainty-broadened equal signs, being unfamiliar and an additional conceptual barrier, are avoided in this treatise.

of the x data points in both upper and lower figures. That first derivative entails a quadrature phase shift, so the second derivative of the x data points (not shown) would experience a second quadrature phase shift and emerge in phase opposition to the initial x data points thereby satisfying equation 5*18. The new harmonics within unfamiliar dimensions have more degrees of freedom than the two-dimensional mode of sinusoids traversing the unit circle on a plane. Their modes can additionally progress axially to the circulatory direction. Each vector n independently references from the origin with each s_n being orthogonal to its respective n. Transpired cycles from the wave's origin are implicitly engrained mathematically in the advancement of n. And enormous number of modes are feasible but only a relative few would be harmonic. These admit interference patterns suggested to be easily broken through slightest interaction with the wave.

Waves represented by particles having energy E = hν are of specific interest. Here h has dimensional units of energy-time/cycle as h = E/ν. Planck's constant ≡ h ≡ E/ν = 6.62606876 x10^{-34} Joule-Seconds/cycle in MKS units. Non-arbitrary dimensional units of energy and time may later be shown as preferable to arbitrarily chosen Joules and seconds **so the numerical value for h may differ in those appropriate units, though the quantitative magnitude represented need not**. The waves described herein have been shown to derive from a population of non-divisible energy-time units. As such, the intervals between integers that depict that population remain inaccessible, indeterminate or void due to the shroud of omitted information concerning each population member's "innards". Population members here may equivalently interpret as a **compounded excess-angle**. Compounding that excess angle (or limit of angular resolution) occurs with each passing that creates the population. Such a population would be devoid of the space between its integers, and that missing information equates to the extent of one population member.

That uncertainty per cycle, one least quantum of action, or a corresponding indeterminate angle θ_{min} within each cycle of spacetime, may limit the otherwise deterministic progression of n from reaching infinity. As earlier noted, each smallest energy-time unit denotes a minimum-angular fixed-cyclic-portion of a temporal wave. For increasing n, when θ diminishes down to that smallest resolvable portion of a cycle θ_{min}, it would correspond to the minimal θ_n energy-time unit depicted between axis integers. The origin angle θ_n of that triangle at some very large n would reach the smallest resolvable portion of a cycle. Progression can advance no further than that "noise level" or "space curvature" within a cycle. Citing figure 1, $2\phi_n$ = θ_n is valid for all n. At very large n

$$\tan \phi_n \approx 1/s_n, \ n\theta_n \approx s_n, \text{ and } n\theta_n\phi_n \approx 1, \text{ so } n\theta_n^2 \approx 2 \approx s_n^2/n. \qquad (5*19)$$

Here 2n becomes unresolvable from 2n+1 terminating the process of n increasing to infinity. The relationship $2\phi_n = \theta_n$ also expresses as the identity

$$2 \tan^{-1}(1/s_n) = \tan^{-1}(s_n/n) = 2 \tan^{-1}[1/\sqrt{(2n+1)}] = \tan^{-1}[\sqrt{\{(2n+1)/n\}}]. \qquad (5*20)$$

Based on the orthogonality axiom the preceding analysis could provide the basis for this theory strictly in terms of energy-time units around the spiral. Unless the mathematics and graphics are incorrect, attributes of the spiral expressed in energy-time terms should elucidate substantial quantum phenomena. Other analysis presented herein such as relating prime numbers to directions ABCD also rest on a strictly mathematical foundation. It however appears that broader descriptions are possible in terms of an excess angle permeating spacetime. Whereas many preceding issues are direct mathematical consequence of the orthogonality axiom, (which will be shown to have several rigorous equivalents) interpreting exact meaning of the mathematics becomes more speculative. Venturing into these unclear territories often involves considerable conjecture but some attempt was deemed justified. In spite of the insecure vantage it seemed warranted that the various possibilities presented be discussed.

The appropriate dimensional units for Planck's constant expresses as $h \equiv E/\nu \equiv$ **(energy)(time)/(cycle)**. This constant is herein treated as dimensionless being the excess-angle encountered per cycle. That is verified through the following dimensional consistency argument. When h is dimensionless in $h \equiv E/\nu$, it makes energy and frequency synonymous, (differing only by a dimensionless constant). Then the small value of h =

(energy)(time)/(cycle) = (frequency)(a minute time-interval of a cycle)/(cycle) =

(cycles/second)(a minute time-interval of a cycle)/(cycle) =

(a minute angular fraction of a cycle)/(cycle) = (cycles)/(cycle) = (dimensionless) (a minute excess angle encountered per cycle) =

(a small angular fraction of 2π radians in each cycle) =

(The fractional excess angle of spacetime that would be encountered with each cycle of every wave) = (dimensionless, independent of frequency or time)

For example, such a constant might signify 1 picosecond of 1 megahertz = $(10^{-12})(10^{6}) = 10^{-6}$ = (dimensionless fraction of a cycle)

Or, 10 picoseconds of 0.1 megahertz = $(10^{-11})(10^{5}) = 10^{-6}$ = (the same dimensionless fraction of a cycle). This demonstrates dimensional consistency for h = $\Delta\theta$ expressed in angle per cycle form with $\Delta\theta$ and h being the same dimensionless fixed fraction of a cycle. They both depict the ratio of: excess space-time radians that prevail in curved space to 2π radians in Euclidean space.

It should be noted that $\hbar \equiv h/2\pi$ of quantum mechanics designates a further division by 2π as $h/2\pi \equiv E/\omega = E/2\pi\nu$. In curved space Planck's constant in the relationship $E = h\nu$ will be shown interpretable as [or equivalent to] a minute "excess-angle $\Delta\theta = h$" of a cycle, within each cycle of any frequency ν. It will be treated as an excess-angle per cycle and dimensionless in that context. That explanation endures being tantamount to an excess angle exceeding a complete conventional rotation within Euclidean space that must be traversed during each cycle. Curved or warped space can express in terms of an excess angle/cycle denominated $\Delta\theta = h$.

Each time the wave traverses $\Delta\theta$ the effect compounds, resulting in multiplication of repeated vectors that minute angle apart. That is the process of compounding triangles (or binomial vectors) that form harmonic waves. That smallest angle between vectors of the-spiral-that-creates-the-wave cannot progress indefinitely to zero as in the exponential function's definition. The outcome portrays a population of smallest repeated constituent "processes" (energy-time units or excess-angle compoundings), requiring orthonormal transition between integers of the accrued population and injecting inter-integer indeterminacy into the system. In physical reality, it suggests an actual miniscule excess angle (or equivalent degree of indeterminacy) $\Delta\theta$ must exist within spacetime obscuring change between some uppermost n_{max} and $(n_{max}+1)$. That smallest unit-of-"discernability" represents a population of excess-angle-increments that accrue in progression of the wave's traversing spacetime. The associated indeterminism equals the angle/cycle and the interval between integers of the population. The compounding entity characterizes as a population for which the space between integers of that population is disallowed. However unfamiliar, **the assumption of some excess-warpage-angle is always more inclusive than neglecting its presence since $\Delta\theta$ can always later be set to zero**. Specifically, when angle θ_n [$= 2\phi_n = \pm \phi_n$] of spiral triangles reach that angle at energy-time n_{max}, the wave seemingly can progress no further obscuring into total uncertainty. There,

$$2n_{max}+1 = s_{nmax}^2 = 4/\theta_{min}^2 = 1/\phi_n^2 \equiv 4/\Delta\theta^2, \text{ and } n_{max} = (2/\Delta\theta^2-1/2) \qquad (5*21)$$

becomes indistinguishable from $(n_{max}+1) = (2/\Delta\theta^2+1/2)$. $\qquad\qquad (5*22)$

The ½'s blur as imperceptible at that uppermost n_{max}, associating Planck's constant in energy-time/cycle, with an excess angle (with an indeterminate equivalent between integers), $h = \Delta\theta = \theta_{min} = \pm\phi_{min}$. This conjecture would yield a maximum "endpoint concerning discernability" of the wave. At some point of ever-increasing n, angle ϕ (extending toward $\pm$ one integer on either side of n_{max} in figure 1) would have diminished down to the intrinsic curvature of spacetime ϕ_{min} . Equation 5*21 and 5*22 would apply within a physically curved spacetime environment possessing indeterminate or excess angle $\theta_{min} \equiv \Delta\theta = h$.

That interpretation would equivalently restrict the maximum number of discernable (energy)x(time) partitions along the original Euclidean axis to

$$n_{max} = 2/\theta_{min}^2 \equiv 2/\Delta\theta^2 = 2/h^2. \tag{5*23}$$

At n_{max} the geometric construction for ϕ_{min} in figure 1singles out where a "straight line" for vector s_n can no longer be distinguished from a curved line on-the-sphere of radius $(n_{max}+1)$. These are all equivalent statements justifying a limit to the spiral at $n_{max} < \infty$. At that limit,

$$\Delta\theta/2 = \phi_{min} = 1/s_{1\text{-}max} = (s_{max} - s_{1\text{-}max}) = h/2. \tag{5*24}$$

Along with other vanishing parameters, the indeterminate gap in energy-time between largest adjacent tangential vectors

$$\sqrt{(2n_{max}+3)} - \sqrt{(2n_{max}+1)} \text{ becomes } 1/s_{max} = h/2 = \phi_{min}. \tag{5*25}$$

This equals the energy-time ambiguity of the uncertainty principle[6], an obscuration wherein $\sqrt{[(4/\Delta\theta^2)+3]} - \sqrt{[(4/\Delta\theta^2)+1]} \approx 1/\sqrt{[(4/\Delta\theta^2)+1]} \approx$

$$1/\sqrt{2}n_{max} \approx 1/s_{max} = \phi_{min} = \Delta\theta/2 = h/2 = \Delta E\Delta t. \qquad (5*26)$$

When representing real-world harmonic solutions of physics, this smallest resolvable angular gap $\Delta\theta/2 = 1/\sqrt{(2n_{max}+1)} = 1/s_{max} = \phi_{min}$ (5*27) **evidently equates to (Planck's half quanta)** $\equiv$ **h/2**, where h has dimensions (energy-time/cycle). **Planck's constant interprets as an indeterminate angle** $\Delta\theta$ **within each cycle of space-time, implying unification of the motional conservation laws.**

Energy, as $h\nu = \Delta\theta\nu = (\Delta\theta)$(cycles/second) $\qquad (5*28)$

is readily seen as the rate of traversing excess angle $\Delta\theta$ with time.

Momentum, as $h/\lambda = \Delta\theta/\lambda = (\Delta\theta)$(cycles/meter) $\qquad (5*29)$

depicts the cycles/meter rate of traversing $\Delta\theta$. Angular momentum or spin, as

$\pm h/2\pi = \pm \Delta\theta/2\pi$ describes the cycles/radian rate of traversing $\Delta\theta$. $\qquad (5*30)$

The motional conservation laws coalesce to become one law, **the conservation of** $\Delta\theta$ **traversal rates**. Moreover, this law would seemingly encompass a conservation of minimum indeterminism equaling the indeterminate angle $\Delta\theta$. Conservation of minimum indeterminism asserts that any experiment attempting to penetrate the veil of uncertainty between integers representing compoundings of excess-angles $\Delta\theta$ will fail, just as experiments will fail to surpass perpetual motion for energy conservation. This applies to physical geometry issues of our curved space environment as well as to cyclical harmonics. The population of smallest energy-time units is seen to derive from an angular curvature of spacetime, a minute angular difference between "complete" rotations in Euclidean space. That "action" as in quantum mechanics, constitutes the "noise level" of all cyclic processes and will be seen to inject a non-commutation into cyclic phenomena related to Planck's constant. A conclusion emerges that maintaining dimensional consistency for cycles in equations of physics is of paramount importance. Planck's constant h has cycles in its denominator even though (as will be shown) the non-divisible entity along the Euclidean axis can represent (energy)x(time), without utilizing the denomination of cycles. [Each unit comprising integer n will be shown as also interpretable to equal energy-time/quadrant, where a quadrant has unity extent being exactly one radian or **the absolute dimensional unit-base of rotation** (unity).] As described herein, constant h dimensionally quantifies per cycle (as the excess-angle/cycle). The variable (energy)(time) [or energy-time/quadrant] depicts the relevant smallest-unit parameter that transfers along or around a Euclidean axis. Within each cycle, a quantity h representing that energy-time, remains indistinct.

For illustrative simplicity, the unified wave has been depicted having its vectors n directed onto a conic surface. Constraining all vectors within a simple surface is unnecessary and represents only one possible mode of vector behavior. Tips of the spiral's vectors could conceivably follow a helical [or any arbitrary surface including an origin-centered sphere] surrounding the Euclidean axis. Examining a cylindrical surface yielding a helix for example, the initial radial vectors near n = 0 could start out with apex angle β close to, or at 90°. Upon reaching whatever cylindrical off axis radius, vector tips would thereafter follow that cylindrical radius for each subsequent cycle, each successive radial vector incurring a reduced angle β as necessary. Whenever accumulated angles θ_n "pass through" the apex angle β = arcsine $(2/\pi)$ the wave could repeat, possessing transient "four-quadrant-symmetry" there. A form of symmetry would reoccur at each apex angle $(1/2)$ arcsine $(2/\pi)$, $(1/3)$ arcsine $(2/\pi)$, $(1/4)$ arcsine $(2/\pi)$, etc. In these circumstances, **extent s_n of respective tangential vectors at n remain unchanged, as well as each θ_n.** The resultant wave would entail "circular polarization", each direction A re-occurring at a different angle relative to the helical axis. The product energy x time (as a portion of frequency x period) thus remains unchanged per cycle, (or for any cyclic segment thereof between radial vectors). This would be true considering any cylindrical radius, or for that matter any path locus taken by the radial vectors in circling the axis or circling the origin. The only requirement in forming the graphical result is that each such triangle of origin angle θ_n defines a plane, but each consecutive plane intersecting the origin can be at any angle relative to all other planes. The summed angles θ_n always accrue in repetitive 4-radian symmetry per triplet triangle independent of the planes those angles are in relative to each other. The resultant (energy x time) or (portion of frequency x period product produced) remains constant regardless of the "orbit" taken by tips of the radial vectors. That is why the interval between radial vectors, can take on intrinsic units of energy x time.

It furthermore indicates how, within the indeterminate shroud created by $\Delta\theta$, energy-time increments of the wave remain invariant irrespective of the amplitude, cyclic rate, or radial vector locus taken. That indeterminacy as to which course the vectors take insinuates non-localization of the wave, there being no definable "trajectory" for radial vector tips while overall consistency of uniform energy-time units prevail. That is why there can be no amplitude for the wave and why its energy proportions to frequency. Non-locality infers that vectors describing these solution-mode quanta "might subsist in multiple places at once". The orthogonality axiom model characterizes what is possible for a monadic population and the full range of angular directions remain "simultaneously feasible". These harmonic-mode-solutions with symmetry and wave properties are in effect "possibility waves". Just as for the equation $aX^2+bX+c = 0$, when $b^2<4ac$ does not exclude two simultaneous solutions based on the orthogonality of $\sqrt{-1}$, **so all orthogonal directions from each n to (n+1) remain "jointly possible" mathematically**. Specific vector locations remain undefined and can be anywhere. It suggests that of the totality of possible vector tip locations, only certain modes of the wave may qualify as "standing wave solutions" [having ABCD overlap]. Other factors not included in presentation here could be influential in when such harmonics occur. In this treatise straight-chord analyses containing approximation errors at small n is included for illustration reasons and to encompass the exponential function's definition at the n $\to \infty$ limit. It also allows demonstration of the four absolute radians resulting at that limit.

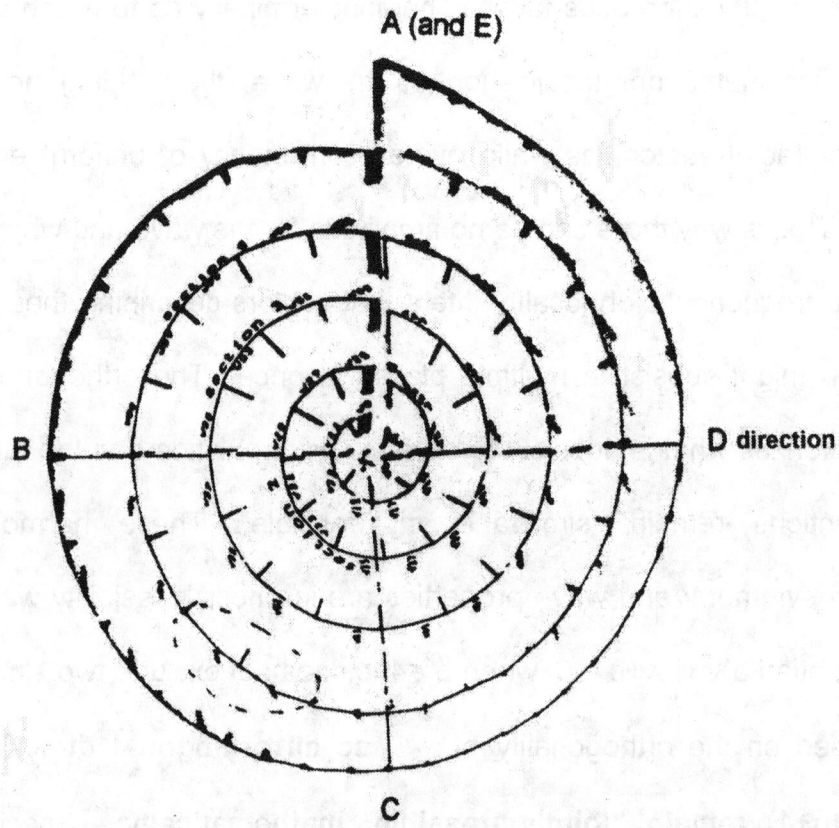

Figure 6. This is an actual photograph of the figure 3 transparency spiraled into a conic surface. The cyclic effect would entail a spin of unity, one rotation about the axis per triplet triangle. Each concatenated right-angle triangle could alternatively exist within any arbitrary plane going through the origin. Than radial vectors and tangential vectors could non-localize randomly, though rendered here to show harmonic-mode solutions on the conic surface. Each such resonance can represent a stationary state harmonic mode and because of the endless right-angle possibilities a multiplicity of such modes remain simultaneously feasible. Right angle triangles appear here with s_n following the curved surface, though calculated as a chord within a plane. That can contribute to approximation errors in the small deviation from four-radians/cycle at small n.

6) MODES AND IMPLICATIONS OF THE NEW HARMONICS

To evidence periodicity, non-localization and other harmonic properties, figure 6 presents an axial view photograph of the cone resulting from a transparency of figure 3 with directions A and E joined. As later demonstrated in figure 49, joining A and E of that same transparency after right-angle folding along directions ABCD yields a tetrahedron-like "*canonical form*" expression of the unified wave. [That canonical form entails curved tangential vectors on a unit-sphere' surface, simulated in initial discussions by the transparency of figure 3 wrapped onto a conic surface. Intersections of all radial vectors n with the surface of a unit-sphere about the origin specify radian angles of each θ_n triangle **by the segment-length on that sphere of the respective great-circle between adjacent radial vectors**.] In figure 7 two embodiments of the transparency are depicted, creased arbitrarily at outermost radial integers. Virtually any topological configuration of folds along radial vectors constitutes a **feasible mode**, which could also include folds at inner vectors. That generalized circumstance might visualize by "cutting out" the spiral strip containing all tangential vectors in figure 2 and randomly folding it along radial vector directions. It would signify how the wave could feasibly "advance all possible outcomes together" within 4π steradians. However, symmetry and interference prevail only when folds at inner triangles register at directions A, B, C, and D synchronously with those at outward triangles.

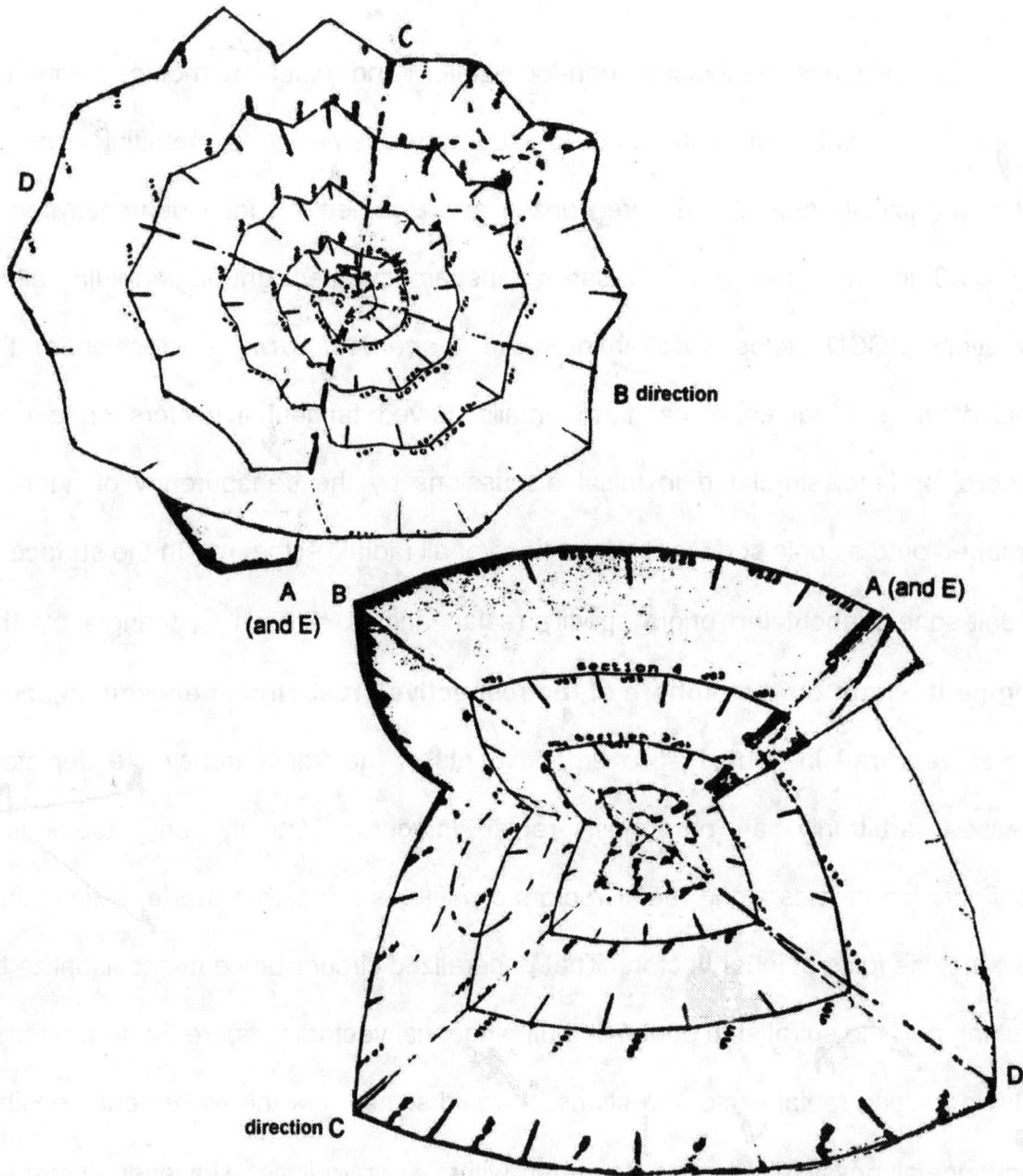

Figure 7. This shows two differing fold-examples of identical transparencies of figure 3 after directions A and E were joined to form a conic surface. In the upper photo arbitrary folds occur at outermost triangles. Folds exist only along directions ABCD for the lower photo. Folds are possible at all radial vectors of the spiral so the concatenated sequence could meander anywhere within a spherical volume (or anywhere within 4π steridians). Conditions allowing synchronous stationary-state standing-waves can only transpire whenever directions ABCD might align. Mode configurations delineating specific symmetries can allegedly represent wave manifestations of physical particles.

Because all of the waves possibly generated incorporate "rotating clock-hand" tangential vectors that "flood" the region, (which vectors allow phase cancellations), the QED sum-over-histories method of analysis[7,8] for progression around the axis can presumably apply toward establishing the probability of a terminal event from the effects of an origin event. Harmonics describing possibility can yield fixed standing-wave representation at specific synchronous conditions. Sequentially serial contingent routes, parallel routes, and concomitant routes to the outcome can all respectively add vectorially as "spinning tangential vectors", whose square root values must be squared to obtain physical probability. The square of any magnitude s_n proportions to the "effective throughput area for vector n", at radial distance n from the origin. Dotted circles in figure 3 indicate how such surface areas associated with each s_n intermesh throughout the entire wave. It is suggested that this circumstance may provide a generalized basis for quantum mechanical interpretation.

Discussion indicates how a monadic increasing population n, "auto-transforms" into a harmonic wave described by n and s_n. Such a cyclic property of natural numbers is inextricable from successive integer values for $\sqrt{(2n+1)}$. That ubiquitous phenomena engrained in the sequence of numbers provides the origin for unified waves. **Indeed, it is suggested the source of spacetime cyclic processes in general base on the four-radian absolute definition of a cycle derived from an inescapable property of numbers under monotonic increase**.

An idealized plane doesn't exist in curved-space physical-reality making traditional circular harmonics Euclidean fiction, while the absolute four-radians subsist no matter what curvature prevails in space. Angles θ_n can all assume different directions following (or irrelevant to) any conceivable curved surface or geometry, but their cumulative sum still always accrues to exactly four radians per each cycle as $n \to \infty$. That generality does not exist with the circle. It demands a plane to define a cycle and breaks down if the plane becomes a deformed "platter". For monads, directions surrounding the energy-time Euclidean axis (or spatial axes when analyzed) emerge as unfamiliar. They are suggested to depict the additional "dimensions" of string theory[9] where up to 12 total dimensions could apply. Again, these degrees of freedom are being called dimensions to minimize alternative semantic difficulties and perhaps aid visualization. The 4x2 = 8 new microscopic dimensions remain obscure in the macroscopic world. If the hypothesis is correct, for smallest non-divisible quanta comprising a population they remain as real as traditional X, Y, Z, and t. Under the orthogonality axiom the spiral forms as appropriate an "axis" for non-divisible entities as the Euclidean axis is for continuous mathematics.

It is conjectured that different harmonic modes of this unified-wave may comprise the coupling bosons of known forces, as well as other particles. What follows entails speculation about further harmonic mode configurations with different topologies and other values of spin. They characterize multiple possible "orbits through time" (and through X, Y, & Z for spatial axes) between an origin event

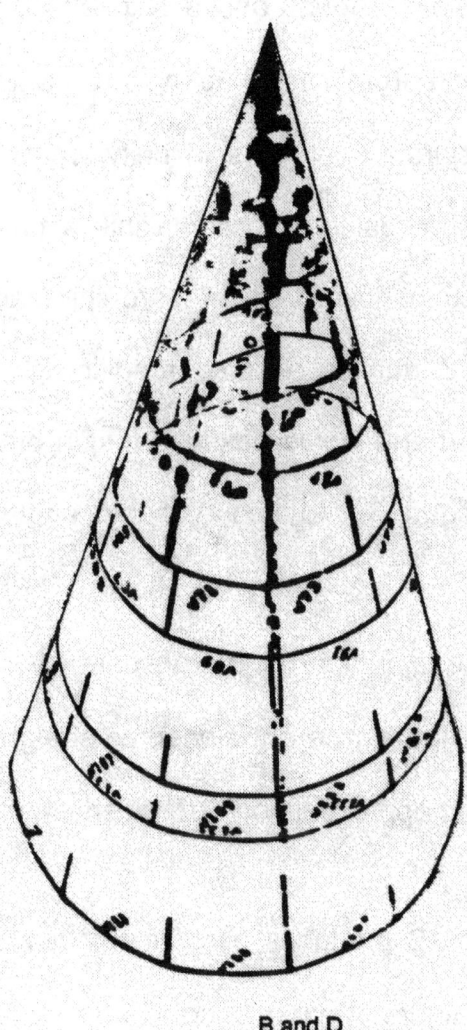

B and D

Figure 8. A transparency of figure 3 was wrapped around twice to form this cone having half the apex angle of figure 6. This harmonic mode would thus exhibit rotational periodicity characterized by a spin of two.

and terminus of that event. Besides the delineated case of spin ± unity illustrated via one cyclic rotation per triplet triangle, generalized embodiments with spin 0, 2, and ½N as described below are possible. Though only a segment of spiral is portrayed in each case, symmetries extend from n = 0 to n_{max} (or toward n = ∞). Two superimposed counter-rotating spirals each of spin unity would form a zero-spin entity because zero net circulation results. Since multiple simultaneous solutions would seem always possible, such "lowest state" zero-spin-pairs might seem a likely-solution condition. Figure 8 depicts a spiral rendition that undergoes two rotations per triplet-triangle-repeat and would possess a spin of 2. This effectively double-wraps a transparency of figure 3, into a cone of apex angle

$$\Gamma = \beta/2 = (1/2)\sin^{-1}(2/\pi). \tag{6*1}$$

It might signify a graviton. [These wraparound transparency simulations utilize curved s_n vectors. They can alternatively be envisioned as straight-line cords within the conic surface introducing slight approximation error at small n.]

The mode model of figure 9 produces a symmetrical wave of spin ½ by cascading two transparencies of figure 3, each "accordion folded" to occupy up to π cyclical radians, rather than 4 radians. Up and down fold-directions are relative to the plane of the page and "depth" of the folds can take on a large range of values. Here, periodic overlaps of respective ABCD directions occur between every **alternate** perfect square of (2n+1). Two sub-embodiments are feasible for repetitions at either odd values of P or even values of P. Higher order incarnations could overlap at any spacing in P. Other fold arrangements allow additional modes

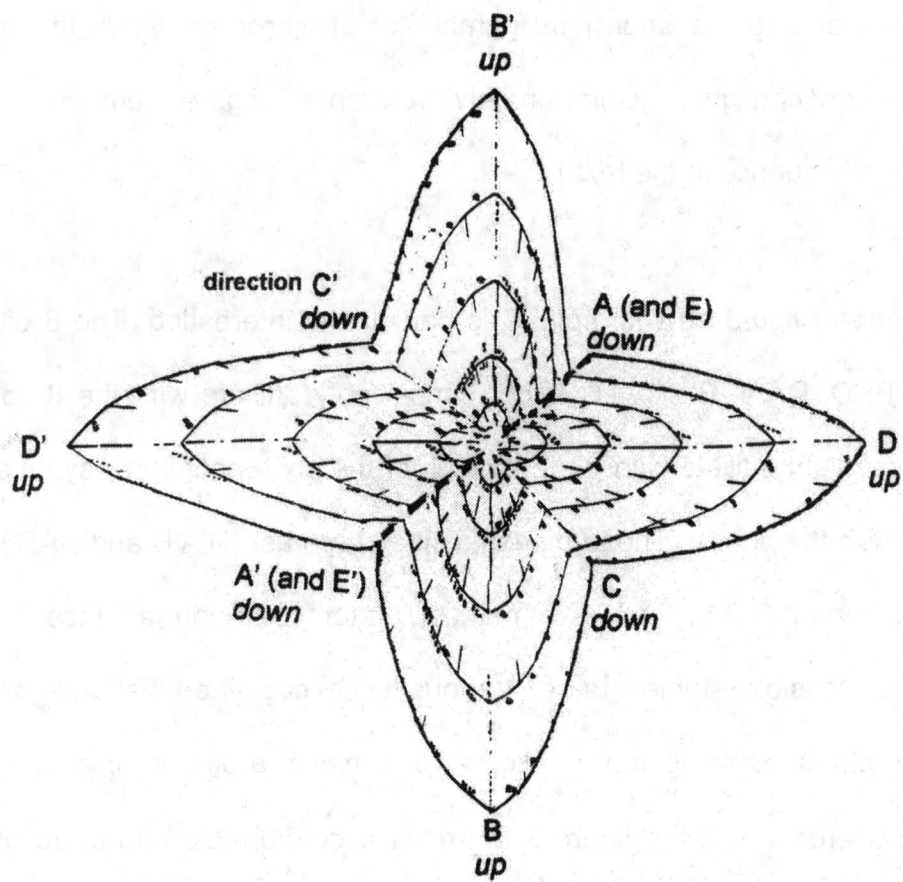

Figure 9. Two transparencies of figure 3 were folded so as to each traverse π radians rather than 4 radians. They were then joined at matched points along directions A-A' to form a single folded transparency with spin of ½. [Each fan of figure 3 would contribute ½ cycle of rotation.] The model could then be "accordion squeezed" or "flattened and expanded" into a smaller or larger angle-at-the-origin figures. It demonstrates the flexibility toward variations of same-spin symmetrical-mode configurations. Like other transparency-figures herein, it was difficult to arrange and photograph axially while permitting its precise symmetries to emerge.

with symmetry. Besides extended-flat-planes leading to sharp folds at ABCD, harmonic modes are feasible exhibiting "more-gradual-directional-changes", including those at non-quadrant locations. Computer generated figures 19 through 31 provide a few examples of such modes. It can be seen how "electron cloud" type distributions are among the enormous number of synchronous ways the unified wave "spiral", can configure. [Unfortunately, referenced figure numbers will no longer remain in sequence in the text.]

Symmetry of figure 10 with spin ½ is particularly interesting. The 8 division directions A, B, C, D, A', B', C', D', might conceivably affiliate with the 8 sorts of gluons. Quarks might affiliate with permutations of the six repetitiously symmetrical planes linked with the figure. These planes display here as: (D'A'B and DAB'), (BC and B'C'), (CD and C'D'). Such symmetrical patterns among all those totally possible signify feasible stable states. Various such cascaded half-integer spin modes might characterize isotopes, etc. To achieve arbitrary spin $N/2$, two cascaded transparencies of the figure 3 fan must accordion-fold into up to (Nx2π) radians of traverse. Figure 10 further displays a condition that might effectuate a longitudinal component wave. Vectors between directions D'A'B, and D A B', remain within the plane of the paper while advance occurs "up and back in the propagation direction" during respective intervals B C D and B' C' D'. This type of electromagnetic mode might portray virtual photons.

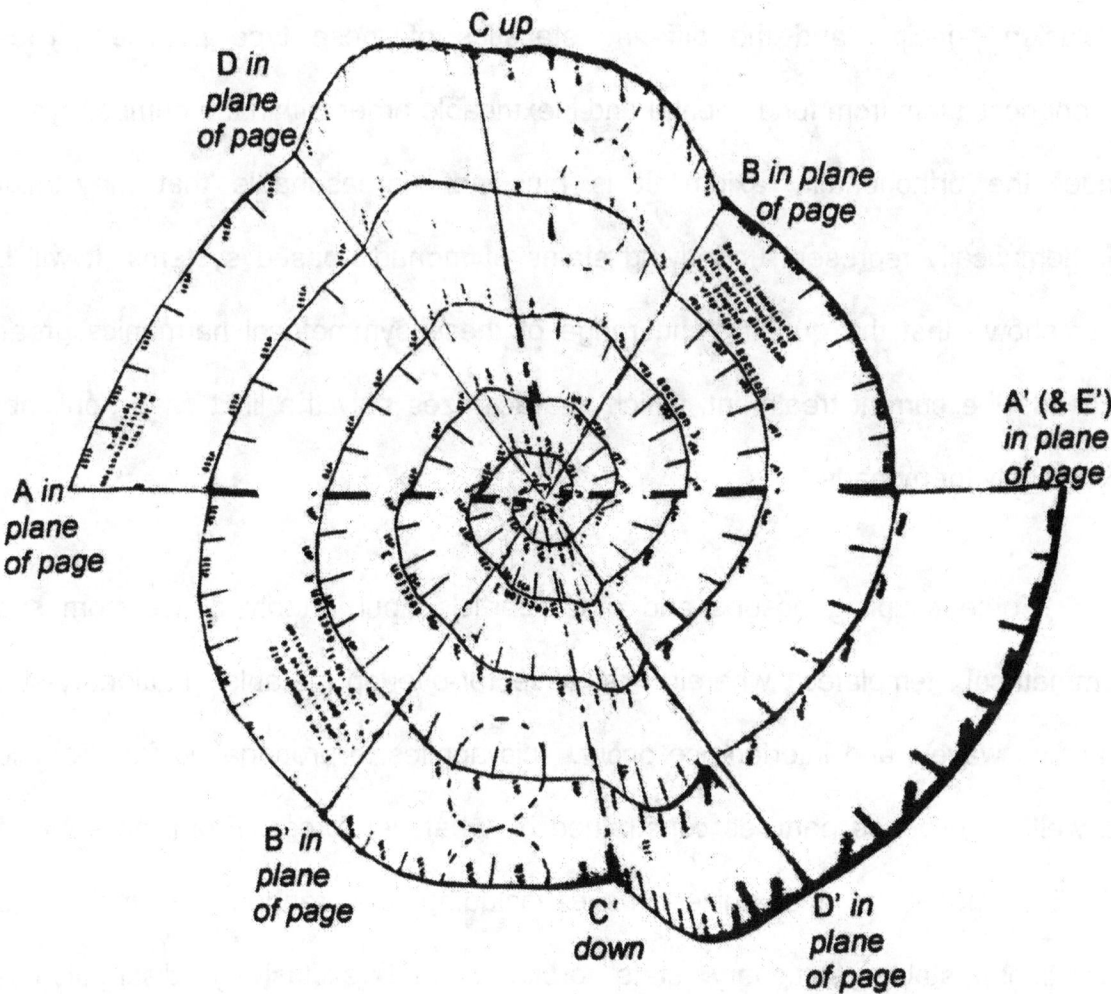

Figure 10. In this mode model, directions D' A' B, and D A B' remain within the plane of the page. Direction C is folded upward from the page with C' folded downward. This could plausibly represent a longitudinal wave with C advancing forward and C' progressing backward. The 6-fold different–embodiment permutations, identifying six separate compartments, partitioned via 8 symmetry lines might speculatively associate with nuclear quarks and gluons.

"Fermions" with half integer spin would comprise from two cascaded figure 3 "fans" coiled about an odd number of times. Integer spin "bosons" by contrast comprise from a single fan coiled about an integer number of times. The two dissimilar-topology categories indicate a basic distinction that might relate to the exclusion principle and the differing statistics of these type particles. These resonances stem from fundamental and inextricable order within the number system under the orthogonality axiom. It is thus not unreasonable that they might mathematically represent underlying states of monadic-based systems. It will be later shown that the quantity and range of these symmetrical harmonics greatly exceeds the current treatment, which characterizes only the limit case applicable exclusively for monads.

Force coupling bosons and other particles purportedly derive from such symmetrical templates wherein radial-vector-overlap, stable stationary-state standing waves, and interference occur. This applies to propagating free particles as well as to bound particles constrained by whatever forces. From the set of all possible modes of the unified wave including the random unsynchronized conditions, stable stationary state orbits should exclusively distinguish at overlapping conditions of ABCD symmetry. If superimposed presence of prevailing potentials, force fields, or energetic restrictions further constrained vector locations, such "standing-wave" overlapping periodicity could sustain only within resulting "potentials" defined by those constraints. Those energetic solutions would represent possible solutions to each respective constraint.

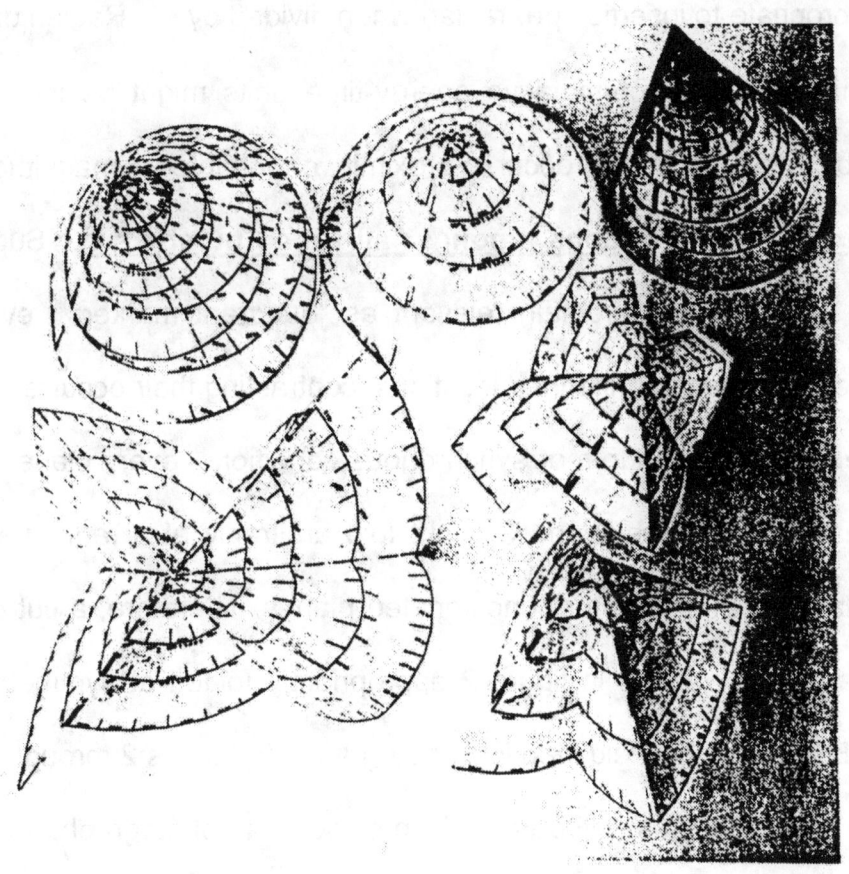

Figure 11. This photograph shows various transparency models used to demonstrate different modes derivable from figure 3. A quasi-infinite number of mode configurations are feasible but only those where directions ABCD of figure 3 can align would be synchronous.

From figure 3, it is seen that each subsequent synchronized quadrant injects one additional energy-time unit into the spiraling process. **In analogy to matrix mechanics, that "deformation" introduces a non-commutation of one energy-time unit per circulatory radian [or ih/2π = iℏ = iΔθ/2π], compared to classical progression in Euclidean space**. Here h and Δθ signify (energy-time/cycle), remaining appropriate to insertion per radian when divided by 2π. Radian dividers A, B, C, D, (where the non-commutative energy-time units might be interpreted to "insert"), will be shown to always occur in proximity of half-integer magnitudes for s_n and jointly **at quadrant partitions amongst all prime numbers n**. Such radian dividers thus mathematically remain relevant as "quadrant markers", even when vectors are non-localized in undefined locations, contrasting their occurrence on the conic surface. For either random or synchronous conditions, every plane hosting a right-angle triangle in figures 2 and 3 might fold arbitrarily at a radial vector in a manner unrelated to prior or later concatenated planes. Therefore, a cut out spiral strip of the tangential vectors in figure 2 appropriately folded at symmetry divider directions A, B, C, D, could yield identical photographs to figures 2 through 10, (plus many other configurations). Figure 11 provides a photograph of various transparency models used to create the presented graphics. The hand cutting, pasting, folding, joining, and photographing of these spiral models makes their exhibition less precise than if they were modeled and printed via 3d-software as in later figures.

7) MODES FROM PARTIALLY MONADIC ENTITIES.

It has been suggested that for monads, integer natural numbers n depict concentric spherical surfaces of radius n about an origin. This manifests under the orthogonality axiom where a right angle from the tip of vector n to the outward spherical surface at (n+1) could occur in any tangential direction to the origin-to-n vector. That right angle is possible anywhere in the two-dimensional plane tangent to the tip of vector n. Previous spiral examples discussed cases where all consecutive tangential vectors were constrained to remain in the same plane. That constraint was unnecessary but chosen for visualization purposes. Consecutive tangential vectors can be in virtually any direction relative to each other. In general, after only a few transitions from the origin out to radial extent n, specific location on that spherical surface of radius n would become totally undefined. Possible locations for the tip of vector n could be anywhere on the spherical surface at radius n because a "zigzag three-dimensional path" through each lesser integer radius is feasible. Only specific modes that progressed through concentric spherical surfaces while having synchronous overlapping ABCD directions would yield harmonic waves. Those synchronous waves portray potential standing wave harmonic solutions in the midst of the totality of all possible "random" routes to any resultant vector n. The natural numbers themselves provide no information, constraint, or clue about how transitions might occur between those concentric surfaces. For monads that possess missing internal information, transit through each "uninhabitable" annular gap between integer-radius spherical surfaces would occur orthogonal to whatever origin-to-n direction prevails. **Each spherically-outward unit-advance thus entails maximum independence from, (or is linearly independent of) the prevailing origin-to-n direction, thereby invoking the orthogonality axiom**. Traversal through each annular transition region between surfaces n and (n+1) would then have minimal directional-component related to n. When enumerating non-divisible quantum entities of this type, passage through the inter-integer gap should exhibit maximum autonomy in relation to n.

But what if the entity whose energy-time was being enumerated were "**partially divisible**"? Postulate population members with some discernable attribute or having a fragmentary shroud. At an extreme of that possibility, suppose the relevant enumerated entity depicted large divisible macroscopic objects. In such cases one would expect the transition route to occur entirely along or very close to the conventional Euclidean axis. In effect, a <u>range</u> of "**inseparability**" exists extending from *totally* **non-divisible monads** to **exceedingly divisible large rest mass objects**. Harmonic modes for the extreme monadic case would entail maximally orthogonal transition directions and likely portray discrete indivisible quantum particles like photons. They would have zero rest mass, propagate at light-speed and have unity spin. The previously derived unified waves could plausibly accommodate such harmonics. Energy-time enumeration of large rest mass objects would by contrast, involve small or zero propagation velocity compared to c. It is thus not unreasonable that a component of the transition along the origin-to-n direction (along the Euclidean axis) might associate with divisibility while the transition component orthogonal thereto affiliated with non-divisibility. Of course the designation "divisibility" is applied somewhat metaphorically here to describe an unfamiliar attribute of a new parameter. The "degree of divisibility" quality could also be presented in terms of degree-of-directional-correlation between transition and n, or the penetrability of the shroud, etc. What is implied is, a ***degree-of-orthogonality*** can apply more generally for transitions; a proportional-type property providing commensurate reduction in the effect of the monadic shroud. Orthogonal transition components maximize when the shroud is impenetrable while they minimize for cases approaching Newtonian continuity. The former yields resonances typical of zero spin, discrete, non-divisible, photon-like particles, whereas harmonics might tend to vanish for the latter case.

Figure 12 develops graphical construction that accommodates **graduated degrees of orthogonality and divisibility**. It elucidates a spiral formation method where transition s_n can have both an **a**xial and **o**rthogonal component respectively designated as s_a and s_o. Axial component s_a will always comprise a fraction of the unit-distance between each pair of integers n and (n+1) and the remainder of that distance defines as $\xi = 1 - s_a$. Each inter-integer advance can be envisioned as progressing diagonally from n away from the origin along s_n to (n+1). Alternatively, by way of components of s_n, it can progress axially from n to $(n+s_a)$ and then orthogonally along s_o to (n+1). Here, component s_a of the transition is being allowed in the axial direction attendant with diminished

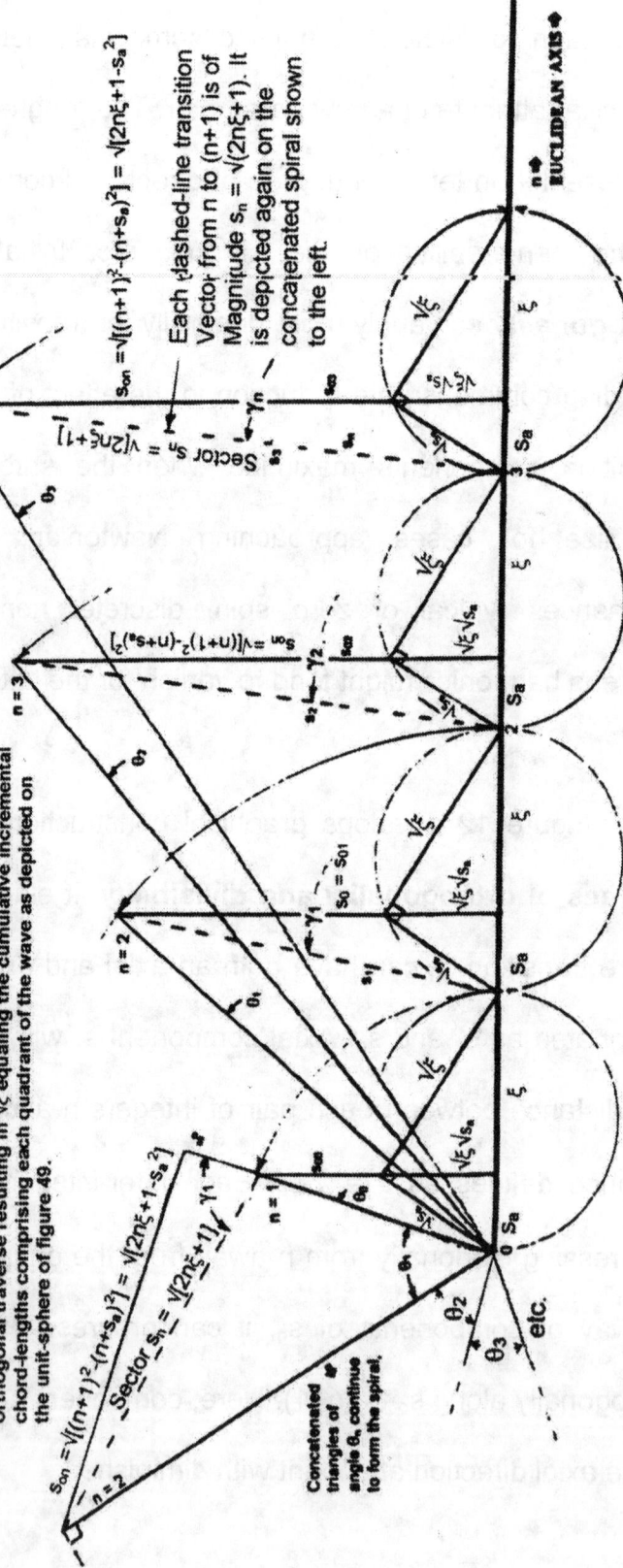

Figure 12. This general-case graphic-spiral-construction depicts **orthogonal progression** s_o to occur at an arbitrary axial point s_a [≡ $(1-\xi)$] between each pair of integers n. A "partial degree" of orthogonality results that can vary from maximum at $\xi = 1$ to a minimum at $\xi = 0$. Every transition-vector progresses between n and (n+1) as the oblique-vector $s_n = \sqrt{(2n\xi+1)}$, [or via its components s_a and s_o]. Here the orthogonal component of each transition s_n equals $s_o = \sqrt{(2n\xi+1-s_a^2)}$. It forms an origin-angle-triangle with sides (n+s_a) and (n+1). The spiral on the left derives from those concatenated origin-angle-triangles constructed on the right for reference. Triangles forming the spiral have consecutive origin-angles θ_1, θ_2, θ_3, etc. Two additional vectors $\sqrt{s_a}$ and $\sqrt{\xi}$ can be drawn inscribed in a hemi-circle between every pair of integers. The inscribed triangles of sides $\sqrt{s_a}$, $\sqrt{\xi}$, 1 so formed, have heights $\sqrt{(s_a\xi)}$ and area $\sqrt{(s_a\sqrt{\xi})}/2$. Expressions are included for the relevant vectors. Physical interpretations and harmonic mode consequences for all these vectors will be developed.

Parameter ξ specifies that portion of the Euclidean axis "bypassed" by orthogonal traverse resulting in $\sqrt{\xi}$ equaling the cumulative incremental chord-lengths comprising each quadrant of the wave as depicted on the unit-sphere of figure 49.

orthogonal component s_o. Accordingly, a triangle of sides, n, s_n, and (n+1) is formed

for each transition where s_n can be oblique to the axis. On the far left of the figure,

two concatenated triangles exemplify how progression from n = 1 to n = 2 would

situate in materializing the resultant spiral. The triangle from n = 1 to n = 2

represents the same (congruent) triangle as the one along the axis having origin

angle θ_1. The spiral forms by concatenation on the left from consecutive triangles

depicted with respective origin-angles θ_n on the right, (as when going from n = 1 to

n = 2 for example). The continued remaining triangles for 2→3, 3→4, etc. would

graphically construct similar to what was previously described. The spiral rotates

angularly on the left by linking triangles shown along the axis having origin-angles

θ_1, θ_2, θ_3, etc. Those triangles have respective vectors s_n labeled s_1, s_2, s_3, etc.

Every axial-component s_a at n has respective $\qquad\qquad s_o = \sqrt{[(n+1)^2 - (n+s_a)^2]}$.

Each triangle's diagonal s_n must therefore be

$$s_n = \sqrt{[s_o^2 + s_a^2]} = \sqrt{[(n+1)^2 - (n+s_a)^2 + s_a^2]} =$$

$$\sqrt{[(2n+1 - 2ns_a + 1]} = \sqrt{[2n(1-s_a) + 1]} = \sqrt{[2n\xi+1]} \qquad\qquad (7*1)$$

In accommodating *degree-of-divisibility*, the spiral incorporates an ξ less than

unity **and <u>ends up with diagonal transition lengths $\sqrt{[2n\xi+1]}$, instead of $\sqrt{[2n+1]}$</u>**

<u>applicable for the totally orthogonal case when ξ = 1</u>. That constitutes an

elegant simplicity engrained within the natural numbers encompassing

fundamental enumeration of "partially divisible" entities as distinguished by

ξ. Figure 12 displays geometric construction for a generalized method of counting similar entities having arbitrary degrees of divisibility. It extends from continuous progression along the Euclidean axis for $\xi = 0$, to totally orthogonal transitions when $\xi = 1$. The case of $\xi = 0$ would represent zero non-divisibility (or total divisibility), while $\xi = 1$ (where transition orthogonality maximizes), depicts maximum non-divisibility. The entire range $0 \leq \xi \leq 1$ covers the gamut of intermediary possibilities. **This "modifier" ξ on the efficacy of the information shroud surrounding each** **population member constitutes an unfamiliar variable inextricable from** **natural numbers and enumeration procedures**. It delineates that fractional amount between all integers where orthogonal progressions begin. It will later be shown that harmonic symmetries intrinsic to the previously examined $\xi = 1$ case preserve for all values of ξ. Values of n where quadrants of the wave occur also remain invariant with ξ. Those quadrant directions ABCD of figure 3, later shown to never associate with a prime value of n, correspondingly retain that feature throughout all ξ. Arbitrary values of ξ do not alter the intrinsic 4 radians/cycle and 1 radian/quadrant synchronization generated by natural numbers. In ξ, an additional identifier emerges that characterizes all basic waves. Its effect on allowable synchronous modes will be later elaborated.

The issue arises as to whether ξ is rational or can be irrational. **If ξ were** **rational, than all irrational values that might be encountered along the** **Euclidean axis would be bypassed by tangential traverse.** All radial extents of relevance would be an integer n plus s_a. The axis-segment ξ between every pair of integers effectively becomes "avoided" by orthogonal progression of s_o. Now ξ's resultant mathematical role in $s_n = \sqrt{(2n\xi+1)}$ manifests exclusively as a multiplier of all even numbers. Its algebraic sum with s_a is always unity suggesting the point of partition it defines is a ratio or rational number. Moreover, since $\xi - s_a = \xi^2 - s_a^2$ because $s_a + \xi = 1$, the difference between their squares would have to be irrational for them to be irrational. Their irrationality could not stem from being an irrational square root of a rational number. While these arguments do not provide definitive proof, ξ will hereafter be assumed rational hinging on later justification. Though the range of ξ displays important properties of numbers and waves, discussions will continue to focus on the $\xi = 1$ case before analyzing waves of divisible entities.

The traditional process of counting is synonymous with a straight-line scale of natural numbers or an axis of possible integer values. That selfsame enumeration process **implemented under the orthogonality axiom** yields unified waves with their interesting attendant traits. No other distinctions are necessary between the two methods of counting. Just as enumeration and natural numbers remain inextricable so unified waves and natural numbers are comparably inextricable. Actual counting procedures merely encompass the range from the familiar linear registration of sequential natural numbers to the unfamiliar unified waves embedded within those numbers. Analysis herein simply exposes natural wave properties of the number system when interpreted under more general conditions.

8) UNIFIED WAVES AFFILIATE WITH POPULATIONS AND UNCERTAINTY

New theories assert contention for scientific minds in accord with the paradigms those theories reshape. Often the new theory overturns a fundamental assumption previously deemed appropriate. It is sometimes difficult to recognize a priori which organizing hypotheses underlie traditional paradigms however. **It should be recognized that mathematics is more encompassing than physics, which occupies only a limited sub-set of the total domain of mathematics**. One must therefore be careful that in the application of mathematics to physical theories, inadvertent encroachment into the excess (non-physical) territory of mathematics does not occur. While mathematically compounded limit processes might legitimately extend to infinity or zero; they may misrepresent reality upon entering that domain. An unverified postulate commonly embedded within the structure of scientific non-relativistic relationships is that space "occupies" *exactly* 2π radians or 4π steridians. Now π can be described to billions of places while still representing exactly zero percent of the information it contains within, the same percentage as if we had assigned $\pi = 3$. A subtle impreciseness in the structure of space may thus permeate certain foundation paradigms, like the presumed applicability of Cartesian Coordinates, Euclidean geometry, $\sqrt{-1}$, etc.

General relativity tells us spacetime curvature exists, but traditional mathematics has not provided a simple substitute for the retrospectively inappropriate rectilinear coordinates often used to represent it. Mathematics within those coordinates may differ from actual physical behavior the mathematics alleges to represent. The premise of "infinite division" down to ***nothingness*** is another refutable substructure corroborating Newtonian calculus, differential equations, Euclidean axes, etc. Though unproven, or even known to be wrong, these conventional modeling tools and their underling assumptions lodge within accepted foundations of physics as described through mathematics. For the circumstance of compounded vectors with respect to actual space, this theory challenges both those suppositions. It asserts the orthogonality axiom, compelling smallest non-divisible compounded entities to require different dimensional echelon then large macroscopic objects. Gradation between those extremes is also possible. Conclusions of the theory then become consequence of that axiom, reshaping traditional paradigms that evolved under assumptions of un-curved space and Newtonian continuity. Variables compounded from indivisible-repeated-constituents employ a different "route" for getting from integer to integer, which route becomes an "orbit around", rather than a straight-line course along the direction of progression. Primary predictions from these ideas should hopefully constitute reduction or elimination of dilemmas hitherto existing under prevailing concepts. The theory is alleged to predict empirical observations aberrant under existing theory, plus a few unfamiliar forecasts.

Non-locality is not often characterized mathematically in descriptions of classical physics. In this theory, it is achieved by the "answer" allowing many-possible "simultaneous-locations". What is called "the answer" may seem unusual because conventional answers under Newtonian continuity pinpoint location exactly. Those conventional interpretations had pre-assumed the validity of infinitesimal continuity for compounded variables, allowing the modeling differential equations to apply down to nothingness. Those describing equations everywhere depict a continuous trajectory or locus of positions for the answer, and that may be acceptable for macroscopic representation. It is unacceptable for variables compounded from smallest indivisible entities for they must somehow progress from integer to integer without directly traversing fractional values in between. We cannot cite or consider the centroid or center-of-gravity of non-divisible articles for example. No sub-properties are definable for them. Their description does not portray accurately as a moving point. That variable's route between integer-units of its minimal-repeated size (or constituent population members) cannot pinpoint onto a one-dimensional path like a Euclidean line axis. Between integers, the route becomes nebulous, ill defined, it assumes non-local properties.

Under infinitesimal continuity, dt or dx allowably reach zero permitting continuous variables to follow each point in t or x along the Euclidean axis. Positions are perfectly definable. Non-divisible compounded variables can be viewed as an ensemble of discrete entities, like conjugate energy-time units whose population number entail specific integers absent fractional values between them.

One might alternatively say either the population increases from integer to integer, or, only integer values of it (e.g. energy-time) are allowed or tangible, with orthogonal progression between those integers. A **dramatic difference exists** between a compounded variable comprised from constituents of zero incremental extent and one whose tiniest sub-division is however minute, **<u>but not zero</u>**. For the former case, the orthogonal component of progression per compounding reaches zero whereupon for population entities however small, but not zero, an entirely new set of dimensions appears. For indivisible entities, the transition path takes on an orthogonal component and leaves the traditional Euclidean axis, "circulating" around it. **An infinite range exists between indivisibility however minute and some divisibility. <u>The indivisible case invariably depicts a higher-order more-comprehensive system because the orthogonal component can always mathematically go to zero at one limit</u>**. The circumstance with minimal-indivisible-increment encompasses that infinite range **including** the limit point of zero size at the end of the range. The exponential of infinitesimal continuity embodies only the zero-size end-point at the limit. The two cases literally describe two different worlds.

Absent a compounding limit process and the orthogonality axiom, the distinction between reaching explicit zero and some smallest size might entail only minor contrast, as for example between finite differentials and infinitesimal derivatives. A differential represents an increment along the directional axis of the variable, like a small zero-offset shift in the variable, which approaches zero under Newtonian continuity. By contrast, the orthogonal transition of a compounded population variable embeds in **exponentiation**, or an **angular** offset increment that minimizes towards zero. It is not a linear translation of addition or subtraction. It produces a compounded (multiplied) minute angle. A population variable as referenced herein invariably signifies **number of angular compoundings, multiplications, or exponentiated products within a limit process**. It will turn out that the 4-quadrant harmonic symmetry of the unified wave allows invariant transformations. The limit process references the previously mentioned binomial vectors; namely right-angle triangles with their origin angles as depicted in figures 1 and 2. As exemplified in the exponential function's definition, [lim n →∞ of $(1+i\omega t/n)^n$], is the limit of multiplied right-angle triangles having sides [1, $\omega t/n$, and $\sqrt{\{1^2+(\omega t/n)^2\}}$]. It provides a <u>compounded *angle*</u> (or triangle) whose incremental size goes to zero as n → ∞. That produces **angular** progression around a spiral just as in the unified wave, a process that only limits at the unit circle, zero increment by zero increment.

By contrast, the smallest increment residue of this theory limits to an angular deviation by amount $\Delta\theta$ of 2π radians, in a "circle" that comprises exactly 4 radians of nutation at the base of a cone. Rather than limiting at absolute zero, Planck's constant will be seen to represent the limit of angular and spatial resolution. Every angular advance equal to $\Delta\theta$ part of a unit-circle then depicts a population member (a compounding). It will be seen there are $2/\Delta\theta$ of them comprising a cycle and $2/\Delta\theta^2$ in the composite unified wave of $1/\Delta\theta$ cycles. For such a compounded population variable, the non-divisible residue increment (the monad) effectively rests in the **exponent** of the describing equation rather than as a linear incremental shift along a directional axis. In de facto curved space that residue increment can not go to zero. The exponential function thus necessitates infinite exponentiation due to $n \rightarrow \infty$, thereby misrepresenting finite physical reality. Unified wave interpretation actually **removes a singularity** harbored within the exponential function and its idealistically angular depiction of space. That singularity is evident from the exponential's definition with infinite exponent n in equation 4*3. Physical and mathematical consequence becomes endless exponentiation instead of limiting to $n_{max} = 2/\Delta\theta^2$, which accommodates the symmetry of curved spacetime delineated via the spiral and $\Delta\theta$.

Human beings generally deal with continuous macroscopically-divisible entities. Our variables and mathematics, particularly statistical interpretations, utilize fractions and decimal notation that allows digits after the decimal point. Such statistical or fractional representations do not describe a real entity but involve a hypothetical ploy (a model) standing for an ensemble of things in some averaging way. The method by which variables "devised-from-repeated-compoundings" progress between integers receives little attention. Rules for functional advance of such variables should take precedence over the academically familiar concepts of continuity. Continuity does not exist between discrete smallest constituents or their integer numbers due to the shroud that hides the innards of each smallest-constituent population-member. That indeterminism, as missing data about how each constituent might be further subdivided, embodies a different information-form than population number. Such distinction between the integer and inter-integer-interval mandates orthogonality and vector expression between integer transitions. The concept of transition between integers is unfamiliar because it is generally assumed all possible fractions of the interval are allowed as consequence of Newtonian continuity. However for monads, inter-integer fractions are not allowed, which brings the transition into relevance.

Often, "linear continuously-causal phenomena" mathematically interpret as totally deterministic within a Laplace-type mechanical-universe. Yet, for the case of populations treated under the orthogonality axiom, classical type solutions are shown here to also bring about non-locality and uncertainty with possible randomness as part of the solution. Those properties become "the **caused effect**" of the monad's information shroud. That missing information eliminates or alters the functionality of available space between integers. **However subtle its presence, indeterminism is inextricable from the interval between integers of a population**. Stipulation of repeated smallest-unit-things invokes unavailable fractional-value information between integer numbers of those things. **Non-divisibility in constituents that comprise a variable places a gap or breach in the inter-integer progression continuity of that variable**. That gap manifests as uncertainty in the behavior of such population variables. If a given set is described by a line boundary, registering the number of **dynamically-changing monadic population members in that set will always entail uncertainty equaling the**

"extent" of one monadic population member. It is an unavoidable requirement. **Ambiguity whether a member in transit is in or out of the set will always exist to the extent of its size, to one integer-width of the population**. The system becomes characterized by the multiplicity of all possible states represented by that band of unit-uncertainty between integers. That in turn becomes the multiplicity of all states describable within concentric-spherical surfaces of integer-radius. The inter-integer interval masking the interior of smallest repeated constituents unavoidably incorporates missing information and requires vector operators to help straddle that missing data.

As the spiral formation shows, for a population variable devised of compounded monads, the information void between integers of the population transforms into an equivalent information void around the axis that could register those integers. Being prohibited from taking a fractional-value course directly along the axis line, population growth traverses a nebulous multiplicity of uncertain routes in "environs" of the axis. This engenders intangible, non-localized "parallel path orbits" surrounding the axis, or surrounding the origin. **Multiple solutions become simultaneously possible**. A small few of those solution modes can experience periodic ABCD overlap with resulting harmonics and interference, while other modes remain random or undefined. Interrogating a suspected harmonic mode to determine its state by external means will typically destroy the self-cadence and interference-symmetry of the wave. It will be seen that cadence depends upon the cumulative sum of an enormous number of different angles θ_n repeatedly reaching

identical rotational positions per cycle. The angular digits of accuracy required to sustain ABCD overlap at large n becomes **unbelievably great** and any perturbation whatsoever can alter the net accrued angle and the harmonic result. That "mathematical insecurity" implies the indeterminism of originally missing information between integers is not surmountable via interrogation methods; the route taken (or the state of the system en route) will inevitably be veiled. Precarious "standing wave" harmonic solutions can vanish if perturbed by interrogation, allowing an outcome to be determined by the process of observation.

The inherent uncertainty of the shroud can manifest in various different ways but will always be present in one equivalent form or another. If the information was missing at the beginning due to the shroud of each population member, there exists no way it can be recovered anywhere, or in any form along the course of events, and such attempts are doomed to fail. The process might be interpreted in the same context as a conservation law, conserving equivalents of that minimum existing uncertainty between integers. One such effect of interrogating the system is collapse from interference conditions to randomness as consequence of disturbing the accrued angles. Indeterminism from the shroud introduces angular accrual through orthogonality, permitting the mathematics to achieve exact but incredibly precarious synchronization. Undefined location of the wave's vectors enable a "mathematical method" to implement "conservation of minimum indeterminism" when interrogation is attempted to establish the system's state. The detailed mechanism by which such conservation occurs may sometimes appear difficult to fathom in the same manner as the predicament to visualize the exact proceedings by which different forms of energy conserve. For the Euclidean energy-time axis one smallest energy-time unit of conserved indeterminism corresponds to: the extent of each monadic shroud, the interval between integers n, the annular space between concentric spherical surfaces, the equivalent non-localized surrounding microzone region bypassing the axial region, the influence of one cyclically traversed excess-angle. Traversal angle $\Delta\theta$ registers as the uncertainty principle or a quantum of action. Each such equivalent constitutes a magnitude related to Planck's constant. Since all of these conserved forms bear congruence, the question of which is the "source" becomes somewhat moot. From the standpoint of a physical mechanism, it would seem that a cosmological excess angle $\Delta\theta$ would most likely constrain the other equivalents and that has been utilized in these analyses.

9) CURVATURE OF SPACE INTERPRETED AS A MINIMAL EXCESS ANGLE

What can a minute excess angle in spacetime physically mean? When zero excess-angle exists, space would be "Euclidean", without curvature. The angle's magnitude specifies the degree of curvature or "warpage", with excess (or deficient) signifying direction. Under General Relativity, Euclidean interpretation of space-time is knowingly inappropriate and a minute excess angle is the here-alleged manifestation (albeit unfamiliar) of prevailing curvature. Such an angle would be angularly compounded during every cycle of every wave. That is what such a simple equation as $E = h\nu$ dimensionally says; (energy)(time) = [(cycles)/(second)](time) = (h)(cycles). It has been shown that compounded entities not sub-dividable below a given size appropriately account for in terms of a population. It was further demonstrated that the interval between integers of that population register that same degree of indeterminism, due to each population member's missing shroud of information. Whatever the magnitude of a smallest population unit, a comparably equal magnitude of uncertainty will infiltrate the system as the interval between its integers. Nebulosity of each population member "makes inaccessible" the interval between integers. Since an excess spatial angle of $\Delta\theta = 2\phi_{min}$ can produce a compounded population; it equivalently introduces comparable uncertainty. It is therefore feasible to express or account for a given excess-spatial-angle/cycle as dimensionless indeterminacy/cycle or space-curvature/cycle.

Like all other variables of nature, there is no reason why variables comprised of cycles should not have a "noise or uncertainty level "associated therewith. The excess angle would constitute that fundamental physical constant articulating monadic irresolution of space-time. In other contexts it might be characterized by a Planck length or a Planck time. Those quantities will be shown to represent smallest distance and time chord-lengths resulting from the same indeterminate angle $\Delta\theta$ on a conic-unit-circle. In so doing, the circumstance of zero-angular-warpage of space remains analytically included by not excluding $\Delta\theta$ from having the assigned value zero. **Nothing is "lost" by this more general inquiry that involves space-curvature while a large new domain becomes encompassed**. It is furthermore likely that any existing angular excess or curvature in spacetime would, by repeated compounding of that angle, convert the variable energy-time (or momentum-position) into a population. That is to say, under an excess-angle that exceeds conventional Euclidean-space radians, each cycle of anything might of necessity assimilate population members with such cycles because compounded units of angular irresolution correspond to units of energy-time. Any non-zero degree of angular "anomaly" effectively "counts" or compounds what each rotation must pass through; just as a "swinging gate" for individuals traversing it might characterize that population by the counted "gate swings". Each population member becomes the compounded angular anomaly (or "angular noise level") of a cycle for which $1/\phi_{min}$ comparable increments are potentially identifiable around each cycle.

The total number of smallest resolvable increments in the entire extended wave, each increment being one population member, would then determine the maximum population. Total resolvable partitions in the entire unified wave derive from: Resolvable angular increments/cycle = $1/\phi_{min}$ = $2/\Delta\theta$ = $2/h$ (9*1)

multiplied by # of cycles/total-wave = P_{max} = $1/h$

The maximum number of discernable angular increments in the entire wave is then $(2/h)(1/h)$ = $2/h^2$. Since one unit of energy-time as a portion of a cycle is equivalent to one unit of angular irresolution ϕ_{min}, n_{max} = $2/h^2$ expresses the maximum number of energy-time units in the entire wave. Each integer n defines one compounding (multiplication) of resolution angle ϕ_{min} = $\Delta\theta/2$, (or one smallest energy-time unit). Here ϕ_{min} equals **Planck's half quanta**, and is not zero as in the exponential function where n_{max} = ∞ forces that limit compounding angle to zero. **It is no "accident" that Planck's half quanta ϕ_{min} will be shown to graphically equal exactly half Planck's constant $\Delta\theta$, since 2ϕ = $\Delta\theta$ <u>track for all n</u>**. Additionally, energy-time/cycle = angular irresolution/cycle = $2\phi_{min}$ = $\Delta\theta$ = h, which provides (the-fractional-resolution-of-a-cycle) proportionality constant   h [= E/ν] between energy and frequency.

These analyses conclude that a fixed angular aberration of magnitude $\Delta\theta$ = h must prevail in spacetime. Every cycle of every thing, no matter how big or small must engage traversal of that variant angle. Nothing described herein prevents the constant $\Delta\theta$ from being interpreted as the fixed prevailing value of an otherwise "possibly" continuous variable. Under the above description $\Delta\theta$ depicts the existent fixed-background excess-angle warpage characterizing spacetime. Because the concept of an excess angle exceeding 2π radians/cycle may be unfamiliar, many redundant descriptions are reiterated below in the hope of improved comprehension. Difficulty to visualize and discuss an excess angle in warped space may produce various speculative dialogues about processes involved. Overall mathematics of the resulting harmonic waves however, stands on its own far more concrete merit. As previously described, energy-time/cycle designates the "angular spatial distortion" or "surplus angle" within each 2π radians of every cycle, being the non-Euclidean-fractional angular-additive of a rotational cycle. It is: $\Delta\theta = 2\phi_{min}$ = h,

or, {Energy} [= frequency] (times a minute increment of {time}/cycle) as {energy}{time}/cycle, $\qquad\qquad\qquad\qquad\qquad\qquad\qquad\qquad$ (9*2)

or, {a minute increment of a cyclic frequency}x{period}/cycle,

or, {angular deviation per cycle from traditionally interpreted space},

The relationship E = h ν, with h = $\Delta\theta = 2\phi_{min}$ expresses as energy-time/cycle so that $\Delta\theta$, $2\phi_{min}$, and h are dimensionless, representing a geometric "excessive angle" <u>per cycle</u>, or a geometric angle divided by 2π. **[Note: Division by 2π usually converts**

cycles to radians; here it converts a minute surplus radian-angle to a dimensionless angle-per-*cycle*.

An important physical constant, Planck's constant = h empirically pertains in the equation $E = h\nu$, whose magnitude applies to several relevant system-parameters. It expresses smallest repeated compounding increments. Each tangential transition s_n, symbolizes inter-integer space and will be shown to characterize **time from the origin to reach vector (n+1)**. Indeterminism in energy-time has the same inter-integer value as the prevailing excess angle $\Delta\theta = h$. After many cycles that indeterminism limits the uppermost population of smallest (energy-time) units to $n_{max} = 2/h^2 = 2/\Delta\theta^2$. Figures 2 to 5 become "continuous" toward $n \to \infty$, graphically simulating sinusoidal advance around the unit circle, (or the infinity circle). **In traversing a cycle, the block of energy-time any wave would require to pass through a given angle remains constant**. Angular segment $\Delta\theta$ would be commensurate with that block of energy-time expended per cycle to traverse such angle. A gamma ray's energy-time advancing through specific angle-of-a-cycle $\Delta\theta$ equals the energy-time a radio wave or any other frequency would expend traversing $\Delta\theta$. The gamma ray has higher energy and traverses angle $\Delta\theta$ in a correspondingly shorter time. Energy and frequency are synonymous. Fixed increment h **depicts an angular segment of a cycle**, that cyclical fraction comprising energy-time value h. It is effectively dimensionless delineating the veiled energy-time/cycle, or the add-on fraction of 2π within each rotation. Ergo $\Delta\theta = h$ represents a smallest physical-angle-of-resolution, the extra-angular traverse above

2π through any cycle in spacetime. Such angular irresolution conveys non-commutation and time's arrow.

The arrow of time remains bidirectional in solutions where multiplied binomial-vectors remain uniform. Time irreversibility (and possibly entropy increase) associates with the differing vector-formed-triangles in a harmonic function's progression. Unified waves possess such time irreversibility. Real world solutions that model unidirectional time require a changing compounded-vector-binomial. **All science and engineering solutions based on exponential harmonic sums will thus be time reversible**. At the limit $n \to \infty$, origin angles for both exponential and unified wave functions approach zero converging toward identical consecutively-compounded-triangles. For that circumstance, the spiral divided by n consolidates the exponential. In preserving n as a functional parameter however, the spiral retains greater generality; it gives time an arrow because the consecutive triangles, (the binomial vectors) always consecutively differ. At the $n \to \infty$ limit utilized for the exponential they would not differ and the unified wave would encompass the exponential exhibiting time reversibility.

10) DIMENSIONAL CONSIDERATIONS, THE EXPANDING

UNIVERSE, & π.

A speculative argument might be made that excess angle $\Delta\theta$ of these analyses associates with one component of the red shift and the expanding universe. The argument is not presented with great emphasis or inferred validity, but it might furnish a possible interpretation for $\Delta\theta$. Its primary feature is visibility for what Planck's constant might actually be. Generality is another quality of the conjecture. It has at least two basic shortcomings. One is that it assumes space is totally homogeneous and isotropic with respect to the excess angle and requires expansion from sub-nuclear to cosmological levels. While that is a very generic assumption it requires an everywhere-minute expansion of the universe like a pre-compressed sponge. Continued expansion of the sponge would allegedly give rise to an excess angle and red shift whose magnitude would appear to increase with distance. Such expansion might conceivably affiliate with an excess-angle. The greater the comparative distance between two points the greater the expansion would appear to be. Homogeneous expansion implies that the average relative velocity between <u>any</u> two objects in the universe roughly proportions to their separation, particularly for objects at greater distances. This highlights the second shortcoming of the conjecture, the difficulty of proving or disproving it at this time. A third consequence of relevance is that dimensional units for energy-time/cycle would become non-arbitrary while the represented quantitative magnitudes within those units could remain invariant. The required dimensional units for energy-time would be set by the angular numerical value of $\Delta\theta$, (the actual excess angle in radians)/(2π radians).

For energy-time in such an expanding universe, an extra segment of angle $\Delta\theta$ becomes encountered during each cycle in correspondence with monotonic expansion. Since $\Delta\theta$ is extremely small, that cyclic rotation would have to be very rapid for any effects to be experimentally noticeable. Effects predominate at high frequencies like those affiliated with elementary particles. An experimenter would be "unaware" of the expansion. In an expanding universe, the actual value of [perimeter/diameter] $\equiv$ "π" is conjectured to increase by the dimensionless factor ϕ_{min}, as $\pi(1+\phi_{min})$. This perhaps expresses a metaphor for an increase in path-length" around each cycle of rotation at diameter $D = 2r$. Then,

$$\text{Perimeter/cycle} = 2\pi r + 2\pi rh/2 = 2\pi r(1+\phi_{min}) = \pi D(1+\phi_{min}) = D(\pi+\pi\phi_{min}) \qquad (10*1)$$

Expansion is a hypothesis made in attempt to justify the mathematics, for which alternative possible physical explanations might apply. The excess angle may be a manifestation of the cosmological constant of general relativity. The empirical value of this excess factor is presumed directly related to h/2. The traversal of this additional factor, ϕ_{min}, might then be expressed explicitly in the various conservation laws. Energy, e.g., could then be articulated as the time rate of traversal of this extra increment in π within each cycle. Momentum, as well as angular momentum may be similarly interpreted as the traversal rates with respect to distance and rotation respectively, as later discussed. Under this conjecture angular curvature of expanding space would relate to the motional conservation laws. Pragmatic determination for the experienced value of π would preclude its depiction as an irrational of infinite digits. The granularity of physical reality necessitated by the factor ϕ_{min}, would plausibly truncate digits of its mathematical expression. The conventional mathematical value calculated for π would obtain only in the absence of spatial expansion, i.e., in traditional Euclidean space.

Unfamiliarity makes it difficult to reconcile equating Planck's constant to a physical angle. A prevailing excess angle/cycle $\Delta\theta$ would make the necessary units of energy and time **become absolute, (explicitly dependent upon that physical angle)** although the magnitude of energy-time/cycle represented by $\Delta\theta = h$ in those absolute units could be the same as the magnitude of h now depicted under MKS or CGS units, etc. A smallest energy and time dimensional unit would be such that their product numerically equaled the physical excess angle as a fraction of a cycle. In those new absolute units, (energy)x(time) would be equivalent to cycles and energy would be cycles per unit time. While we have become accustomed to arbitrary energy and time units like those under MKS or CGS, physical articulation of an existing universal excess angle would stipulate that energy and time be **characterized in absolute dimensional units based on the prevailing excess angle**. Since the physical angle remains unknown at this time, with only the magnitude of its energy-time equivalent $h = \Delta\theta$ being known in relative dimensional units, the absolute new dimensional units are also unknown. Given that science gives no credence to a minute excess angle per cycle now, it matters only academically now what its numerical value is. **Suppose someday it is determined the absolute excess angle/cycle was 10^{-30} of a cycle.** **Then the required absolute dimensional units would necessitate that <u>6.63 x 10^{-34} Joule-seconds/cycle equals 10^{-30} in the new absolute dimensional units of energy and time</u>.** Then energy-time would be in appropriate dimensions such that the energy-time/cycle of physical constant h would exactly equal the prevailing excess angle/cycle of physical space. It is alleged that those absolute dimensional units of energy-time that cause numerical match to the physical angle be the appropriate units to utilize in physics.

The question emerges as to how $\Delta\theta$ interacts with whatever absolute dimensional units are necessary. When a fundamental smallest-element-process physically governs a phenomenon, dimensional units would typically express in terms of that smallest element. For processes where things are made of indivisible quarks for example, quarks would characterize the process. The analogy of time being defined by earth's rotation in the absence of a more stable clock might be another example. Then, earth's rotation would become the "physical source of time" without an alternate method to check its absolute periodicity. If we did not know earth rotation slows minutely over extended intervals, the physical phenomenon (the day) would set the standard dimensional unit of time, rather than an arbitrary dimensional unit for time. The picture changes when awareness of the drift exists. Such changes can then be compensated for and factored into defining a standard time unit based on whatever process allowed improved resolution of the drift. If the drift were not known or discernable, the physical standard of actual rotational earth periods would prevail thereby unknowingly changing time's "absolute" dimensional units, though the magnitude of what is called "one day" would remain the same. The unit of time would follow the earth's rotation rate even if divided into a billionth or a trillionth of a day increments. **When dimensional units depend exclusively upon a physical parameter without availability of a greater-resolution alternative source, that phenomena becomes the absolute basis for dimensional units** whether it changes minutely in an unmitigated sense or not. A similar phenomena would exist regarding changes in (or invariance of) physical angle $\Delta\theta$ where an alternative greater-resolution source is unavailable. If or when that phenomenon drifts, the absolute dimensional units set by it would "drift" correspondingly, even if unknowingly when a greater precision source is absent.

At some limiting value of n, the spiral operationally converges into the circularity of the exponential harmonic. That limiting convergence represents the attainment of 4 radians defined through adjacent integer pairs whose sum is a perfect square; i.e., $n + (n+1)$ = perfect square. As discussed above, these radians may be analyzed or interpreted as rectilinear quadrants on a cone or however, thereby providing <u>**an absolute definition for the unit-radian in terms of the**</u> <u>**natural numbers n**</u>. **It should be noted no irrational values need be involved; the relevant number becomes 4 rather than 2π, (or unity if one considers a quadrant or radian as the base). Furthermore, <u>throughout all values of n</u>,**

those quadrant-marker integers will be shown as never being a prime number.

Thus, the natural numbers themselves provide: (10*2)

♦ An absolute rational definition for unit-radians;

♦ A definition of and demarcation of quadrants for all cyclical waves;

♦ Symmetrical partitioning of all prime numbers as coincident with those quadrants;

♦ A discrete representational mechanism for all harmonic type waves;

♦ An alternative limit process for delineating exponential-type harmonics applicable in Euclidean spacetime;

♦ "Rational polarization" based on the odd number sequence of perfect squares of form $s_n^2 = (2n+1) = $ (odd integer)^2;

♦ A direct relationship between integer values of n and transpired cycles P since the origin, as $n = 2P^2 + 2P$;

♦ A direct relationship between ξ, each transition magnitude s_n, and integer values of n as $s_n = \sqrt{(2n\xi+1)}$

♦ A hierarchically-ordered spiral structure for visualizing angles and vectors within the natural numbers;

♦ Possible expressions relating physical variables and universal expansion, including;

{a} A maximum encounterable sequence of numbers n, limited to n_{nax}

{b} A limit to the diminution in angle θ_n, $[\geq\Delta\theta]$

{c} A maximum resolvable accumulation of P consecutive cycles

{as $n_{max} = 2P_{max}^2 + 2P_{max}$};

Several additional features like root mean squared summations will be developed later. By contrast exponential harmonics converge to 2π radians/cycle at $n = \infty$, zero increment by zero increment.

The greater the rate of universal homogeneous expansion, the larger the numerical value of $\Delta\theta$ and the smaller would be the maximum integer n_{max} that could depict physical reality. [The largest encounterable integer n along the spiral would end up further from $n = \infty$, since $n_{max} = 2/\Delta\theta^2$.] With only one chance in infinity of being at that ∞ point, it depicts the "odds of a perfectly static universe", or of achieving exponential harmonics where effectively $n = \infty$. Considering the reality of homogeneous expansion or contraction, some however-minute excess-angle $\Delta\theta$ would be expected to algebraically influence

perimeter/diameter = $\pi(1+\phi_{min}) = \pi(1+h/2) = \pi(1+\Delta\theta/2)$. 　　　　(10*3)

Excess angle $\Delta\theta$ may be the consequence of a Cosmological Constant or other mechanism. Trying to establish its "root cause" might be impractical or premature. The question emerges however, as to how to establish the excess-angle/cycle precisely. It might not be possible without knowledge of the precise receding velocities of stars at known distances and what those velocities are due to. What fraction of that velocity was due to the big bang ballistic aftermath and how much if any to homogeneous expansion would similarly have to be determined. Or, more specifically, if $\Delta\theta$ is not the consequence of homogeneous expansion, then the issue becomes how to determine the numerical value for the prevailing excess-angle/cycle $\Delta\theta$. Such determination remains one of the difficulties of this conjecture. Homogeneous expansion/contraction would seemingly preempt arbitrary distance units and earmark associated suitable units, with $\Delta\theta$ being the "automatic-coupling" means. Were this scenario valid, an unavailable degree of freedom has been extended to analysis when employing arbitrary units like MKS or CGS. Such use entails an assumed zero-value for a physical parameter, namely setting the excess-angle/cycle to zero, whether due to homogeneous spatial expansion or whatever. It treats the universe as if $\Delta\theta$ was non-existent and $n = \infty$.

The more general circumstance should always allow possibility for an excess or deficient angle, wherein that angle could, (if later desired) be analytically set to zero. Inclusion of a possible value for excess angle $\Delta\theta$ is always more universal. It would then seem appropriate to retain the same **magnitude** for Planck's constant by adapting absolute dimensional units that set $h = \Delta\theta$ to the specific numerical value of prevailing excess-angle/cycle. Only the freedom to employ arbitrary dimensional units need be relinquished. That waive should be anticipated however, when such units must satisfy an absolute physical property, analogous to the example of the day being established by earth's rotation. Freedom to utilize arbitrary dimensional units surrenders when a physical process instituted the units and alternative measurement mechanisms with greater precision are unavailable. Were $\Delta\theta$ numerically large presumably due to increased homogeneous expansion, that angle/cycle becomes **compounded for precisely an inverse number of cycles**

P = 1/Δθ. This analogues compounding of n in the exponential function, lim n → ∞ of $(1+i\omega t/n)^n$. For very large n, [or large 1/Δθ] the resultant outcome from more compounding of a smaller angle becomes almost independent of 1/Δθ (or n). In the compounding sense the role of 1/Δθ is thus seen as bearing some analogue to n [$\approx 2/\Delta\theta^2$] in the exponential's definition, **but it would not go to infinity and vanish, except for zero homogeneous expansion or contraction.**

Therefore conclusion-wise, it is not highly consequential at the present time whether MKS, or CGS, or Δθ specified, or whatever similar units are employed. If all such systems yield very small numbers for Planck's constant h, than P = 1/h or $n_{max} = 2/h^2$ represent extremely large numbers of compounding. As analogy, consider compound interest type growth. If the principal in your bank account is compounded at the same interest rate every microsecond, nanosecond, picosecond, or infinitely as with exponential growth, it makes very little difference in the result. Similarly, the result would not differ greatly if "presumed" angles were made larger or smaller and then respectively compounded the inverse amount, about 10^{52} or 10^{66} times per total wave. So long as the actual physical angle involved in that compounding procedure is not employed for other ultra-precision calculations, that actual excess-angle would make little difference. In absence of this theory the excess-angle value has been assumed zero implying infinite compounding with interpretation that the real-world's perimeter/diameter ratio was (mathematically calculated π). Societies have survived without adverse consequences using ratios as divergent as 22/7 ≡ π, so whether the actual ratio discerns to roughly 26 or 33 digits, or whatever, is not highly relevant.

This train of thought suggests a "reason" why distinguishing the basic-process-units for energy and momenta has remained so elusive, though these variables are unquestionably of prime importance. Energy, or work, possesses equivalents and designative dimensional units, **but not a "phenomenological-descriptive-process-unit".** **We haven't known what a smallest unit of energy is** and yet these units are conserved with mass being one specific form. A basic though unfamiliar energy or momenta unit seemingly depends upon the rate of traversal of an excess spatial angle whose existence has not previously been validated. The conventional (but arbitrary) Joule or Erg as currently utilized thus inadvertently inhibited involvement of any natural (or supposedly correct) action unit. Energy is here characterized by the reciprocal time to traverse a smallest discernable angle. With the existence of such angle unknown, deciphering variables dependent upon it becomes improbable. Distinguishing the phenomenological-descriptive-process-unit for energy is formidable when what it depends upon remains obscure. With $\Delta\theta$ numerically larger under increased homogeneous expansion for example, the time rate of traversing that angle (or energy involved) changes, requiring a commensurate scale factor on dimensional units. **A plausible conclusion of this theory is that the excess-angle and Planck's constant are interpretable as an additive increment to the fixed value of mathematical-π. That would circumvent irrational values in real-world situations**. The motional

conservation laws would coalesce. De broglie waves and a basis for quantum phenomena emerge. Though there is only the logic of the orthogonality axiom to substantiate the basic conjecture, the bonuses are not trivial and inclusion of $\Delta\theta$ is more general than its omission.

Mathematical π is the ratio of perimeter to diameter, for a perfect circle on a perfect plane in Euclidean space. If the space is not Euclidean and/or circulation about the center is not perfect, the experienced value of perimeter/diameter does not remain the same. Were space curvature greatly exaggerated to simulate in two dimensions by earth's surface, the value of perimeter/diameter could vary from $\pi = 2\pi r/2r$ (for a very small diameter 2r taken about a point at the North Pole), to **2** (for a diameter increased to equal that of the earth). When r becomes a longitudinal line-length from the North Pole origin point to the equator, than perimeter 4r around the equator equals 4 such longitudinal line-lengths and twice diameter. On that curved surface there would be 4 radians/cycle, as measured around the equator from a North Pole origin.   (10*4)

We can by this analogy recognize (or interpret) the ***experienced value*** of π (contrasting Euclidean mathematical-π) as involving "dimensions" dependent upon the degree of curvature present. Those dimensions effectively influence what constitutes a cycle. In that simulated curved space, a cycle expressed in radians for very small r centered at the North Pole differs from a cycle around the equator centered at the North Pole. It should thus be recognized that in any generalized

113

case including curved spacetime, units of dimensional consistency for cycles should not be discarded. Such units must maintain dimensional consistency to stipulate what a cycle is. Dropping dimensional units for cycles might only retain validity for the specific case of Euclidean spacetime. That condition will always be less comprehensive than inclusion of dimensional consistency by accounting for cycles, because such accounting does not induce any harm in Euclidean analysis.

Thus a fundamental error is enacted in dropping the denominator dimensional unit of cycles in Planck's constant = h = E/ν = energy-time/cycle. It will be shown that h can interpret in units of energy-time/quadrant, where a quadrant is one mathematical radian. Being an absolute standard, the unit-radian in the denominator might simply express as unity allowing h to represent conventional action units of quantum mechanics. Discard of denominator cycles immediately constrains the analysis to only apply in perfect Euclidean space the antithesis of what non-commutation parameter h phenomenologically represents. Just as π can appear a dimensionless constant in Euclidean space or a variable with "dimensions" in curved space, Planck's constant h can also appear as a constant having dimensions energy-time in Euclidean space, or dimensionless when accounted for as energy-time/cycle in curved space. Such "dimensional unit" representation depicts wave energy in cycles-traversing-$\Delta\theta$/second being multiplied by time-per-cycle to yield a dimensionless parameter for Planck's constant. Alternatively, h can be seen as energy-time (or momentum-position), being a minute excess-angle segment-of-a-cycle per full cycle, or a minute excess-number-of-degrees-due-to-curvature per 360 degrees, both interpretable as dimensionless. The non-obvious association between Planck's constant and π share similarity with respect to having dimensional-unit ambiguity in common, and both encompassing the crux of what constitutes a cycle. Issues coupling π and h can only be resolved by treating the higher-order system possessing curvature, which is in fact the essence of what Planck's constant allegedly represents. Pre-anticipation of seeming dimensional-unit disparity between "dimensionless" mathematical-π and Planck's constant's action-units should be postponed until the end of this treatise. The historic viewpoint from the vantage of Euclidean space needs to accommodate the more general perspective of spacetime warpage.

11) PRIME NUMBERS IN RELATION TO QUADRANT AXES OF THE WAVE

Various other attributes of the spiral of figure 3 are significant. For example, the four quadrature axes A,B,C,D will be shown to never be occupied by a prime number value for n. Figure 13 shows an axial cone view of the primes out to n = 480. This sketch of the vector nodes circles all primes within that mapping and they never lie on the quadrant axes (except for n = 1, 2, 3). This tendency continues out to n = ∞. **The A,B,C,D axes therefore divide microzone space into compartmentalized partitions defined by exclusively non-prime (sub-divisible) integers**. Directions A,B,C,D form true axes of symmetry since **they cleaver all divisible and indivisible integers by methodical partitions of assuredly divisible numbers**. <u>**These quadrant partitions also constitute the peaks and zero-crossings of all harmonic waves**</u>. The quadrant dividers constitute respective separation points between alternately interchanging energy forms in all waves; for example between the electric and magnetic or potential and kinetic energy. The unified wave harmonics possess a unique symmetry that embeds within the intrinsic structure of the number system. That symmetry will be shown to sustain for all ξ, over the range $0 \le \xi \le 1$. The articulated partitioning properties are totally additional to conventionally acknowledged "sinusoidal properties". <u>**However unfamiliar, harmonic frequencies are alleged to more generally possess all of the additional attributes categorized here through the unified wave**</u>!

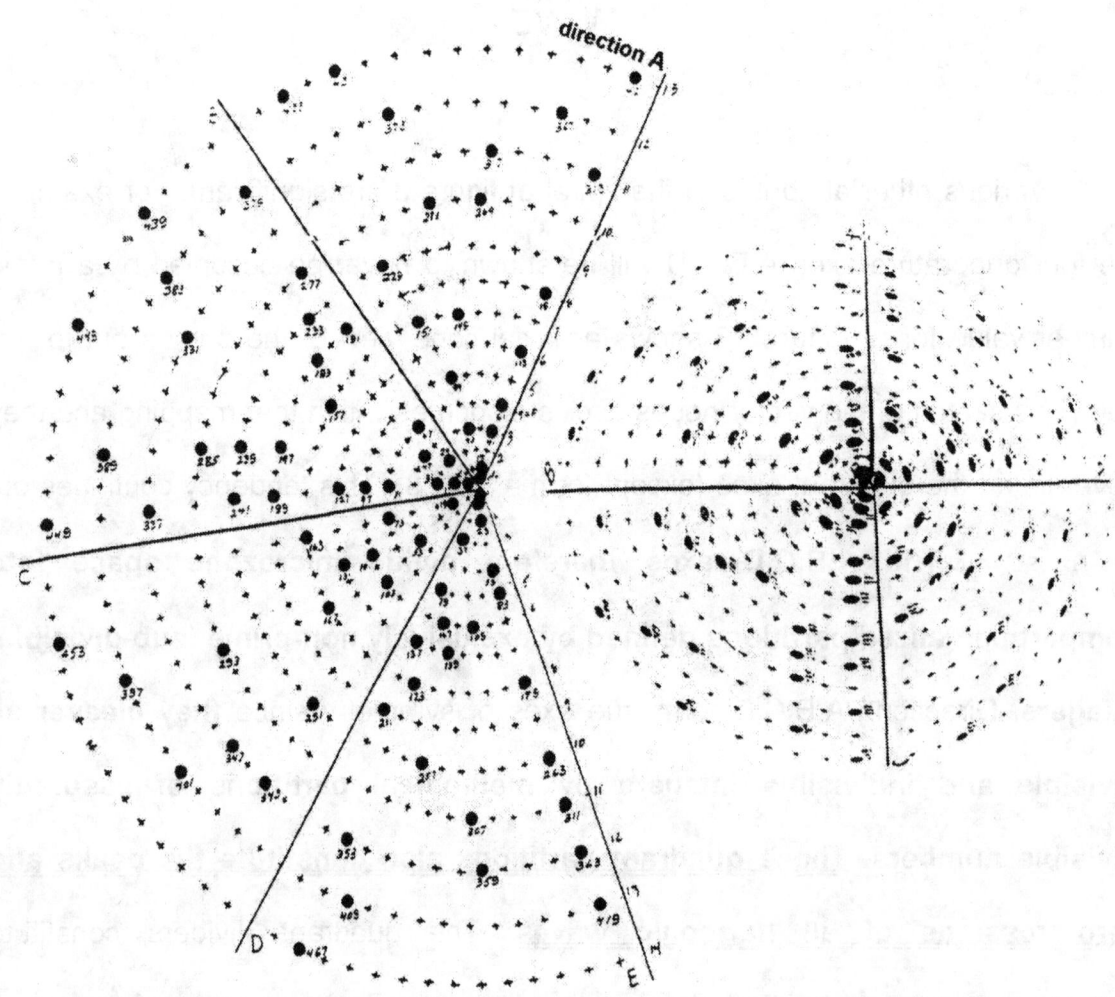

Figure 13 depicts an end-view conic-axis photograph of a coiled transparency displaying the nodes of a unified wave spiral. Prime values for n below 480 are circled and labeled. Except near the origin at n = 1, 2, 3, prime numbers never fall on quadrature axes A, B, C, D, and the tendency continues to n = ∞. Axes A and C always yield even numbers while B and D alternate odd-even between consecutive cycles. One integer less than each n value along axes B and D (except n = 5) always appears to be non-prime, presumably to infinity. The radian quadrants of every wave thus partition all integers above 3 into closed sets of primes and non-primes with specific location propensities. Even in its simplest terms, the harmonic can resemble a rotating "four toothed gear" allowing and disallowing engagement with particular sub-periodicities. The four symmetrical radians that comprise all cyclic phenomena therefore contain these numerical-composition interrelationships.

As will be elaborated, axes values at directions A and C of the unified wave can never be prime since they always have n = (even number). For each axis B and D, sequential alternation occurs between n being odd-even-odd-even out to infinity. Moreover, although only checked as far as n = 5500, figure 13 displays that one integer less than the n value associated with axes B and D are also never prime (except for n = 5). They alternate even-odd-even-odd in intermeshed fashion with the axis integers. In conjunction with non-primes at axes A and C, and the perfect squares, that sequence [like the pitch of a propeller blade], jointly provides inherent indication of rotation direction within the wave. Such clues inside the harmonic are consequential because they occur by virtue of the summed θ_n angles. They potentially apply even when the wave takes on different non-localized, random, or indeterminate forms, in addition to applying to wave modes as displayed on the conic surfaces in figures 6 to 11. During such circumstances, **the wave's potential harmonic nature would remain encoded within summed θ_n angles even if the wave did not exhibit "sinusoidal shape"**. Figure 3 easily verifies that upon traversing every four consecutive axis-directions of the spiral, [every 4 quadrants established by the summed θ_n angles] the number of inter-axis partitions must increase by unity. As shown below, (where each complete cycle at direction A articulating a triplet triangle is signified by underline), **the values of n along all A,B,C,D directions contain no primes**, (other than 1, 2, 3). Specifically from the tabulation below, directions A (underlined) and C (not underlined) are always seen to have an even value of n and are never prime.

A, **B**, C, **D**, etc. ≡ 0, **1**, 2, **3**, 4, **6**, 8, **10**, 12, **15**, 18, **21**, 24, **28**, 32, **36**,

40, **45**, 50, **55**, 60, **66**, 72, **78**, 84, **91**, 98, **105**, 112, **120**, 128, **136**,

144, **153**, 162, **171**, 180, **190**, 200, **210**, 220, **231**, 242, **253**,

264, **276**, 288, **300**, 312, **325**, 338, **351**, 364, **378**, 392, **406**,

420, **435**, 450, **465**, 480, **496**, 512, **528**, 544, etc. to infinity.   (11*1)

The tabulation shows the **D** axis (in bold) having the consecutive-cycle-sequence as n = 3, 10, 21, 36, - - -etc. is factorable by (a)(2a+1); [where a = 1, 2, 3, --- etc.]. The next higher **B** axis (in bold) having the sequence n = 6, 15, 28, 45, - - - is factorable by (a+1)(2a+1). Accordingly, **being either even or factorable, none of the axes values at** A, **B**, C, **or D** **can contain a prime value for n**. [The difference between any n at **B** and the proceeding **D** will be (2a+1) = (consecutive odd numbers) = (2P+1), relating quadrant divider directions **D** and **B** with transpired cycles P.] The n values along **above-highlighted** directions **B** and **D** can be expressed algebraically below as

N_P at P cycles.

$N_P = N_{P-1} + (N_0+4P)$	For **B** direction	For **D** direction	(11*2)
	$N_0 = 1$; n =	$N_0 = 3$; n =	
@ P = 1 $N_1 = N_0 + (N_0+4P)$	6	10	
@ P = 2 $N_2 = N_1 + (N_0+4P)$	15	21	
@ P = 3 $N_3 = N_2 + (N_0+4P)$	28	36	
@ P = 4 $N_4 = N_3 + (N_0+4P)$	45	55	
" " "	"	"	
@ P = P $N_P = N_{P-1} + (N_0+4P)$	etc. to ∞	etc. to ∞	

One less than the n value at each B and D axis is also seen to never be prime, except for n = 5. Directions A,B,C,D therefore **partition all indivisible natural numbers (as well as the divisible numbers) into associative radian quadrants**. In the form of values for n, non-primes manifest in collated regularity at each radian within the quadrant structure of waves. That literally becomes what could be an alternate definition for a radian; the symmetrical angular partition of primes numbers to infinity. It underscores that the n → ∞ limit of the spiral defining a standard unit-radian also distinguishes the non-prime quadrant and radian partitions. **The condition of wave peaks and zero crossings never associating with a prime (non-divisible) number provides an alternate definition for pure-harmonic-periodicity.** Furthermore in figure 13, radial-direction and P-cycles-of-outward-radial-extent both retain simple relational significance with n and s_n. **These are in addition to other unified wave features like the correlation of tangential vectors with half-cycles, perfect squares, quadrants, rotation rate, energy-time units, non-commutative insertion points, consistent reference to the origin, irrationality, RMS sums, and angles $\Delta\theta$ and ϕ.** Associations between and amongst all vectors comprising the wave are highly ordered in terms of both harmonic qualities and the number system. Such qualities tend to be different manifestations of invariants. Regardless of the shape or form a mode of the wave takes on, each integer number of radians from the origin [4P] is gauged, which associated number n is always factorable into some sub-multiple array of lessor integers (radial vectors), with special recursion every four radians [P].

Non-divisibility is synonymous with both prime numbers and the orthogonality-axiom-foundation applicable to quanta. That axiom provides the basis for the spiral that partitions prime and non-prime numbers. The sequence for vectors n that make up non-divisible quanta embodies both primes and non-primes. It will subsequently be shown that enumerating even large divisible objects of arbitrary ξ retain the same harmonic symmetry as non-divisible entities. Indivisibility used to explain one extreme condition provided a useful essence for understanding the issues. It justified treatment as a population with attendant "integerization" and emerging microzone dimensions. Encoding related to spiral vectors remains part of the harmonic wave structure and of the underlying natural numbers. These aspects of excess-angle compounding are inferred to apply to **all waves of physical reality**, (in the real world of excess-angle space and $\Delta\theta$). The concept seems foreign only because the association has been unrecognized. Wherever the unified wave harmonic replaces the exponential harmonic, (**and allegedly it should do so in any physical representation**), indivisibility features emerge and take on relevance. **Only at the non-physical infinite-limit-process of mathematics (or in idealized Euclidean space) does the issue of divisibility vanish**. Only at the n = ∞ fictional-limit within the exponential's definition does endless splitting of the compounding angle cease to be germane, specifically being erased by that infinite partitioning process.

Unified waves probably apply for all spacetime descriptions of objective reality following an origin event. They just have not been analytically brought to bear in treatment. A contention of this thesis is that these resonances portray all particles and force carriers and thereby pertain to all possible wave descriptions in Nature. It is speculated, and will be later elaborated for example, that rungs in the nucleotide ladder of DNA may utilize this divisibility in encoding gene-makeup from the double-helix embodiment of the conic spiral. [This can be envisioned from figure 3 by conceiving directions A and E as "helical twisting railroad tracks" {for β close to $\sin(2/\pi)$} between which lie consecutive strings of four serially-connected radians or two half-cycle "ladder rungs". All features of the unified wave, **<u>like unending reference to the origin and half-cycle tracking in s_n, then remain embedded in the encoding process.</u>]**

The spiral constitutes the "applicable axis" in the microzone. All of its attributes, namely **the associative features interrelating numbers n within the figure 3 configuration apply to such axes**. These features include both radial and tangential-direction grouping of all integers to infinity. For monads both n and $s_n = \sqrt{\text{(odd integer)}}$ **base entirely on integers**. That affiliation, plus an undefined degree of angular freedom allegedly delineate properties of microzone spacetime. That manifests no differently than features of Cartesian axes applying to Euclidean space. <u>Besides the linear algebraic sequence of unit-increasing n in analogue to Euclidean axes, unified waves interrelate:</u> **half cycles signified by s_n, proximity to half-integer magnitudes for s_n articulating axis dividers A,B,C,D, the quadrant partitioned prime numbers, the odd perfect-square values of (2n+1), radians, intermeshed odd-even successions, irrationality, quadrants associated with**

inserted energy-time units, symmetry, rectilinearity, the RMS sum of transition magnitudes equaling integer n, and Planck's constant h equaling an excess angle $\Delta\theta$. It would seem an odd accident of Nature if such a web of inter-relationships and fundamental associations inherent within the A, B, C, D, axis structure were **not** the basis of some physical processes. These inseparable interconnections within the harmonic denote unified wave potency in contrast to exponential harmonics. The four quadrants **of all waves** purportedly possess these associative affinities. **<u>Every cycle could not occur without engaging the</u>** $\Delta\theta$ **<u>excess-angular-curvature of the universe</u>**. Integers of the number system are the fundamental organizing agent binding this web as compelled through indivisibility. Through these alliances it is alleged that entangled quanta of quantum mechanics mathematically link-up. Intricacies embedded within natural numbers avail like a hidden variable theory throughout all of Nature's domains. <u>Portrayed in spiral format for illustration, the sequence of integers furnishes a substructure for cosmic order</u>.

An important distinction between mathematics and physical reality stems from limit processes in mathematics that proceeds to infinite divisibility. One cannot divide the indivisible and that truth broadens the conceptual framework of the physical world. A partition detaches the two domains between divisible and indivisible. On the one hand, traditional mathematics provides the ultimate tool of quantitative science within the realm of divisible entities and classical physics. However, when inadvertently applied below the threshold of physical indivisibility, that mathematics leads into an analytical abyss of non-reality. It is at this dividing juncture between severable and non-severable that logic, physics, and mathematics bifurcate. A new logic, an altered physics and changed mathematics emerges for treating non-sub-divisible entities. The new paradigm purportedly becomes the orthogonality axiom where to quantitative degrees direct encroachment into the interval between integers is impossible and variables become populations. Two categorically different domains of existence emerge. The transition is totally logical though totally unfamiliar. Consequences become the unified wave spiral, non-locality, quantum mechanics and eight additional "dimensions" with seemingly attendant dilemmas when interpreted within traditional dimensions and logic. The results are not so enigmatic when viewed in light of the traversed indivisibility threshold. The mathematical culprit was the infinite limit process, with an important explicit oversight in the exponential harmonic's definition. A much more general harmonic applicable to both sides of the indivisibility threshold emerges when the orthogonality axiom pertains. The orthogonality axiom allows two worlds, contrasting the familiar macroscopic world of X, Y, Z, and t. It grants harmonics of eight microzone dimensions having non-locality for indivisible entities, and indeterminism from the shroud of each population member. Intersections of those two worlds occur experimentally when observation into the microzone pierces and annihilates the consequence of an origin event, when the possibilities terminate. Allegedly this is synonymous with collapse of the wave function, the measurement problem of quantum theory viewed under macroscopic logic.

12) AN OVERVIEW OF UNIFIED WAVES

Advocating eight new dimensions where conventional measurements cannot occur might violate a byword of physics. Theories of physics supposedly require acts of observation or measurement to confirm validity. However, no stipulation specifies exactly where in the course of events, phenomena must be observed. Certainly, we do not see or measure electromagnetic radiation everywhere along its propagation route. That route is only deduced after observation at some terminus position. We do not observe or measure irrational values traversed along a variable's presumed progression. No assurance really exists that those values were actually traversed. We do not everywhere measure the presence of force fields like gravity, only its consequence at specific locations. That result then allows us to attribute via a standing paradigm, an interpolated gravitational field or space curvature. The tenet requiring observation for validity thus is often inadvertently bypassed. Inference prevails that if intervening measurements were made in the unobserved regions, results would be consistent with expectations of the standing paradigm. Usually, descriptive equations like Maxwell's serve as the vehicle to straddle the unmeasured regions, it being assumed such relationships predict what would have occurred in those regions without actually making the measurements. Equations for a suitable paradigm of understanding justify representation of the phenomena, for example the method by which radiation propagates. Inability to measure except at the terminus of an experiment is endemic to quantum mechanical descriptions.

However, as in Godel's Proof[13] for mathematical relationships, the paradigm utilized can never prove its own validity. Unexplored degrees of freedom, hidden variables, unrevealed applicable dimensions, indeterminism, etc., external to (and neglected or unknown in) the situation may always exist. They could alter validity of the representation if recognized and invoked. What Godel described for mathematics, applies for mathematical physics and physics. There can always be an implied assumption whose validity was overlooked. These considerations extend to the eight extra dimensions for indivisible particles. The wave/particle remains unobserved for the microzone portion of its route prior to the particle's terminus. An analytical characterization like the unified wave straddles description of what is presumed to have occurred in the unmeasured region. Under the prevailing new paradigm, portrayal in this area must interpolate satisfactorily with results from the terminus measurement. It should further be possible to duplicate the experiment with a new terminus measurement at an intermediary stage of the particle's route. Results from each such experiment need consistently comply with the conjectured new paradigm. If so, the new analytical procedure utilizes observation and measurement for verification to the same degree as applied for Maxwell's Equations.

An excess angle in space exemplifies a neglected or unknown condition suggesting unexplored territory requiring new physical modeling. That angular distortion is extrinsic to all problems treated via Cartesian Coordinates. The actual condition is outside the existing paradigm that presupposes those coordinates portray de facto space. Rectilinear coordinates do not portray real space. When the bounds of applicability for the mechanism analyzed, (and/or the depicting equations) become inadequate we begin to search for a more comprehensive facsimile to describe the processes in question. Often, newly applicable more-inclusive models dictate significant revision of the old thinking, although quantitative computational methods of the old situation may retain their precision. Epicyclical analyses of celestial bodies circling an earth-centered cosmos provide one historical example of such an older computational model. The model may provide essentially the right quantitative values under the wrong paradigm. It is suggested that a similar modeling circumstance exists in quantum theories that can benefit from conceiving microzone dimensions for propagating indivisible entities. Unfamiliarity with orthogonal progression between integers makes it difficult to visualize variables rotating like a corkscrew, rather than progressing along a straight axis. That is why de Broglie waves of motion seem so out of context, even when we know they exist. They are entirely logical under the orthogonality axiom framework, applicable wherever fractional values for a changing population become impossible or improbable. A population of indivisible entities may only change momentum-position and energy-time along a "spiral-type" axis because they must progress at right angles between integers.

This also allows them to be non-local since consecutive right angles need not correlate to each other. Orthonormal transitions earmark the antithesis of taking a direct route to each next higher integer. Non-locality within the wave manifests from, and is tantamount to, inter-integer indeterminism. That intrinsic uncertainty derives from and is self-same with the impenetrable shroud of each population member via the "conservation of minimum indeterminism". Non-locality thus emerges as a fundamental property associated with smallest indivisible entities. The shroud of population members due to their unknowable interior supplants each definitive X, Y, Z, t, macrozone degree-of-freedom with two nebulous degrees-of-freedom in the counterpart microzone. By contrast, infinitesimal continuity signifies continual association from one increment to the next (as via a linear Euclidean axis), while monadic entities possess individual or quantized attributes, (discrete and less dependent on each other). Though non-locality is foreign to the macro world, it is a salient feature of the microzone.

A rather general misconception in modeling physical reality is the assumption that sufficient piecewise linear approximations can always extend to adequately approximate systems that are more complex. An example occurs in modeling de facto excess-angle-spacetime with rectilinear space. The higher order system of extraneous-angle space will always be more general. No matter how many lower order iterations are attempted, subtleties can obscure under piecewise approximations. Familiar expectations of the lower order system may not appear at all in the higher order one, which in turn may exhibit unanticipated new phenomena. Interpreting those new phenomena under the paradigm for the lower order system often yields unresolvable enigmas. For discrete indivisible objects, that is the case with respect to issues of non-locality, extra dimensions, and the indeterminate shroud between integers. For empirical results to appear reasonable, one must view the situation in context of the higher order paradigm involving both sets of dimensions, the macrozone and microzone.

Determinism is often considered the necessary criterion that makes a physical theory have one-to-one correspondence with each element of physical reality. However, **any actual system containing smallest repeated constituents will always possess indeterminism equaling the impenetrable interior of each constituent, and as the interval between their integer numbers**. With that understanding then, theories with identifiable quanta <u>must all violate one-to-one correspondence with confirmed physical reality</u>. Indeterminism stemming from treating population variables is **unavoidable**; it should preferably be considered **a causal consequence** of those smallest constituents. **Inter-integer uncertainty is the "caused effect" from constituent-indivisibility** (of quanta), in an analogous way that acceleration is a causal effect of applied force, or of any other rules producing cause and effect.

Indeterminism is the consequence of missing information; a condition created by truncated decimal digits, incomplete initial conditions, background Gaussian noise, or any system unknowns. No actual system can have all of its parameters known to infinite degree so **indeterminism always exists, whether mathematically accounted for or not**. Accounting for expansion of the universe

131

for example would not be typical in a localized problem. This theory indicates it may cause or correspond to excess angular $\Delta\theta$ in spacetime, producing indeterminism to degree h energy-time in every cycle. That is a causal effect of the impetus phenomena just like any other causal effect. A smallest non-divisible entity will always exist that voids the space between integers or truncates digits. Classical account of a clockwork-type deterministic universe has simply been an erroneous orthodox interpretation since Newton. No such condition can exist just as actual tangible variables do not take on irrational values. Such values would have infinite information content, which is impossible to describe. Though mathematics can express magnitude to infinite precision, it remains infinitely more encompassing than physical reality. Actuality is more realistically limited by spatial angular curvature $\Delta\theta$ to unknown, (but in MKS units perhaps about 33) decimal digits, or whatever in non-arbitrary units. **The extra digits of precision implied by classical limit processes misrepresent the physical situation**. This theory attempts to supplant such mathematics with analytic methods to accommodate $\Delta\theta$. It effectively truncates the unrealizable mathematical digits via non-local harmonics of the microzone. That computation and the produced motional de Broglie waves allegedly characterize the actual situation, in contrast to Newtonian continuity with determinism.

No justification exists to assume that an infinite limit process invoked in mathematics actually possesses an analogue in the physical world. Resorting to a limit operation in analysis can breach a threshold boundary wherein the

132

mathematics modeling reality deviates from physical reality. Nothing is "wrong" with the process as a mathematical operator. It is a logical invention of the human mind. However, physical reality need not comply toward uniformly approaching an idealized point or line. Those entities themselves represent limit-process abstractions, there being no assurance the approach procedure must remain invariant, even if the object might elsewhere represent suitably as a point. We may for example consider an electron a point object when far away. That does not mean it remains like a point in arbitrarily close proximity. The range of applicability of a mathematical limit process is inadvertently presumed infinite upon enacting that limit operation. Reality nonetheless, typically possesses some smallest repeated constituent invalidating the assumption of functional invariance under endless approach. Whenever real objects are not idealized points or lines, phenomena involved when approaching them may change, while the mathematical limit procedure does not change.

When a coalesced signal formed from two same-frequency amplitude-components A and B have totally random phase, their energy proportions to A^2+B^2. When the two phases overlap exactly, the affiliated energy proportions to

$$|A+B|^2 = A^2+B^2+2AB. \qquad (12*1)$$

The change from overlapped to random phases represents breakdown-in-interference and different combined influence of the wave components. The unified wave possesses multiple possible modes. One possible mode may entail repeated overlap of directions A, B, C, D and four-quadrant sinusoidal type symmetry. That could depict a state of an indivisible elementary particle. Interrogating the state of the system by any means would typically alter that wave by injecting some component however small affiliated with the interrogation. Altering any phase accumulation of the different θ_n angles from the exact prescription necessary can destroy the wave's symmetry. Upon interrogation the resultant wave would no longer accrue its angles precisely so symmetry would dissipate. That constitutes cadence loss for repeated overlap of directions A, B, C, D. The wave can thus loose interference properties through interrogation (interaction). For the conic spiral, all of the separate possible non-local routes from the origin to an outcome, (for example through slits A and B by analogy) can add as possibilities Ψ_A and Ψ_B. That may be looked upon to represent geometric fractions of the "flooded" open-aperture-slits to the surrounding non-throughput area. When the fraction of waves passing through the slits remains un-interrogated, the two sub-components overlap and interfere, having net vector effect as $|\Psi_A + \Psi_B|^2$. Squaring is needed to convert throughput tangential vectors s_n into rational values, or to associate areas with "tangential radius" s_n. Any successful external interrogation will add a component that destroys perfect overlap and interference, the combined effect of the two non-overlapping (randomized) sub-components then proportions to $\Psi_A^2 + \Psi_B^2$. In this manor, (by destroying overlap through interrogation) the inherently conserved system-indeterminism mathematically prohibits external determination of the state of the system en route. In doing so, the mathematics implements a "conservation-of-indeterminism law".

A theory is unattractive if it generates answers, however accurate, through abstract mathematics devoid of a logical association to the related conceptual entities of the problem. Heaviside operators initially were an example where computations provided correct answers without a fundamental foundation for why. Eventually, reasons for the operator's predictive attributes were understood. Quantum theories are also very much in this category. The meaning of a possibility wave Ψ or of a probability wave $\Psi\Psi^*$ is academic, non-realistic in the sense that it does not connote what transpires within the physical reality to cause the answer. It does not tell why, based on known conceptual entities. As previously mentioned, determinism need not be the mandatory criteria for an appropriate theory to have one-to-one correspondence with physical reality. Other criteria may be preferable. What's more appropriate for correspondence of theory with reality is one-to-one logical identification of entities, processes, phenomena, dimensions, and variables between math of the theory and the respective objects of reality. One of those entities will inevitably be indeterminism, interpretable as a causal effect of a logical process.

Under such rational, the unified wave helps fill-in the physical phenomena that bring about quantum mechanical results. It identifies the occurring physical mechanisms corresponding to the mathematical principles and analytic processes employed. This theory serves to help explain why QM works. It also indicates oversights, like times arrow, the coalesced conservation laws, the eight microzone dimensions, the wave comprised of ray vectors, the orthogonality axiom, and the

135

curvature of spacetime $\Delta\theta$ = h. It extends QM to prime numbers, to reveal harmonic wave/particle interactions linking force fields, and possibly to the cosmos of the red shift and expanding universe. In accord with Occams razor it provides insight for these processes without introducing any arbitrary constants. It requires only a paradigm shift for quanta, from linear progression along Euclidean axes, to bypassed irrationals via orthogonal advance between integers. The missing one-to-one correspondence between depiction and reality has been consequence of that unrecognized axiom. In retrospect, the axiom is totally logical. Its mathematical inclusion clarifies enigmas and couples intrinsic properties of ordinal numbers into an explanation for physical phenomena. Waves generated from natural numbers model an inherent property of space and conjoin the theory, the numbers, the dimensions, the variables, the mathematics, and reality.

Intervals between Euclidean axis integers n can each stand for a smallest possible unit of energy-time. As radial vectors from an origin those integers describe a population of least energy-time units. Always forming a right angle transition to the next-higher-integer (n+1), transverse vectors s_n straddle the space between population members. Extent along s_n will be seen to connote time from the origin and affiliates with energy-time units. For monads the configuration forms a sequence of tangible, radial, ordinal-integer-vectors n each followed by a tangential orthogonal-transition-vector s_n to reach radial population-vector (n+1). Radial vector magnitudes are invariably rational with tangential vectors always irrational and intangible, except at triplet triangles. For the solution mode where every vector s_n

remains within the same plane, the spiral rendition of figures 1, 2, and 3 result. More general modes entail arrangements wherein consecutive vectors s_n are not within the same plane but might meander as influenced by prevailing fields, potentials or motion relative to an observer. One such illustrative mode involves a conic spiral of apex angle β = arcsine($2/\pi$), which displays four-quadrant "four-radian" symmetry whenever directions A, B, C, D, overlap. Radial vectors n and tangential vectors s_n diverge from the Euclidean axis with periodicity in accrued-origin-angles, irrespective of how they are portrayed. The cone serves as a pictorial metaphor to show angular and vector spacing under conditions of quadrant vector alignment. **Individual angles θ_n, their sum, and all vector magnitudes, remain identical, whatever collective method is used to display them**. However standing wave solutions (stable states) emerge only when conditions of synchronous periodicity prevail and that requires overlap of the ABCD vectors. Each consecutive vector n delineates one greater unit of energy-time. The four quadrant division-vectors of each cycle that partition each perfect square magnitude s_n are always occupied by a smaller divisible value of n, (except for n = 1, 2, 3). All prime values for n partition into four symmetrically closed sets angularly converging at large n to precisely one radian each.

The object of a physical theory is to explain and predict observable order in the universe. Here, order is revealed within generic harmonics through the unified wave theory. The analysis should furthermore apply to a relatively large ensemble of processes, all phenomena amenable to harmonic description. Particular utility

137

should emerge for characterizing microscopic non-divisible building blocks. Since macroscopic happenings may always decompose into those more elemental constituents, the theory's applicability should in concept encompass most situations within the cosmos. The limiting exponential harmonic case (where n = ∞), is familiar and utilitarian but may prove problematic, except under circumstances where elemental indivisibility and excess-angle spacetime can be ignored. The microzone might perhaps interpret as a "medium" that could supplant the ether in the sense that it supports the distribution of waves outside the macrozone. It could conceivably also relate to the sea of quantum in vacuum. It is the medium where waves can exist and promulgate. That would imply electromagnetic waves (and fields) are likely states of the microzone medium, not states of the macrozone. Further elaboration on this point will later be made regarding RMS sums of transition magnitudes. Figure 12 infers an origin of those states through graphical presentation. Empirically knowing electromagnetic waves propagate at light-speed, which process alters classically perceived macro-space and time, provides further justification for delineating two differing media.

Summarizing, orthogonal transitions between population integers create cyclic phenomena in the form of a unified wave. A much richer field of harmonics becomes available than the subset of exponential sinusoids. The apparent "non-uniform" interval between population integers effectively injects time irreversibility currently missing in Newtonian, Einsteinian and Quantum Mechanics. That orthonormal traverse and missing information consequently creates non-locality, possible non-separability, with uncertainty in the solution response. A question then emerges regarding whether dynamics of populations having unavoidable missing inter-integer space should be considered "causal processes" that non-the-less yield indeterminate (and seemingly non-causal) effects. If, in otherwise "linear systems" the impetus dynamics of smallest-member populations remain analogous to and as "causal as" similar dynamics for continuous variables, then causality should prevail for either of such proceedings. In processes related to population variables, indeterminism emerging within the solution would simply be an unavoidably "caused effect" of each population member's shroud. That shroud designates the inability to establish what is within (or what partitions are possible within) each smallest indivisible entity. For monads the veil remains inextricably present between integers in the very definition of a population.

Interestingly, Einstein, in his objection to quantum mechanics was unwilling to relinquish causality. Perhaps his vision that QM was "incomplete" and that a form of causality would prevail in its ultimate explanation, was in retrospect a reasonable interpretation. Under that interpretation indeterminism can be one "causal result" of the initially missing information affiliated with members of a population and it manifests within obscure dimensions. **Non-linearity has historically provided one explanation for non-causal results. Non-linear equations allow bifurcated solutions with criteria for bifurcation undefined. It appears that treatment of compounded population variables (contrasting continuous variables) can likewise introduce indeterminism into the system. This occurs by virtue of** missing information regarding the composition of each smallest discrete entity, (or what constitutes the interval between population integers.) Modeling equations for those population variable systems may appear linear or non-linear, depending upon interpretation and how microzone dimensions are construed. In these unified wave analyses, deviation from perfect periodicity is negligible for straight-line chords vs. curved-arcs, except for very small n. The distinction is thus often overlooked in subsequent discussion although only the circumstance of circular-arcs on a unit-sphere or cone yields exact periodicity and radians for all n.

TABLE 1. A tabulation of the initial 10 perfect square cycles for radial vector n, tangential vector s_n, and s_{On} the closest half integer to s_n over the four quadrants.

Definitions for Partitions A, B, C, D in figure 3 where n=radial extent s_{On} = tangential extent to closest(integer/2) [forQuadrants] s_n = actual tangential extent for d-radians at n	Quadrant Q = 1 Radial extent n, where the sum of consecutive numbers = n + (n+1) = perfect square so $\sqrt{2n_1+1}$ = odd integer.	Quadrant Q = 2 Calling n_2 the ¼ quadrant to the next perfect square.	Quadrant Q = 3 Calling n_3 the midway quadrant to the next perfect square.	Quadrant Q =4 Calling n_4 the 3/4 quadrant to the next perfect square.	Calling n_5 the next perfect square so s_{n5} = next odd number after s_{n1}	Average radial extent between quadrants $\delta n = P_n+1$ "cycles"	Average change in radial extent between perfect squares	Fractional deviation from exact radians per d-radian [(d-radians-radians)/radians] = $\Re$ =
$s_{On}=\sqrt{2n+1}+Q/2-1/2$ $s_n=\sqrt{2n+1}$	$s_{On}=s_{n1}$ & $s_{m1}=\sqrt{2n_1+1}$	$s_{O2}=$ $\sqrt{2n_1+1}+1/2$	$s_{O3}=$ $\sqrt{2n_1+1}+1$	$s_{O4}=$ $\sqrt{2n_1+1}+3/2$	$s_{O5}=$ $\sqrt{2n_1+1}+2$	$\delta n=$ P_n+1	n_5-n_1 = total δn	
n→→→→→→→→→→ $s_{On}=\sqrt{2n+1}+Q/2-1/2$→ $s_n=\sqrt{2n+1}$→→→→→	$n_1=0$ 1 1	$n_2=1$ 1.5 1.732	$n_3=2$ 2 2.236	$n_4=3$ 2.5 2.645	$n_5=4$ 3 3	1	4	
n→→→→→→→→→→ $s_{On}=\sqrt{2n+1}+Q/2-1/2$→ $s_n=\sqrt{2n+1}$→→→→→	$n_1=4$ 3 3	$n_2=6$ 3.5 3.6055	$n_3=8$ 4 4.123	$n_4=10$ 4.5 4.582	$n_5=12$ 5 5	2	8	0.181
n→→→→→→→→→→ $s_{On}=\sqrt{2n+1}+Q/2-1/2$→ $s_n=\sqrt{2n+1}$→→→→→	$n_1=12$ 5 5	$n_2=15$ 5.5 5.5677	$n_3=18$ 6 6.082	$n_4=21$ 6.5 6.557	$n_5=24$ 7 7	3	12	0.0433
n→→→→→→→→→→ $s_{On}=\sqrt{2n+1}+Q/2-1/2$→ $s_n=\sqrt{2n+1}$→→→→→	$n_1=24$ 7 7	$n_2=28$ 7.5 7.5498	$n_3=32$ 8 8.062	$n_4=36$ 8.5 8.544	$n_5=40$ 9 9	4	16	0.0188
n→→→→→→→→→→ $s_{On}=\sqrt{2n+1}+Q/2-1/2$→ $s_n=\sqrt{2n+1}$→→→→→	$n_1=40$ 9 9	$n_2=45$ 9.5 9.5394	$n_3=50$ 10 10.049	$n_4=55$ 10.5 10.5535	$n_5=60$ 11 11	5	20	0.0105
n→→→→→→→→→→ $s_{On}=\sqrt{2n+1}+Q/2-1/2$→ $s_n=\sqrt{2n+1}$→→→→→	$n_1=60$ 11 11	$n_2=66$ 11.5 11.532	$n_3=72$ 12 12.041	$n_4=78$ 12.5 12.53	$n_5=84$ 13 13	6	24	0.00871
n→→→→→→→→→→ $s_{On}=\sqrt{2n+1}+Q/2-1/2$→ $s_n=\sqrt{2n+1}$→→→→→	$n_1=84$ 13 13	$n_2=91$ 13.5 13.527	$n_3=98$ 14 14.035	$n_2=105$ 14.5 14.525	$n_5=112$ 15 15	7	28	0.00465
n→→→→→→→→→→ $s_{On}=\sqrt{2n+1}+Q/2-1/2$→ $s_n=\sqrt{2n+1}$→→→→→	$n_1=112$ 15 15	$n_2=120$ 15.5 15.524	$n_3=128$ 16 16.03	$n_4=136$ 16.5 16.522	$n_5=144$ 17 17	8	32	0.00341
n→→→→→→→→→→ $s_{On}=\sqrt{2n+1}+Q/2-1/2$→ $s_n=\sqrt{2n+1}$→→→→→	$n_1=144$ 17 17	$n_2=153$ 17.5 17.521	$n_3=162$ 18 18.027	$n_4=171$ 18.5 18.52	$n_5=180$ 19 19	9	36	0.00261
n→→→→→→→→→→ $s_{On}=\sqrt{2n+1}+Q/2-1/2$→ $s_n=\sqrt{2n+1}$→→→→→	$n_1=180$ 19 19	$n_2=190$ 19.5 19.519	$n_3=200$ 20 20.025	$n_4=210$ 20.5 20.516	$n_5=220$ 21 21	10	40	0.00202
n→→→→→→→→→→ $s_{On}=\sqrt{2n+1}+Q/2-1/2$→ $s_n=\sqrt{2n+1}$→→→→→	$n_1=220$ 21 21	$n_2=231$ 21.5 21.517	$n_3=242$ 22 22.022	$n_4=253$ 22.5 22.516	$n_5=264$ 23 23	11	44	0.00166

13) INTERRELATED RADIAL AND TANGENTIAL VARIABLES

Here we elaborate characteristics of the four quadrants located between all perfect squares of the number system. For each perfect square of two summed consecutive numbers $[n_1 + (n_1+1)] = (2n_1+1)$, four distinguishable quadrants or partitions at n_2, n_3, n_4, will exist to the next higher perfect square $(2n_5+1)$. **Those four quadrants interspersed between consecutive odd perfect squares of the number system [i.e., between where $(2n+1) = 1, 9, 25, 49, - - -$ etc.], represent peaks and zero-crossings of all waves!** (13*1)

The cycles and quadrants can be most readily visualized in figure 3 by noting the ½ integer increment-magnitudes of s_n, which increments denote cyclic quadrants. A full cycle transpires between each consecutive interval where the tangential vector affiliated with any radial vector n_1 is a perfect square; i.e., wherever $s_{n1} = \sqrt{(2n_1+1)}$ = odd integer. A cycle occurs for each increase by 2 in the magnitude of s_n, and each quadrant consists of an increase by ½. The tangential vector thus provides a scale for periods or time **making periodicity literally governed by consecutive intervals between odd numbers**. Certain analysis employing straight chord-lengths necessitated approximations that can be circumvented using equivalent circularly-curved line-segments. The previous breakdown discussed large n where quadrants between A, B, C, D, of figure 3 became so close to an absolute radian the distinction was neglected. Inclusion of small n now illustrates using Table 1. In this tabulation each n_5 represents radial extent at the next higher perfect square above $(2n_1+1)$, and therefore becomes

143

p	n	n+1	2n+1	(2n+1)^.5	Angle(deg)	Angle(rad)	Cuml Angle	Cuml.Angle .2(pi)	Angle÷2(pi)-3rad	θ	Cuml.Angle .2(pi)	(n)^.sinB	x=.cosB	y=.sinB	z=.cosA	Sum(AngleCycl)
0	0	1	1	1.7320508	90	1.5707963327	1.570796633	1.84159265		0	1.842	0.637	-0.170	0.613	0.771	
	1	2	3	2.2360680	60	1.047197551	2.617963366	1.617963388		1	3.460	1.273	-1.209	-0.398	1.542	
	2	3	5	2.6457513	48.189686	0.841068671	3.459006255	1.41186500		2	4.671	1.910	0.303	-1.886	2.314	
1	3	4	7	3	41.409622	0.722734248	4.181796960	1.29353057		3	6.165	2.546	2.529	-0.300	3.085	4.181706796
	4	5	9	3.3166248	36.659898	0.643501109	4.825207911	0.02889927		4	7.004	3.183	2.193	2.307	3.856	
	5	6	11	3.6055513	33.557310	0.585085543	5.410983545	0.87100371		5	7.965	3.820	-0.423	3.796	4.627	
	6	7	13	3.8729833	31.002718	0.541096520	5.952080297	0.82040769		6	8.791	4.456	-3.592	2.637	5.398	
	7	8	15	4.1231056	28.955024	0.505360510	6.457440807	0.79075667		7	9.502	5.093	-5.030	-0.799	6.169	
	8	9	17	4.3588989	27.266044	0.475882250	6.933325573	0.76126041		8	10.344	5.730	-3.477	-4.654	6.941	
	9	10	19	4.5825757	25.841933	0.451026812	7.384352573	0.73642498		9	11.080	6.366	0.536	-6.344	7.712	
	10	11	21	4.7958315	24.619077	0.429609668	7.81405221	0.71500783		10	11.795	7.003	5.021	-4.852	8.483	
2	11	12	23	4.7958315	23.556464	0.411137862	8.22519007	0.69653603		11	12.492	7.639	7.616	-0.571	9.254	4.043393278
	12	13	25	5.1961524	22.619865	0.394781120	8.61908119	0.58505656		12	13.077	8.276	7.222	4.042	10.025	
	13	14	27	5.3851648	21.786789	0.380251207	9.00023240	0.57051665		13	13.647	8.913	4.195	7.664	10.796	
	14	15	29	5.6677664	21.039470	0.367206021	9.36744042	0.55747348		14	14.205	9.549	-0.644	9.528	11.568	
	15	16	31	5.7445626	20.364135	0.355421202	9.72286162	0.54560654		15	14.750	10.186	-5.861	8.331	12.339	
	16	17	33	5.9160798	19.749023	0.344701160	10.06756260	0.53490662		16	15.285	10.823	-9.870	4.440	13.110	
	17	18	35	6.0827625	19.186136	0.334896158	10.40245366	0.52516160		17	15.810	11.459	-11.399	-1.172	13.981	
	18	19	37	6.2449980	18.671718	0.325882957	10.72834192	0.51614840		18	16.327	12.096	-9.654	-7.014	14.652	
	19	20	39	6.4031242	18.194872	0.317540429	11.04590235	0.50782587		19	16.834	12.732	-5.473	-11.406	15.424	
	20	21	41	6.5574385	17.752780	0.308844840	11.35574699	0.50011008		20	17.335	13.369	0.746	-13.348	16.196	
	21	22	43	6.7082039	17.341443	0.302665274	11.65841228	0.49293077		21	17.627	14.006	7.305	-11.950	16.966	
	22	23	45	6.8556546	16.957426	0.295962921	11.95437518	0.48622836		22	18.314	14.642	12.590	-7.478	17.737	
3	23	24	47	6.8556546	16.657426	0.289686994	12.24406218	0.47995244		23	18.704	15.279	15.255	-0.654	18.508	4.018072103
	24	25	49	7.1414284	16.260205	0.283794109	12.52785620	0.42640319		24	19.220	15.915	14.835	6.784	19.279	
	25	26	51	7.2801099	15.642399	0.278246823	12.80610311	0.42044590		25	19.641	16.552	11.632	11.775	20.051	
	26	27	53	7.4161985	15.542471	0.273012622	13.07911573	0.41571170		26	20.057	17.189	6.113	16.065	20.822	
	27	28	55	7.5498344	15.268088	0.268063123	13.34717685	0.41076220		27	20.468	17.825	-0.841	17.806	21.593	
	28	29	57	7.6811457	15.000185	0.263373416	13.61055227	0.40607250		28	20.874	18.462	-8.084	16.599	22.364	
	29	30	59	7.6102497	14.835112	0.258921542	13.86947381	0.40162062		29	21.276	19.099	-14.410	12.535	23.135	
	30	31	61	7.9372539	14.592551	0.254686054	14.12416187	0.39738714		30	21.673	19.735	-18.742	8.181	23.907	
	31	32	63	8.0622577	14.361512	0.250656662	14.37481753	0.39335474		31	22.066	20.372	-20.315	-1.523	24.678	
	32	33	65	8.1853528	14.141110	0.246808933	14.62182646	0.38950902		32	22.455	21.008	-18.784	-9.408	25.449	
	33	34	67	8.3066239	13.930565	0.243134044	14.86476051	0.38583313		33	22.841	21.645	-14.283	-16.204	26.220	
	34	35	69	8.4261498	13.729133	0.239616596	15.10437008	0.38231785		34	23.224	22.282	-7.365	-21.018	26.991	
	35	36	71	8.5440037	13.536203	0.236251308	15.34063058	0.37895039		35	23.603	22.918	0.631	-22.880	27.762	
	36	37	73	8.6602540	13.351164	0.233022126	15.57365261	0.37572121		36	23.978	23.555	8.628	-21.543	28.534	
	37	38	75	8.7749844	13.173551	0.229921841	15.80357435	0.37262092		37	24.351	24.192	17.167	-17.045	29.305	
	38	39	77	8.8881944	13.002824	0.226944209	16.03051645	0.36051645		38	24.721	24.828	22.749	-9.946	30.076	
4	39	40	79	8.8881944	12.638568	0.224075285	16.25459173	0.36677437		39	25.087	25.465	25.439	-1.156	30.847	4.010529556
	40	41	81	9.1104336	12.680363	0.221314442	16.47590617	0.33547371		40	25.42280067	26.1014107	25.011	7.465	31.616	
	41	42	83	9.2195445	12.527905	0.218653106	16.69454037	0.33261248		41	25.75561913	26.7380304	21.717	16.598	32.380	
	42	43	85	9.3273769	12.380769	0.216085705	16.91054508	0.33032497		42	26.08586410	27.3746502	16.854	22.317	33.161	
	43	44	87	9.4339811	12.238758	0.213606564	17.12425166	0.32776585		43	26.41362995	28.0112700	8.007	26.842	33.932	
	44	45	89	9.5393920	12.101482	0.211210860	17.33546254	0.32537015		44	26.73900010	28.6478898	-1.016	28.630	34.703	
	45	46	91	9.6435508	11.968746	0.208894017	17.54435656	0.32305328		45	27.06205338	29.2845095	-10.276	27.423	35.474	
	46	47	93	9.7467043	11.840274	0.206651784	17.75100823	0.32081103		46	27.38286441	29.9211293	-18.709	23.279	36.245	
	47	48	95	9.8488578	11.715852	0.204480109	17.95548952	0.31863948		47	27.70150387	30.5577401	-25.680	18.563	37.017	
	48	49	97	9.9498744	11.595273	0.202375696	18.15786421	0.31653465		48	28.01803862	31.1943588	-30.175	7.908	37.788	
	49	50	99	10.0498756	11.478341	0.200334642	18.35810905	0.31449411		49	28.33253293	31.8309608	-31.777	-1.851	38.559	
	50	51	101	10.0498756	11.364877	0.198354522	18.55655357	0.31251379		50	28.64504672	32.4676084	-30.262	-11.762	39.330	
	51	52	103	10.1488916	11.254713	0.196431792	18.75208538	0.31059106		51	28.95563708	33.1042282	-25.714	-20.849	40.101	
	52	53	105	10.2489500	11.147691	0.194563912	18.94754928	0.30873218		52	29.26436095	33.7408479	-18.512	-28.209	40.872	
	53	54	107	10.3440804	11.043566	0.192748325	19.14029760	0.30690759		53	29.57126854	34.3774877	-9.297	-33.096	41.644	
	54	55	109	10.3440804	10.942690	0.190982634	19.33120024	0.30514190		54	29.87641044	35.0140877	1.095	-34.997	42.415	
	55	56	111	10.5396538	10.844053	0.189264586	19.52054483	0.30342398		55	30.17983431	35.6507073	11.711	-33.672	43.186	
	56	57	113	10.6301458	10.748236	0.187592105	19.70813594	0.30175137		56	30.46156668	36.2873270	21.567	-29.183	43.957	
	57	58	115	10.7238053	10.654905	0.185963182	19.89410012	0.30012245		57	30.78170612	36.9239488	29.744	-21.879	44.728	
	58	59	117	10.8166538	10.563065	0.184375970	20.07847609	0.29853524		58	37.5605666	37.5605666	35.404	-12.373	45.499	
5	59	60	119	10.9087121	10.475314	0.182828717	20.26130480	0.29698768		60	31.37723134	38.1971863	38.160	-1.478	46.271	4.006713073
	60	61	121	11												

Table 2 . A computer tabulation is presented for some of the more relevant variables from n = 0 to n = 60.

144

the next n_1 at the subsequent Quadrant 1 column a triple-row below. Each horizontal range between n_1 and n_5 constitutes a perfect-square-cycle and one cycle of wave comprised of its four quadrants. Table 2 tabulates comparable values to n = 60.

Discussing **cases with <u>straight-line chord approximations</u>** simplifies by using another name for angular partitions of a wave very close to, but not exactly one radian. Angular separations between n vectors of figure 3, where tangential-vector s_n magnitudes become ***closest to*** multiples-of-½, are now called ***derived radians***, or "***d-radians***". Angles $\theta_n = \cos^{-1}[n/(n+1)]$ between directions ABCD sum to those d-radians. For straight-chord analyses, d-radians provide a **derived degree of rotation** very close to a radian. They are mathematical (rather than physical) and monotonically approach an absolute radian as n $\rightarrow \infty$. Errors of straight-chord approximations would vanish if sin θ_n everywhere equaled θ_n in radians. **Than d-radians would equal absolute radians**. Angular traverse between two n vectors that both result in consecutive perfect squares of (2n+1) is called a **quad-radian** or a cycle. The four summed d-radians intervening consecutive odd perfect squares of (2n+1) will be very close to <u>the 4 absolute radians of actual unified waves</u>. As n $\rightarrow \infty$, quad-radians asymptote into 4 exact radians with each d-radian evolving to one absolute radian. A d-radian thus represents a quarter-cycle or quadrant, which designates as a quad-radian/4. All cumulative angular amounts between where the magnitude of s_n is integer or differs from integer by ostensibly ± 1/2, represent one angular quadrant of the wave. Those conditions therefore depict the four quadrant partitions of time per each incremental increase by 2 in the magnitude of s_n. The four quadrants of each cycle thus distinguish those intervals between where the magnitude of s_n is   odd, odd +1/2, odd +1, odd +3/2, odd +2.   (13*2)

From equation 5*11, for each consecutive cycle then, transpired cycles

$P_n = (s_n-1)/2$ increase by unity and s_n increases by 2.   (13*3)

For the "spiral fan" plotted within a plane as in figure 3, angular radial-vector-directions for n close to A, B, C, D, readily mark the four d-radian partitions between odd perfect squares. Though in figure 3 the triangles illustrate on a plane, the accumulated-sum angles of θ_n between many n vectors would remain **unaltered** if concatenated-triangle directions changed into or out of that plane. **Cumulative summed intervening-triangle-angles θ_1 + θ_2 + θ_3 + + + etc. determine odd perfect square spacing and transpired cycles, <u>not the surface on which the spiral is rendered</u>**. The compiled numerical sum of individual triangle angles remains the same regardless of what relative direction each consecutive θ_n heads in. Table 1 shows those angles divide each cycle of consecutive perfect squares (between n_1 to n_5) into the 4 quadrant regions of figure 3. Those quadrants approximately take place between A & B, B & C, C & D, and D & E (or D & A). The average radial change during each quadrant is δ_n and each d-radian delineates the angular sum of δ_n distinct origin angles (triangles) in figures 3.   Comparative directional planes of those triangles remain irrelevant to the compiled sum of the individual angles θ_x. Each d-radian always "begins and ends" were the magnitude of

s_n has increased by an increment of ½. Directional planes of the triangles have no effect on the magnitude of each s_n.

Every quarter of a quad-radian (or quadrant) designates a conventional quadrature extent of a composite harmonic wave. **Odd integers of s_n at perfect squares of (2n+1) indicate "start-and-finish" zero-crossings of each cycle.** Quadrants between those zero-crossings denote the quadrature components of the wave. Just as four quadrants occur in the X – Y Cartesian Coordinate planes of rectilinear Euclidean space, four d-radians analogously divide each repetitive region between perfect squares in this unified wave "Phase space". It might be called phase space because relative phase angles of the vectors within have relevance. It is repeated that the additive sum of consecutive θ_x values between any pair of n vectors remain invariant, even if those successive concatenated triangles did not remain within a plane or on a cone. For arbitrary or randomly oriented triangles, the **_net_** resultant angle between that vector pair **_can be any value_** less than the accumulated sum angle. That net resultant angular location may be indeterminate (lack predictability) and not be of consequence, but that sum must be traversed in the separate triangles as

$$\theta_n + \theta_{n+1} + \theta_{n+2} + \theta_{n+3} + \text{- - - -} = \theta_{sum} \qquad (13\text{*}4)$$

regardless of relative triangle orientations. **The gross sum angles θ_{sum} constitute each cycle, not the undetermined net resultant angle.** Since the 4 quadrants A, B, C, D of figure 3 occur wherever magnitude s_n is closest proximity to integer multiples of ½, that qualification serves as a marker. Those half integer markers of

147

s_n delineate where tangential magnitude "passes through", $(s_{n1}+1/2)$, $(s_{n1}+1)$, $(s_{n1}+3/2)$, $(s_{n1}+2)$; where s_{n1} depicts any odd integer as listed in Table 1. Between n_1 and n_5 the three numerical locations n_2, n_3, and n_4 (each a half-integer apart in s_n), define four quadrants partitioning each pair of odd perfect squares. **Quadrants of the wave thus enunciate solely from properties of the natural numbers articulated by odd perfect squares of (2n+1).**

Table 1 can help clarify these nomenclature definitions. Starting from n = 0, d-radians occur associated with changes in n by amount δ_n. Thus the first four d-radians apportion as n = unity apart. The first quad-radian sum of those four d-radians ensues between n = 0 [where (2n+1) = 1], and n = 4 [where (2n+1) = 9]. Both those values [1 and 9] represent consecutive odd perfect squares and the next odd perfect square would be at n = 12, [where (2n+1) = 25]. For these extremely small values of n, the angle of respective d-radians differs moderately from radians, but as n increases, convergence toward radians occurs very rapidly. [The right hand column for $\Re$ indicates the convergence rate, which error **varies as $1/P^2$ =** $(1/P)(1/P) \approx$ [1/(chord-radians)][1/(chord penetration into cone)]. The four quadrants between n = 0 and n = 4 articulate partitions **closest to** where magnitude s_n equals increments of ½, i.e., between where s_n =1, 3/2, 2, 5/2, and 3. The four quadrants where P = 0 constitute the zero'th prior cycle of the wave. In the next quad-radian, P = 1 with an increment of 2 = δ_n per d-radian (or quadrant); namely quadrants occur at n = 4, 6, 8, 10, 12. There quadrants are between s_n = 3, 7/2, 4, 9/2, and 5. Angular traverse amid those n vector values comprise d-radians of the cycle with

P = 1. That cycle from n = 4 to n = 12 also establishes 4 quadrants. Each quadrant expressed through every ½ integer increment in s_n constitutes a quarter cycle. The process continues to n → ∞ as indicated by the table.

Radial vector n as energy-time units might interpret as a vehicle for determining s_n, which will be shown to indicate time, or numerically, half-periods from the origin. Once determined, the closest extents of s_n to increments of 1/2 delineate a tangential marker scale for the 4 "quadrants" per cycle. **Integers of s_n mark zero-crossings of the wave, the odd ones delineating a full cycle synchronous with perfect squares**. The magnitude of n (called n_1 where s_{n1} is an odd integer), begins each cycle that ends at n_5 with s_n greater by 2. The initial odd integer of $s_{n1} = \sqrt{(2n_1+1)}$ for instituting tangential quadrant markers derives from n at n_1. Tangential vector increments of magnitude ½ add to s_{n1} to demarcate consecutive quadrants until $s_{n5} = (\sqrt{(2n_1+1)} +2)$ (13*5)

coincides with radial vector n_5. The ½-integer tangential extents **so defined as s_{Qn}** form quadrants to grant a **"linear-tangential-scale for time orthogonal to energy-time"**. Table 1 indicates how that linear half-integer scale bears similarity to, and intermeshes with the scale for radial vector n. **Integers δ_n apart between n_1 and n_5 define the quadrants redundant with the four half-integer partitions of s_n.**

Table 1 makes visible inherent relationships that embed within consecutive perfect squares of the number system. A partial listing is as follows:

$(n_5-n_1) = s_{Qn2} + s_{Qn4} = 4\delta n$ $\qquad$ $s_{Qn3} = 2\delta n$

$(n_5-n_1)/2 = \sqrt{(2n_3)}$ $\qquad$ $2(P_n+1)^2 = n_3 = 2\delta n^2 = (n_4-n_3)s_{Qn3}$

$n_2+n_4 = (n_4-n_2)^2$ $\qquad$ $(n_2+n_4) = s_{Q3}^2$

$(\delta n)s_{Qn3} = n_3$ $\qquad$ $(n_2-n_1)s_{Qn1} = n_2$

$10\delta n = s_{Qn1} + s_{Qn2} + s_{Qn3} + s_{Qn4} + s_{Qn5}$ $\qquad$ $(n_3-n_2)s_{Qn1} = n_2$

$(n_5-n_4)s_{Qn3} = n_3$ $\qquad$ $(s_{Qn1}+s_{Qn5})\delta n = 2n_3$ $\qquad$ (13*6)

These equations interweave integers and half integers associated with directions A, B, C, D in figure 3. **They constitute "identities" under "*orthogonality-axiom trigonometry*"**, with the spiral being the fundamental circulatory axis of that number system. A host of further expressions can be compiled with these and other formulations derived herein. While these relationships exhibit simplicity, it is important to recognize that they all apply to an infinite array of circumstances between all perfect squares from n = 0 to n = ∞. Each subscript refers to an occurrence within a cycle and each consecutive cycle **embodies repeated applicability of all equations**. The equations represent intrinsic periodicity within natural numbers and a generic basis for waves. It is not by chance that the orthogonality axiom brings forth this chain of consistency woven into its numerical fabric. **<u>For non-divisible entities, orthogonality between tangential and radial progression from an origin is virtually as basic as the condition of linearity, just less familiar</u>**. Natural harmonic architecture affiliating

150

tangential advance between integers with the cadence of perfect squares has simply been absent from our repository of basic concepts. **In that format the number 4 in forming harmonic-quadrants supplants the 2π of a circle, conveying "rectilinearity and rationality" to curved space, rather than the discordant irrationality of Cartesian-Coordinate Euclidean-space. Being the foundation of mathematics, integers n directly link <u>radial direction n, tangential advance s_n, quad-radian angle, cycles P, spacing per quadrant δ_n, quadrature wave components about an origin, the non-prime quadrant dividers, and as will be shown, summed RMS transitions equaling n, the energy-time/cycle of Planck's constant $\Delta\theta = h$</u>.**

Numerical symmetry within the entire number system materializes relevant to those half-integers of s_n, with its odd and even integers signifying the wave's "zero crossings". Referring to Table 1, the four partitions characterize 4 divisions between five radial values, n_1, n_2, n_3, n_4, n_5, with n_1 and n_5 representing any pair of consecutive perfect squares in (2n+1). At n_1, tangential vector $s_{n1} = \sqrt{(2n_1+1)}$ = an odd number, and the tangential vector magnitude at n_5 is

$s_{n5} = \sqrt{(2n_5+1)}$ = (the next higher odd number) = $[\sqrt{(2n_1+1)} +2]$.   (13*7)

These odd integers of s_n at perfect squares everywhere **mesh exactly** with the 4 angular quadrants they create. Portrayed within a plane, the fractional deviation error between d-radians and exact radians as, [(d-radians - radians)/radians] $\equiv \Re$ is shown in Table 1 for magnitudes of s_n between 1 and 23

151

(about 10 cycles). Toward larger n, ratio $\Re$ diminishes to miniscule value in proportion to

$1/(1+s_n)^2 \approx 1/s_n^2$, and equates to zero as $n \to \infty$.   (13*8)

Deviation between d-radians and absolute radians will functionally vary as the effect of (summed curved-chord-lengths on a conic or unit-sphere surface)÷ (summed straight chord-lengths within a conic surface). That factor varies as $1/P^2$. The tabulation of Table 1 interrelates: tangential progression time as s_{Qn} and s_n, radial progression as energy-time units n from the origin, the associated perfect square cycles, plus the numerical integer advance of quadrants. It portrays the intrinsically symmetrical periodicity of ordinal numbers (as values of n and s_{Qn} in the table) between all consecutive perfect squares throughout the number system.

As successive perfect squares of (2n+1) advance with increased n, the number of consecutive integers of n that comprise the 4 partitions, [i.e. δn in the table] increases by unity. Corresponding tangential extent at n_1 is

$s_{n1} = (2δn-1) = 2P_n +1$, where P_n = transpired cycles at n_1.   (13*9)

Magnitude s_n apportions to half-cycles traversed, and (2n+1) expresses the repeated perfect squares that signify those cycles. Radial increase between quadrant partitions, $δn = P_n+1$ emerges as another useful parameter besides the variables previously discussed throughout this work. During each "cycle of perfect squares", growth in n characterizing each traversed d-radian (e.g., between n_2 and n_3, or n_3 and n_4, etc.), exceeds the number of transpired cycles by unity, [by (P_n+1)].

Each d-radian thereby signifies how many cycles it is from the origin by that integer increase in n ["by the number of sample data points'] during the course of that d-radian. <u>The unified wave thus carries an inherent "message" of where that d-radian is relative to the *origin*, a message missing from sinusoids</u>.

(13*10)

14) GRAPHIC PORTRAYAL OF INTERCONNECTED VARIABLES

Figure 14 reiterates the structural procedure used to devise figures 1 and 2. The insert at the upper left duplicates figure 1 illustrating graphical construction topology for the initial 5 triangles of the 13 triangles shown below and to the right. Abscissa increment n represents fixed discrete units of (energy)(time), with the ordinate being time in half-periods (using the same scale factor). The ordinate direction is likewise s_n since that variable describes tangential extent associated with every n. The situation depicts the earlier discussion for monads with orthogonal transitions along a Euclidean axis for non-divisible energy-time units. The ½ increment quadrants along s_n profile an harmonic wave, a dotted-line segment being delineated for reference at each perfect square of (2n+1). Each sequential n vector would add

$$\{\sqrt{[2(n+1)+1]}-\sqrt{(2n+1)}\} \text{ to the prior magnitude of } s_n. \qquad (14*1)$$

Those segment lines **are drawn broadened** in the main and upper left sketch. Between each cycle of successive perfect squares of (2n+1), the sum of those increases in s_n totals 2. Considerable discussion herein relates those broadened-lines to great-circle extents on a unit-sphere, which equivalence results in perfect harmonic periodicity throughout all values of n. Rather than straight-chords previously analyzed for understanding, when subsequently interpreted as curved arcs around any 4-radian perimeter on a unit-sphere, those chord-extents produce exact periodicity for all n. Perfect squares of (2n+1) occur at n = 0, 4, 12, etc.,

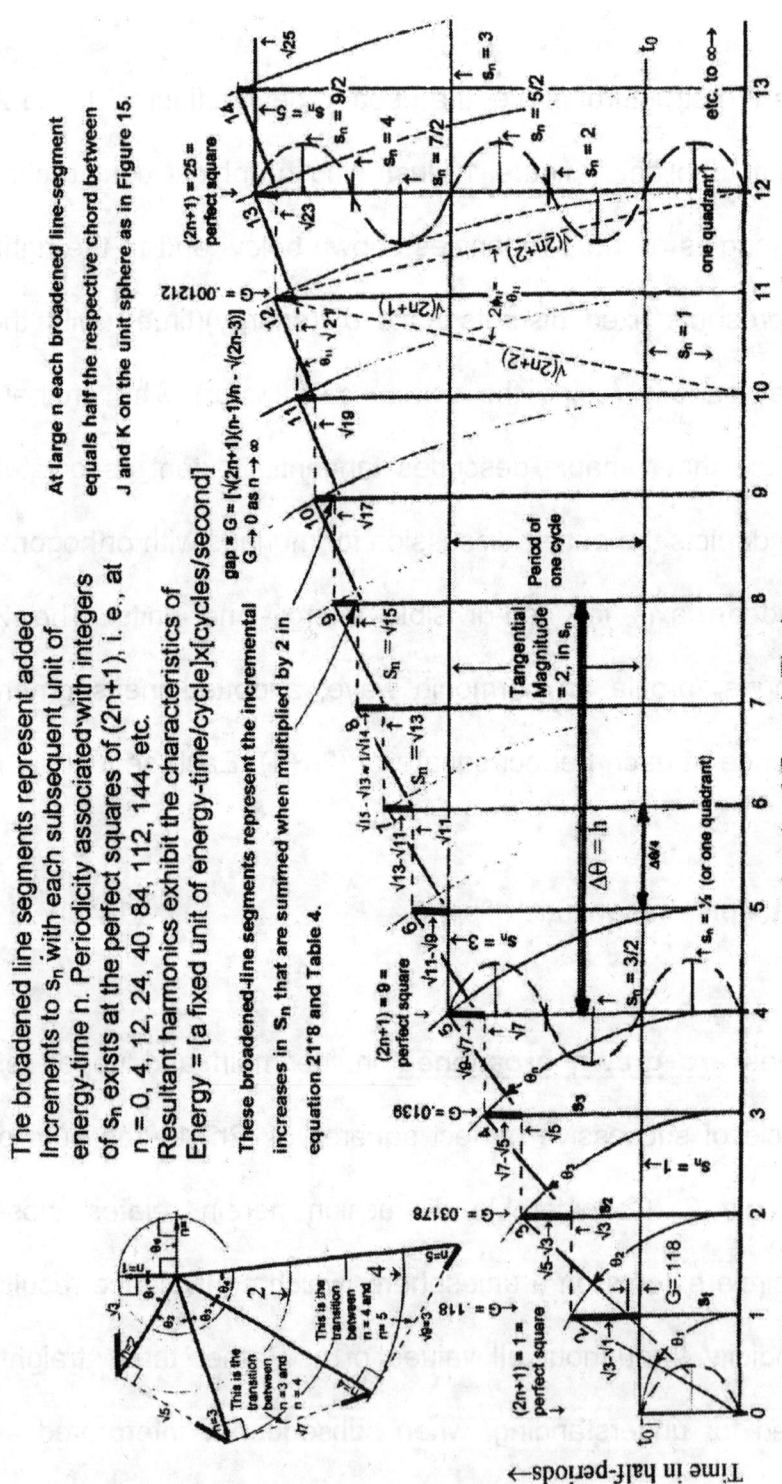

The broadened line segments represent added increments to s_n with each subsequent unit of energy-time n. Periodicity associated with integers of s_n exists at the perfect squares of $(2n+1)$, i. e., at n = 0, 4, 12, 24, 40, 84, 112, 144, etc. Resultant harmonics exhibit the characteristics of

Energy = [a fixed unit of energy-time/cycle]x[cycles/second]

These broadened-line segments represent the incremental increases in s_n that are summed when multiplied by 2 in equation 21*8 and Table 4.

At large n each broadened-line-segment equals half the respective chord between J and K on the unit sphere as in Figure 15.

Figure 14. This shows detail in the construction of figure 1, which is duplicated in the upper left. Each origin angle θ_n (and triangle) of the spiral derives through an orthogonal transition s_n (shown vertical) from a lower to a higher integer (n+1). A periodicity prevails associated with incremental increases of 2 in s_n at perfect squares of $(2n+1)$ with quadrants of that periodicity straddling increments of $s_n = \frac{1}{2}$. Waves of that periodicity are portrayed as dashed sinusoids synchronous with those perfect squares. Throughout this plot, even when extended toward n → ∞, fixed units of (energy)(time) everywhere comprise from (energy)(time beyond t_0 @ $s_n = 1$). Thus, (energy-time/cycle) remains a fixed unit allegedly equaling Planck's constant as in h = E/ν. Broadened vertical lines at the end of each s_n vector portray the (extension in time-from-the-t_0-origin) contributed by that respective s_n. As n increases to the right, s_n continues to increase by 2 for each perfect square in $(2n+1)$, i. e., at n = 12, 24, 40, 60, 84, 112, etc.

and the process continues to n = ∞. Off to the right in figure 14 the sum of all incremental increases in s_n between any two perfect squares will always be 2, characterizing one complete cycle in time. Each such cycle would exhibit synchronous overlap of ABCD directions when the equivalent spiral in the figure's upper left would plot around a cone of apex angle $\beta = \sin^{-1}(2/)\pi)$.

Figure 14 is useful to illustrate how fixed units of (energy)(time)/cycle result. That intrinsic characteristic elucidates the constancy of Planck's constant in the relationship E = hv. Consider two consecutive perfect squares as n_{PS1} and n_{PS2}. Using $n_{PS1} = 4$ and $n_{PS2} = 12$ as exemplary, $\Delta n_{PS} \equiv (n_{PS2} - n_{PS1}) = 8$ represents the energy time intervals between those perfect squares. During that total interval, s_n increases by 2 granting four quadrants of tangential magnitude ½ each. Were the next interval to the right between n = 12 and n = 24 examined, s_n would similarly increase by 2 during one cycle. Such increase by 2 would apply for each and every interval between perfect squares of (2n+1) out toward n → ∞. Now consider the vertical region beyond where n = 0 and s_n = unity in figure 14. Let $s_n = 1$ constitute the reference ordinate t_o starting value for time. Phenomena of interest begin after t_o where $s_n = 1$. The energy-time between n = 0 and n = 4 represents the first pair of perfect squares. During that interval one cycle beyond t_o occurs. The energy-time/cycle there would be $\Delta n_{PS}/1 = 4$. During the next perfect square interval (where n increases from 4 to 12), two cycles transpire after t_o. Those differential $\Delta n_{PS} = 12 - 4 = 8$ energy-time units associate with two cycles since t_o, so the energy-

157

time/cycle there would be 8/2 = 4. For the next perfect-square interval n increases from n = 12 to n = 24 in three associated cycles since t_o. There Δn_{PS} = 12 in three cycles since t_o and energy-time/cycle = 12/3 = 4. **In this manner the energy-time/cycle remains invariant as 4 throughout all perfect square intervals to n** $\rightarrow \infty$. Each full cycle represents an s_n increase of 2, so writing the relationship in s_n rather than cycles,

$$(\Delta n_{PS(1to2)})/(s_{PS2} - 1) = 4/2 = 2 = (\text{energy-time/cycle})/(\text{unit-of-}s_n). \qquad (14*2)$$

This description utilizes the "same" unit scale for abscissa and ordinate but simulates the relationship $h\nu = E$. When traversal through the four quadrants occurs faster, timescale diminishes and energy must increase to maintain the same fixed units of (energy)(time) along the scale. For frequencies that are higher [when timescale is short], the energy variable must be correspondingly greater. Thus, the orthogonality axiom intrinsically yields a non-localized wave where energy can proportion to frequency instead of amplitude squared, with a fixed energy-time/cycle constant of proportionality between them. The earlier part of this work showed that the magnitude of that constant is Planck's constant, the minute excess-angle per cycle $\Delta\theta$. In the figure, Planck's constant in $E = h\nu$ represents the triple horizontal line that would always extend 4 integers of n, being the intrinsic energy-time/cycle of the wave. The portrayal is valid whether all vectors n are represented along one axis as in figure 14, or as a circulating spiral in figures 1 to 4 (or however radial vectors are depicted). Earlier discussions distinguished between smallest energy-time units n and Planck's constant in the relationship $h = E/\nu$ = energy-time/cycle.

The latter per cycle unit happens to be equivalent to a separation of 4 in the sequence of n smallest energy-time units. **Thus, smallest energy-time units and the energy-time/quadrant should be identical**. Moreover, each quadrant represents one absolute radian on the cone. Perhaps a simpler statement stems directly from the periodic repetition every two-fold increase in s_n. Time/cycle around the cone = time/cycle between perfect squares of (2n+1) = sum of the broadened-line intervals between perfect squares of (2n+1) in figure 14 = constant = $1/\nu$ for any fixed ν. Since energy decreases as time/cycle increases, multiplying this by energy allows energy-time/cycle to be valid for all frequencies.

A subtlety emerges in establishing the wave energy. It indicates the necessity of treating h as energy-time/cycle in E = hν. Dividing energy-time (n_{PS2}) by time (S_{PS2} − 1), [rather than dividing $\Delta n_{PS(1to2)}$] as previously employed for the (energy-time/cycle) computation, yields an ever increasing result for energy. Namely, $(n_{PS2})/(S_{PS2} - 1)$ results in the non-constant consecutive results:

4/2 = 2, 12/4 = 3, 24/6 = 4, 40/8 = 5, etc. Using time at consecutive perfect squares, [as n_{PS2}/s_{PS2} rather than $n_{PS2}/(s_{PS2} -1)$], would equal 4/3, 12/5, 24/7, 40/9, etc. and is similarly non-uniform. Constancy for the ratio of (energy-time) to (time) can be satisfied only by using $\Delta n_{PS(1to2)}$ in terms of per cycle dimensional; units, i.e., as (energy-time/cycle) rather than (energy-time). While $\Delta n_{PS(1to2)}$ signifies the (energy-time) of a full cycle, n_{PS2} would reference the origin instead of the earlier perfect square. **Such an (energy-time) variable would therefore avoid cyclic units in its denominator and be inappropriate**. Since n advances faster than s_n,

159

(energy-time) units divided by (time) would be inconsistent with a constant energy. The ratio could not satisfy h in analogue of the equation E = hν = (energy-time per cycle)/(time per cycle) =

(energy-time/cycle)(cycles/time)   (14*3)

The extra half cycle encountered when using time t_0 referenced at s_n = 1, [instead of at s_n = 0], adds deception to the picture. Though offhand seeming problematic, it is appropriate since $s_n = \sqrt{9}$ associates with n = 4, $s_n = \sqrt{7}$ with n = 3, $s_n = \sqrt{5}$ with n = 2, $s_n = \sqrt{3}$ with n = 1, $s_n = \sqrt{1}$ with n = 0.   (14*4)

The gap between n = 0 and n = 4 is the interval between s_n = 1 and s_n = 3. Were h not in units (energy-time/cycle), constancy for the ratio (energy-time)/(time) = n/(time from a reference) might then only be obtained by referencing two time events one cycle apart, namely at $(S_{PS2} - 1)$ and $(S_{PS2} + 1)$. That is, the only relationship that would allow the referenced time to "track" with (energy-time) is in the form $[(n_{PS2})/(S_{PS2} - 1) - (n_{PS2})/(S_{PS2} + 1)]$. That would be the only possible way for (n_{PS2}) to yield a constant energy result <u>by virtue of the everywhere valid identity</u>:

$[(n)/\{\sqrt{(2n+1)} - 1\}] - [(n)/\{\sqrt{(2n+1)} + 1\}] = 1$ (14*5)

However, the E = hν expression does not entail any subtraction between two periods or frequencies so h could not depict (energy-time) dimensional units without including dimensional cycles (or quadrants, or radians) in its denominator! Ratios involving (n_{PS2}) could remain constant (as unity) only by referencing two separate terms for time. It would also require reference to **a full cycle's separation in time** as between $(S_{PS2} - 1)$ and $(S_{PS2} + 1)$. **In the physics of natural numbers the**

equation E = hν can occur only with constant h involved on a per cycle basis.

Cycles in the denominator relate to traversals through Δθ = h normalized per cycle, reemphasizing that energy must be the rate of passing through Δθ. Units of (energy-time/cycle) are equivalent to a fixed angular segment of a cycle per cycle, as would result from an additive increment to π ≡ perimeter/diameter. The energy-time/cycle of Planck's constant represents the energy-time associated with each additive period of the wave as depicted in Table 1 or any figure herein.

Figure 14 can clarify dimensional units. The illustration portrays a wave wherein each vertical-line at integer-n delineates advance of one smallest unit of energy-time. Each such unit was not originally specified with per-cycle dimensions. It appears to have per quadrant or per conic radian dimensional units, where each quadrant will later be shown to associate with unity. Each quadrant represents one d-radian. An important characteristic of this wave exemplified in the photon energy equation E = hν is as follows. Throughout the entire wave a constant k exists wherein energy E = kν. Dimensional units for this k would be:

k = [(the number of energy-time units n between perfect squares)/(number of cycles since the start of the wave)] =

(energy since start)(time of one cycle)/(cycles since start) =   (14*6)

(energy/cycle)(time of one cycle) = (energy)/(cycle/second)

Thus k retains the dimensions: energy-time/cycle in the relationship E = kν [or E = hν]. Now it turns out this constant k can "coincidentally" associate with a "distance" along the energy-time axis of the figure. Because of its definition and constancy, k will everywhere correspond to four units of energy-time n as exemplified by the longer horizontal triple line with arrows. That is the (energy-time) between perfect squares)/cycle = (energy)/(cycle/second).

The constancy property of k takes on dimensional units of energy-time/cycle though it emerges as equivalent to **four energy-time integers** along the abscissa. Constant k with dimensional units of cycles in its denominator has the same numerical value of four consecutive energy-time integers, each of which contributes **one unit of energy-time originally specified without cycles in the denominator**. The distinction clarifies between h = E/ν = [energy-time/cycle] in the relationship characterizing composite properties of the wave, and each abscissa unit-of-n originally expressed as energy-time. The differentiation analogously applies to h = E/ν = energy-time/cycle in the photon energy equation and a unit of action in quantum mechanic equations. The original energy-time unit is the same as energy-time/quadrant, where a quadrant is unity as one absolute radian. Constancy and dimensional units between energy and frequency also verify algebraically from Table 1. They derive from the expression, δ_n = (P_n+1), making δ_n/(P_n+1) = the constant unity throughout the entire wave. Four times δ_n (14*7) represent the per-cycle energy-time increment $\Delta n_{PS(1to2)}$ and (P_n+1) is the accumulated number of Prior cycles from the origin to the respective second perfect square. The constant k = [4δ_n/(P_n+1)] will always numerically equal four energy-time units along the abscissa and has dimensional units [energy-time/cycle].

Multiplication by cycles/second ν effectively interchanges the scales of time and energy under mutual exclusion. As time/cycle decreases, energy reciprocally increases. That confirms the fundamental nature for a monadic unified wave of maximum orthogonality.

Each integer n in figure 14 equals $\Delta\theta/4$ = Planck's constant/4. While frequency is recognized as the cycle after cycle rate, **quadrants actually delineate smallest wave elements. They are segmented by non-prime axis-values-of-n in figure 13 and the tabular rows of Table 1**. Thus θ_{min} represents two energy-time units and in E = hν, Planck's constant h = $\Delta\theta$ represents the energy-time of four quadrants. <u>**A wave comprised of n consecutive (energy-time) units never encounters a prime number for n at the peaks and zero-crossings of the wave**</u>. Missing primes along A,B,C,D, quadrant axes indicate (energy-time) will be divisible there by small-prime n-values that previously occurred "off the quadrant axes". Quadrant repetition points always associate with n values that are divisible, which <u>**physically links numerical divisibility with wave properties**</u>. It grants unification and invariant properties to the wave. **Quadrants cleaving perfect squares in (2n+1) of the natural number sequence n distinguish all waves, even for exponential harmonics definable as n → ∞. The process and its associated phenomena remain valid throughout all values of n to ∞, which limit case encompasses exponential harmonics.** <u>Recognizing the role of non-primes and perfect squares within waves provides an example of attributes gleaned through higher order treatment that are not discernable from the lower order (exponential) structure.</u>

163

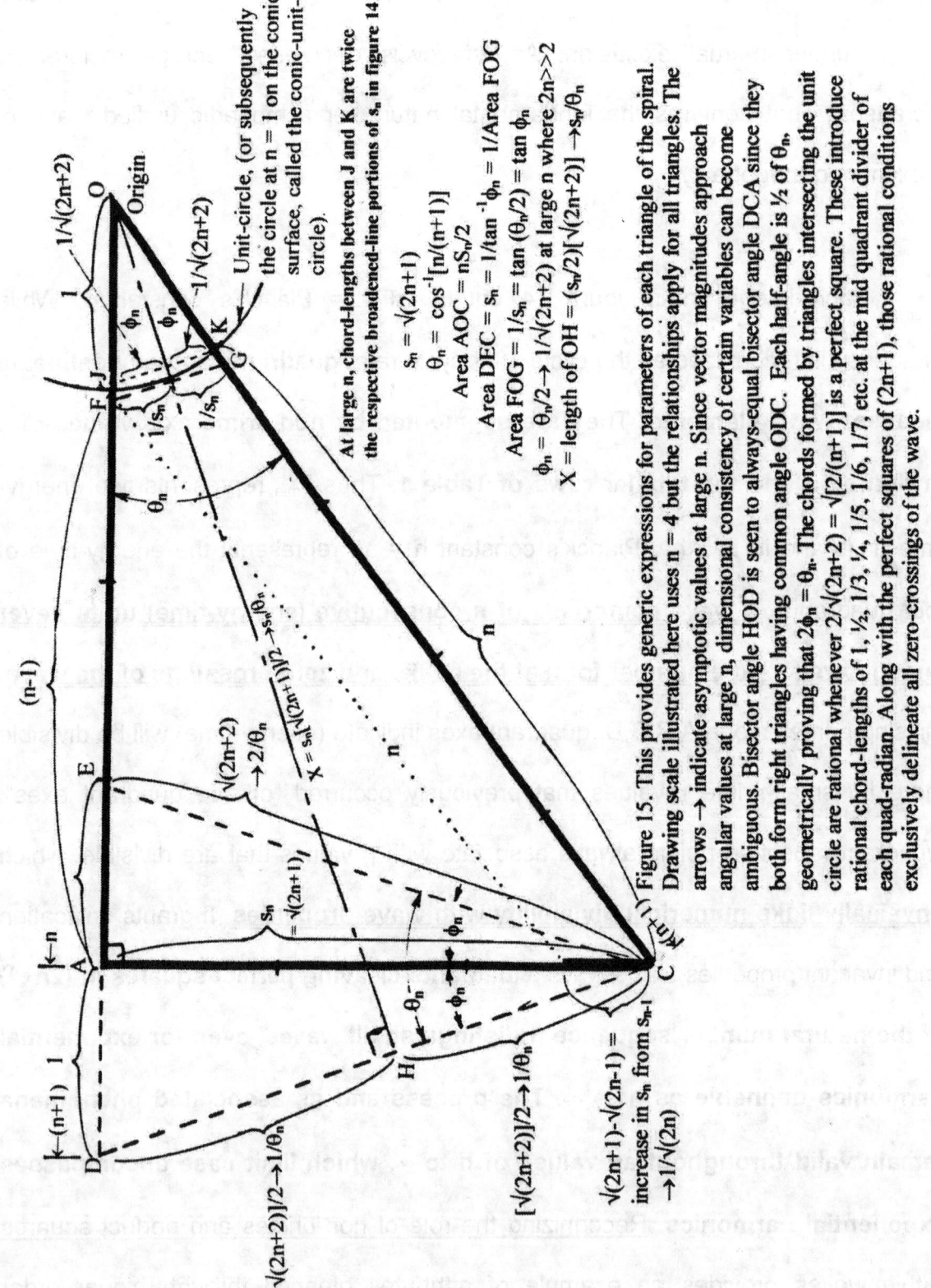

Figure 15. This provides generic expressions for parameters of each triangle of the spiral. Drawing scale illustrated here uses n = 4 but the relationships apply for all triangles. The arrows → indicate asymptotic values at large n. Since vector magnitudes approach angular values at large n, dimensional consistency of certain variables can become ambiguous. Bisector angle HOD is seen to always-equal bisector angle DCA since they both form right triangles having common angle ODC. Each half-angle is ½ of θ_n, geometrically proving that $2\phi_n = \theta_n$. The chords formed by triangles intersecting the unit circle are rational whenever $2/\sqrt{(2n+2)} = \sqrt{[2/(n+1)]}$ is a perfect square. These introduce rational chord-lengths of 1, ½, 1/3, ¼, 1/5, 1/6, 1/7, etc. at the mid quadrant divider of each quad-radian. Along with the perfect squares of (2n+1), those rational conditions exclusively delineate all zero-crossings of the wave.

Within the figure:

Origin

O

ϕ_n ϕ_n

1 1/√(2n+2)

-1/√(2n+2)

K

Unit-circle, (or subsequently the circle at n = 1 on the conic surface, called the conic-unit-circle).

F J

G

1/s_n

1/s_n

θ_n

At large n, chord-lengths between J and K are twice the respective broadened-line portions of s_n in figure 14.

$s_n = \sqrt{(2n+1)}$

$\theta_n = \cos^{-1}[n/(n+1)]$

Area AOC = $nS_n/2$

Area DEC = $s_n = 1/\tan^{-1}\phi_n = 1/$Area FOG

Area FOG = $1/s_n = \tan(\theta_n/2) = \tan\phi_n$

$\phi_n = \theta_n/2 \rightarrow 1/\sqrt{(2n+2)}$ at large n where 2n>>2

X = length of OH = $(s_n/2)[\sqrt{(2n+2)}] \rightarrow S_n/\theta_n$

(n-1)

E

√(2n+2) → 2/θ_n

X = $s_n[\sqrt{(2n+1)}]/2 \rightarrow s_n/\theta_n$

n

n

↓n

A

↑(n+1) 1

D

$s_n = \sqrt{(2n+1)}$

ϕ_n

θ_n

ϕ_n

H

[√(2n+2)]/2 → 1/θ_n

[√(2n+2)]/2 → 1/θ_n

√(2n+1)-√(2n-1) = increase in s_n from s_{n-1} → 1/√(2n)

C

The exponential function gives no clue that its peaks and zero crossings always associate with non-prime values of energy-time. Being part of the family of unified waves, that property must correspondingly apply as n → ∞. Additional interpretations emerge from this analysis. **Time per cycle is always constant, precisely equaling each s_n increase by 2**. Frequency and units of (energy-time/cycle) remain invariant. It furthermore may "justify" the prime values [where n = 1, 2, 3] on the axes of figure 13. Points less than P = 1 or n = 4 avoid the repetition criteria for subsequent quadrants of the perfect squares.

Figure 15 furnishes expressions for relevant vector lengths and angles of every spiral triangle. The radical term **√(2n+2)/2**, [or its reciprocal] emerges to describe various vector magnitudes, and at large n, angles. Those magnitudes are rational only when the radicand is a perfect square. **Thus for example the only rational-value-chord-lengths between J and K where each triangle would intersect the unit conic circle (or unit sphere), arise whenever 2/√(2n+2) = √[2/(n+1)] comprises a perfect square in the radicand**.   (14*8)

Triangles satisfying that condition occur only where

(n+1) = 2, 8, 18, 32, 50, 72, 98, 128, etc. with respective chord lengths

√[2/(n+1)] = 1, ½, 1/3, ¼, 1/5, 1/6, 1/7, 1/8, etc. to 1/∞.   (14*9)

Likewise, integer rationality for vector lengths occur wherever the reciprocal √[(2n+2)]/2 similarly appears in figure 15. These values (n+1) all characterize "midpoints" between the perfect squares of (2n+1), i.e. along direction C of figure 3. Circular markers on the spiral of figure 17 also depict such points, which continue to n → ∞. They signify the n_3 mid-quadrant "zero-crossing points"

165

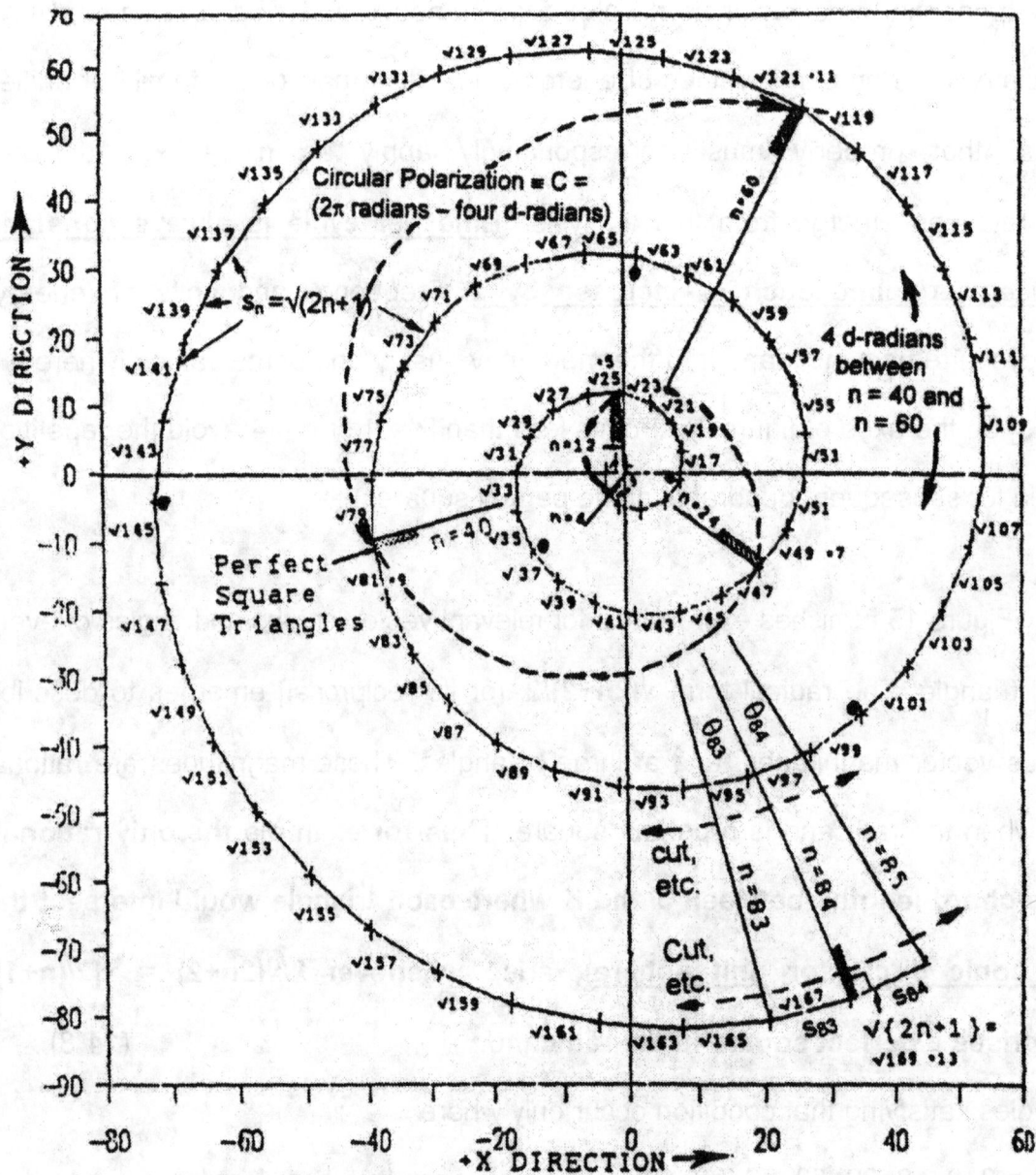

Figure 17. This is an identical plot to figure 2 with added inference to the degree to which the wave would accumulate circular polarization if the vectors were within a plane, (within a conic surface of apex angle α = π/2). On that plane where α = π/2, circular polarization angle C plus the four d-radians (four quadrants) between perfect squares, sum to 2π radians. For a cone of arbitrary apex angle χ, circular polarization would diminish with diminishing χ, approaching zero as χ → [β = arcsin (2/π)], but the 4 d-radian angle remains unchanged as the sum of the intervening θ$_n$'s. Polarization issues are not particularly consequential to the periodicity of the wave.

between perfect squares of (2n+1) in Table1. Equation 11*1 further delineates them as the non-underlined values between bold entries of the sequence. A large-scale recursion relationship for each four-fold increase in n_3 would be:

At (P+1) = 2^k cycles, $n_3 = 2 \times 2^{2k}$, where k = 0, 1, 2, 3, 4, 5, 6, —etc. to ∞.   (14*10)

Four-fold increases in midpoint n_3 respectively occur at

2, 8, 32, 128, 512, 2048, 8192, etc. to ∞. It is therefore relatively straightforward to examine inter-perfect-square regions of ultra-large n, where k might extend to several hundred four-fold increases, or greater. A perfect square of (2n+1) should occur above and below each n_3 by respective amounts $\pm 2\delta_n = \pm 2(P+1)$.   (14*11)

The square of arbitrary consecutive odd numbers s_n however large also specify consecutive perfect squares of (2n+1), as $(s_n{}^2-1)/2 = n_{PS}$ {= n_1 and n_5 in Table 1}.

Referring to figure 14, time between n vectors, $(s_{n+1} - s_n) = \sqrt{(2n+3)} - \sqrt{(2n+1)}$ approaches $1/\sqrt{[2(n+1)]}$ at large n.   (14*12)

From equation 14*8, this is half of $\sqrt{[2/(n+1)]}$, the respective unit-circle chord lengths of equation 14*9. Therefore for large n, <u>half the unit-sphere or conic-unit-circle chord lengths of figure 15 coalesce with the broadened-line portions of s_n in figure 14</u>. <u>Those incremental chord-lengths expressed on the unit-sphere or unit-conic-circle can thus represent each added increment of time Δt</u>.   (14*13)

Equation 21*8 and Table 4 will show that for all n, twice those incremental chord-extents also represent exact radians on a 4-radian perimeter! <u>The unified wave specified by chord-lengths on the cone's n = 1 circular base, or around 4-radians on a unit-sphere, can thus interpret in terms of progression around that conic-unit-circle analogous to exponential harmonics</u>. That is, from figure 15, on the n = 1 circular base of the conic-surface, [or on the unit-sphere] progression around chords of length $2/\sqrt{[(2n+2)]}$ analog cyclic traversal on a "conic unit circle". Radius of that circle at the cone's base is not unity but depends upon the apex angle of the cone. <u>Distance from the cone's apex to the circle is unity</u>. That would characterize circular-progression in "discrete-jumps". When the jumps approach zero size as $n \to \infty$, the situation analogues the exponential. Taking the cone-side equaling unity as "normalizing radius" (instead of its circular base of radius $r = 2/\pi$) the wave could construe as formed on a "<u>conic unit circle</u>" or a <u>unit sphere</u>. As later shown, changes in apex angle of that cone alters the size of the circular base as a function of ξ, namely as $r = (2\sqrt{\xi}/\pi)$. [Figure 37 and 49 depict how each quadrant of summed chord-lengths equaling one radian for $\xi = 1$ on the unit-sphere would diminish to ξ radians.] The circular base with radius $= (2/\pi)$ on a cone or on a unit-sphere can be in terms of a limit process of those consecutive chords as $n \to \infty$. For $\beta = \sin^{-1}(2/\pi)$ where $\xi = 1$, radius of the circle at the unit cone's base would be $r = 2/\pi$. (14*14)

As progression variable, time would increase by advancing around the chord-lengths of the circle on the unit-sphere or cone's base. The circumstance is analogous to time's progression around the unit-circle in the exponential definition, $\exp(i\omega t) = \lim n \to \infty$ of $(1+ i\omega t/n)^n$, but discretely along chord nodes rather than continuously. **That the chord-lengths around the conic-unit-circle can represent a harmonic wave's progression in time verifies by noting the per-cycle sum of half the chord lengths approaches two, [full chord-lengths sum to four]. {These summed straight-line chord lengths remain valid for all possible arrangements of consecutive triangle planes, not only around a conic surface or unit-sphere.} I.e.,**

$$\sum_{n_{ps1}}^{n_{ps2}} 1/\sqrt{(2n+2)} \to 2, \text{ between } \underline{\textbf{all}} \text{ perfect squares of } (2n+1). \qquad (14*15)$$

The sum of $1/\sqrt{(2n+2)}$

E.g., from:	terms equals:	Deviation from 2 is:
n = 0 to n = 3	1.9689086-- -	.03109153---
n = 4 to n = 11	1.9987980-- -	.00120193---
n = 12 to n = 23	1.9997883-- -	2.116455×10^{-4}
n = 24 to n = 39	1.9999356---	6.430293×10^{-5}

etc., with deviation approaching zero as $n \to \infty$

With only P = 3 prior cycles, deviation approaches .003% and diminishes rapidly thereafter. When triangle angles become very small at large n, an average taken between each J-K chord $[2/\sqrt{(2n+2)}]$, (and tangent $2/s_n$ straddling F and G in figure 15) can converge to 2 per cycle even faster than the chords alone. [They can provide an Archimedes-type inner and outer polygon definition for π or the circle as $n \to \infty$. (see Appendix B).] For monads, summed chord-lengths between n-vectors around the conic-unit-circle always approach 4, [unity per quadrant or per radian] and can represent time/quadrant or time/radian, (or time/cycle), or if accumulated continually, time from the origin. Each increase by unity for accumulated chord-lengths-mapped-onto-a-cone of apex-angle $\sin^{-1}(2/\pi)$ [or however accrued between perfect squares] signifies a quadrant of the wave coincident with quadrants formed in figure 14. [Later discussion in connection with figure 37 will elaborate this.] The wave always possesses an inherent time base of **unity per quadrant** or per radian, [4 per cycle] on the conic-unit-circle or the unit sphere, {even when orientations of triangle-planes are random}. **<u>That unity-in-the-denominator obscures the per-quadrant or per-radian dimensional-unit basis for the original energy-time units that integers along the Euclidean axis signify.</u>**

These synchronous repetitions delineate basic periodicity throughout all numbers. <u>They define and interrelate, quadrants, one or four radians, rectilinearity, time-scale, energy-time/cycle, chords of the unit circle, perfect squares, prime numbers, the perpetually referenced origin through n, $\Delta\theta$, P, s_n, and RMS summed transitions,</u> **(14*16)**

It can therefore be understood how the limit of this unified wave as n → ∞ **visualized on the conic-unit-circle or unit-sphere converges to, encompasses, <u>and would be the preferred definition for exponential-type harmonics</u>.** The exponential's n → ∞ limit for the summed chords renders a mathematical definition that would have infinite energy-time constituents, **<u>each of zero "size"</u>**. <u>Chords would become of zero extent</u>. Allegedly, such condition does not exist in physical reality where the minimum energy-time unit is more appropriately $\Delta\theta/4$, the energy-time/quadrant of the wave. For a complete cycle, that minimum is otherwise called a Planck time, which would be chord-lengths on the unit-sphere for angle $\Delta\theta$. Before (and at) that convergence toward infinity, the 4-radian mathematical periodicity supplants conventional 2π radians/circle, to unify four-quadrant coordinate rectilinearity with ordinal numbers, quadrature components, prime numbers, RMS sums, and harmonic resonances.

Elegant symmetry, unification and invariance properties intrinsic to all perfect squares exist within natural numbers. They form the basis for all waves. The process described would remain valid for any value of $\Delta\theta$ however small or large, and provides a more general definition for space, time, and waves, than under neglect of these concepts. **<u>Disregard of an excess-angle as treated under contemporary Newtonian and Euclidean analysis presumes Planck's constant is zero</u>**. The smallest physical radial-angle to specify a conic-unit-circle chord would be $2\phi_{min}$. Extending figures 14 and 15 to large n would portray that minimal discernable origin angle as $\theta_{min} = 2\phi_{min}$. Equations 5*19 to 5*22 indicate that such smallest origin angle $\theta_n = 2\phi_{min}$ would associate with the uppermost $n = n_{max} = 2/\Delta\theta^2$ = $2/h^2$. From equation 14*8, **the time associated with such a half-chord advance would derive from n_{max} substituted into the half-chord**

extent $1/\sqrt{[2(n+1)]} = 1/\sqrt{[2(n_{max}+1)]} \approx 1/\sqrt{[2(2)/\Delta\theta^2]} = \Delta\theta/2.$ (14*17)

Time to advance a full chord length on the conic-unit-circle or a unit-sphere would be twice that or $\Delta\theta$. Instead of experiencing infinitesimal increments around the unit-circle as implied by the exponential, cyclic traverse more appropriately limits to and comprises of **polygon-like consecutive chords, all of extent $\Delta\theta$. <u>Each such chord must then subtend angle $\Delta\theta$ representing one smallest unit of energy-time. This confirms the original premise of $\Delta\theta$ = h as a smallest angle, engendering all the subsequent consequences outlined herein. It provides self-consistent mathematical corroboration of both the orthogonality axiom and the smallest compounded excess angle $\Delta\theta$. That minimum chord-length would signify a Planck time, which emerges mathematically from the unit circle chord of angle $\Delta\theta$.</u>**

When corresponding analyses apply for momentum-position along the distance axes, each chord-length on the unit-sphere from angle $\Delta\theta$ would signify minimal distance and represent a <u>**Planck length**</u>. Just as we conventionally describe a sinusoidal wave along distance X at a "snapshot in time", and a sinusoid in time when passing one position X, so a unified wave describes in momentum-position for X and in energy-time for time. It can thus be recognized how **an excess spacetime angle can effectively produce both a minimal time interval (a Planck time) and a shortest distance (Planck length) from the same minimal angle $\Delta\theta$**. They are both chord-lengths of that angle on the conic unit circle or unit-sphere. **Planck's constant portrays one smallest chord of the "highest resolution polygon" around the perimeter of a conic unit circle**. Appendix B demonstrates that from n = 0 to ∞, no single chord-length on the conic-unit-circle would fit as the element of a **regular polygon** around that circle. There will always be ambiguity related to maximum partitioning of a cycle though quadrants are synchronous and periodicity exact. This affirms an effective "real world" increase in and obscuration of mathematical-π, as $\pi' = \pi(1 + \Delta\theta/2) = \pi(1 + \theta_{min})$, again validating equation 10*1. It further suggests, a preferable dimensional-unit-choice for the velocity of light might be c = unity. Then respective ratios of energy-time and momentum-position axis-integers enjoy the same smallest units around the unit circle and would have the same scale-factor and excess angle $\Delta\theta$. Then c could be a (Planck-length/Planck-time) = 1 and variables of length and time would result in integer numbers without the irrationality of Newtonian continuity..

To reiterate using figure 15, tangents to the unit circle formed at triangle mid-angles have extent $(2/s_n)$. Rational values for those tangent lengths between F and G only result for triangles with perfect squares of (2n+1), and those tangent magnitudes follow the sequence $2/s_n$ = 2/3, 2/5, 2/7, 2/9, 2/11, etc. The chords J to K of the unit circle are rational wherever $\sqrt{[2/(n+1)]}$ (14*18) becomes a perfect square with respective chord lengths 1, ½, 1/3, ¼, 1/5, 1/6, 1/7, etc. to 1/∞. In analogue of exponential harmonic cyclic progressions around the unit circle, those rational chord locations delineate quad-radian conditions on the unit circle. This analogues how repeated Y-axis crossings delineate 2π rotational locations for exponentials. Referring to figure 14 and since

$$(\sqrt{X})[\sqrt{(X+1)} - \sqrt{(X-1)}] \approx 1 \qquad (14*19)$$

for large X, each broad-line-segment incremental increase in s_n of figure 14 approximately proportions to $1/\sqrt{(2n)} = [\sqrt{(2n+1)} - \sqrt{(2n-1)}]$,

[or $1/\sqrt{(2n+2)} = [\sqrt{(2n+3)} - \sqrt{(2n+1)}]$, for n > ≈2. $\qquad (14*20))$

E.g., by n = 10, $\sqrt{21}$ - $\sqrt{19}$ differs from $1/\sqrt{20}$ by only 7×10^{-5}. **The broadened-line-lengths of figure 14 can accurately approximate as $1/\sqrt{(2n+2)}$**. Also,

$2\phi_{11} = \theta_{11}$, and similarly each $\phi_n = \theta_n/2$, express as the everywhere valid identity

$$2\phi_n = 2\tan^{-1}(1/s_n) = \theta_n = \cos^{-1}[n/(n+1)] = 2\tan^{-1}[1/\sqrt{(2n+1)}]. \qquad (14*21)$$

It will be shown that for any n (when $\xi = 1$), **the radians traversed around any 4-radian-perimeter-per-cycle on the unit-sphere will exactly equal twice the magnitude of each respective broadened-line segment in figure 14;** Radians-traversed @ each n = $2[\sqrt{(2n+3)} - \sqrt{(2n+1)}] \equiv \Delta s_n$. The rate of change of radians-traversed with n will be $\Delta^2 s_n \equiv -2[\sqrt{(2n+5)} + \sqrt{(2n+1)} - 2\sqrt{(2n+3)}]$. $\quad (14*22)$

These interrelationships between all straight-line vectors of the spiral, and their exclusive emergence as rational at all perfect squares, characterize the periodic origin of harmonic waves. Quadrature divisors of waves delineate their peaks and zero-crossings. **It should be noted that nodes of the spiral divided by n as utilized in figure 5 are also equivalent to radial-vector intersections with the unit-sphere or conic unit circle**. This analysis utilizes a variable comprised from monadic constituents in time (and space for momentum-position scrutiny) after an origin event. The variable is quantized into discrete units so the process involves population accrual rather than being continuous. Growth of a population essentially describes succession or counting, a basic phenomenon indigenous to natural numbers. The sequence of counted integers is extremely inclusive to all mathematics, numeration, and physical things. While one might take issue with interpretive meanings underpinning certain material presented here, there can be little argument with details produced by the mathematics.

15) FURTHER EXPONENTIAL AND UNIFIED WAVE COMPARISONS

Rational and irrational numbers are elements of mutually exclusive sets. They differ as day and night without continuity between them. A given value is either in one set or the other, but never in both. Irrational numbers occur throughout man's construct of mathematics but only tangentially in the domain of physical reality. A scientific school of thought exists that believes precisely the contrary, that measured values depict truncated irrationals. Possibly, this is so because measurement accuracy often limits prior to encountering a system's intrinsic indeterminism. In addition, mathematically calculated values designated through symbols **are usually presumed to constitute actual real values** as with the length of the diagonal of an idealized unit square on a plane. The fact that you can measure the ten fingers on both hands as a rational number stated with absolute accuracy confirms that rational values occupy the domain of measurable things, not the other way around. The counting or statement of constant-amounts (e.g., the number of fingers) differs from the accrual of dynamical variables, (e.g., the number of ball bearings entering a set like a box) since discernability the size of one ball bearing (if not sub-divisible) will always associate with dynamical variable enumeration. Algebraic and differential equations are generally of the dynamical variable type. Dynamical variables affiliated with a smallest-size-entity must always cope with the added indeterminism equaling the shroud of that smallest constituent. By contrast, no measurement ever yields an irrational value. They yield either the integer constant if static, or the obscuration of the one smallest unit if dynamic, which obscuration would hide irrational values. If one employs "non-physical" mathematical-units (like those involving π) within the basis for measurement, that error at the onset can make it appear measurements yield irrational numbers. Whenever we substitute an **actual** numeric value for the symbol-π, it will however not represent perimeter/diameter of a circle, but invariably something simulating a summation of chords. The circle on a plane is an abstraction of the human mind while algebraic equations can plausibly represent straight-line chords. Mentally, in substituting the numerical value used for π, we assumed trailing digits of the symbolic quantity π were simply truncated for expedience. More realistically, a physically realizable chord summation represented the value employed, and its resolution petered out somewhere due to the excess angle. The de facto model applied to derive numerical answers never expresses what the symbol π exactly stands for mathematically, but the substituted value used effectively establishes a specific chord configuration. Final answers entail solutions to actual algebraic equations though they are often expressed and symbolized as transcendental along the way. We become so familiar with the procedure and symbolism to eventually believe π is actually subsumed in the process. It is not immediately clear that transcendental numbers even exist.

Transcendental numbers like π and e remain outside the realm of solutions to physical-model equations. They are not actually compatible with expressions that model physical things comprised of countable entities. Similarly, $i \equiv \sqrt{-1}$ is devoid of physical meaning though as a mathematical operator (as in the harmonic exponential) it produces a right-angle transition. Moreover, no physical evidence exists to the contrary of time progressing orthogonal to radial direction from the origin. **The exponential harmonic and the orthogonality axiom both constitute embodiments of that postulate**. They apply for smallest entities with the exponential being a continuous variable and the unified wave derived from discrete action units. However common the use of π, e, i, and ∞,   (15*1) such constants are artifacts of mathematical manipulation rather than being attributes of the physical world. They were operationally defined by mathematicians and then glibly applied to physical reality. Planck's constant $\Delta\theta$ on the other hand is an attribute of both the mathematical and physical world though absent from the repository of mathematical foundations. **A variable that represents an excess spacetime angle is a perfectly legitimate mathematical entity and more general than a priori assuming that angle is zero**. The symmetry of harmonics produced by such angle in conjunction with the natural numbers confirms its legitimacy. It accommodates spacetime curvature of the environment under analysis. Spacetime curvature may be a metaphorical equivalent for the information shroud necessitating the orthogonality axiom and the missing gap between integers. As $\Delta\theta$ it might parametrically assume any of the values: excess-angle = $\sqrt{(2/n_{max})}$, where n_{max} is integer [and plausibly $\sqrt{(2n_{max}+1)}$ is integer.] **Planck's constant depicts one particular embodiment of spacetime curvature** wherein n_{max} allegedly derives from the empirically prevailing excess-angle in our world and $n_{max} = 2/h^2 = 2/\Delta\theta^2$. In unified wave solutions Planck's constant thus supplants mathematical abstractions like π, e, i, and ∞ with only one parameter that has both

179

mathematical and physical significance. In the Occams razor sense of a preferred model, Planck's constant eliminates a host of messy irrational and imaginary constants, in effect arbitrary to the real physical system under analysis. Also, unit circle functions like e^{ix}, sin x, cos x, are not algebraic expressions. They involve transcendental values. By contrast, up to some $n_{max} < \infty$, Table 1 and figure 12 demonstrate that tangible integers n can characterize waves of physical reality. No postulate extends transcendental and imaginary values to real-world situations however commonplace they may have become in our mathematics and models. Distinctions between algebraic vs. non-algebraic equations typically depict systems that are respectively countable vs. non-countable, the latter case being applicable for mathematics but not for the physical world.

The more attributes a macroscopic entity possesses, the greater can be the ambiguity in counting, classifying, or measuring a group of them. The system of natural numbers thus most appropriately describes smallest, discrete, indivisible **objects of only one attribute where the orthogonality axiom would apply**. Those are objects for which counting can be done to within one unit. A measured quantity is invariably the result of a comparison with the standard dimensional units employed in the measure. If possible, standard units should preferable be taken as a multiple (or simple divisor) of some absolute property of the prevailing environment. The original basis standard for time for example divided the year into four seasons defined via the solstices. They comprise cycles and quadrants in a yearly description of time. Days and hours similarly derive from earth's rotation. For increased precision, such dimensional units later became linked to less-divisible-increment standards like cesium 133 oscillations. Being unaware of relativistic dilation, our ancestors based time's standard on what they believed were absolute heavenly periods. Analogously, the dimensional unit for an important environmental property i. e., Planck's constant should not vest in unrelated arbitrary units like MKS or CGS, but from what Planck's constant actually signifies, namely the excess angular numeric per cycle. Only then can the proper numerical excess-angle rationally modify the unphysical Euclidean-interpretation of π, while consistently satisfying the *magnitude* of energy-time/cycle that angle represents in traditional, arbitrary dimensional units.

These analyses indicate that rather than 2π radians around a planar unit-circle; **a perimeter of 4 radians encircling a cone at unity from its apex [or a circle on a unit sphere] establishes the four quadrants of harmonic oscillation. [The conic-unit-circle formed on a cone's base everywhere unity from its apex would be an identical circle to the intersection of that cone with a unit sphere.]** Oscillatory processes emerge from more complex phenomena than "time progressing around an idealized planar-circle" as in exponential harmonics. Instead, waves affiliate with conic-unit-circle chords or chords fitted within the surface of a unit sphere. Rather than experience 2π radians per revolution, summed Planck-length chords as "unequal-sided polygon-sides" undergo four symmetrical rectilinear radians commensurate with each quadrant of the wave. **The resultant harmonic references its causal origin event by "counting" excess-angle**

increments thereafter as consecutive integer n [or counting successive chords of the conic-unit-circle.] **The disturbance begins at that origin, not at - ∞ in time or distance, and does not extend to +∞, as sinusoids do**. **Those sinusoids and their root-source harmonic exponential violate causality by producing a response before the impetus that causes it**. Why should such an unphysical instrument be employed so prolifically in modeling reality? **Peaks and zero crossings of the unified wave's rectilinear quadrants never ally with prime numbers n. Unvarying divisibility of a wave's peaks and zero crossings is what perfect periodicity means! Those locations always represent divisible points of the harmonic sequence**. Those peaks, valleys, and zero crossings unite with odd integer [and ½ integer] chord-lengths derived from the sum of two adjacent population values, (from {n+(n+1)}). The harmonics obtain from quantized integer-multiples of Planck's constant, the de facto excess spacetime angle. Each complete cycle registers a tangential magnitude as a consecutive odd number signifying temporal extent, in synchronization with wave periods. Although traditional exponential solutions simulated by planar-unit-circle traversals can closely approximate some characteristics of this more general unified wave, they omit many features, particularly those of discrete phenomena. Another trait avoided by the unified wave involves negative numbers. Everything occurs after the origin event that it references, so possibilities always occur in positive time or distance relative to that origin. In forming the spiral, negative numbers are never encountered in the wave generation process.

A simple circumstance allows "polygon-like" chords within the conic-unit-circle [or on the unit-sphere surface] to replace the circle itself. At large n, that condition simulates the exponential function by time-advancing as chord-length progressions around a circle one-unit from the cone's [or sphere's] origin. Toward n_{max} of the spiral for example, a tangential transition $s_2 = \sqrt{(2n_{max}+3)}$ would have prior $s_1 = \sqrt{(2n_{max}+1)}$. **The difference between them being**

$$\sqrt{(2n_{max}+3)} - \sqrt{(2n_{max}+1)} \approx 1/\sqrt{(2n_{max}+2)}. \hspace{2cm} (15*2)$$

From figure 15, <u>this is identical to $1/\sqrt{(4/\Delta\theta^2)}$ or ½ each chord length around the conic unit circle or on the unit sphere</u>. Therefore as shown in figure 14, the broadened-line incremental-time "added" by s_2 (in excess of the time registered at s_1) would be $1/\sqrt{(2n_{max}+2)}$. **That extent is equal to half the chord length at the unit circle. Thus the <u>time variable produced by spiral-triangle tangential-magnitudes depicted in figure 14 is identical to time progression along polygon chords of the conic-unit-circle</u>.** The circle formed in the plane unity-from-the-origin on the cone has radius $2/\pi$. The harmonic wave generated by the spiral in figure 14 can equally well be delineated as the wave formed by proceeding along chords of the "polygon" inscribed in that circle, (and at the n→∞ limit along the $2/\pi$ radius circle, thereby mimicking the exponential function). **The otherwise ad hoc "circulating time" in exp(iωt) becomes "logically justified" by the orthogonality axiom and the equivalence between chords of the $2/\pi$ circle and the broadened-line increments of s_n in figure 14.** At large n, both broadened line increments of s_n in figure 14 and chords around the $2/\pi$ radius circle of figure 37 are mathematically equivalent.

Unified wave development can provide logic and mechanism whereby chord-lengths summing to 4-radians/cycle are consecutively traversed. The method employed a spiral that identifies four-quadrant symmetrical rectilinearity of the wave, one radian for each quadrant, (rather than 2π radians). The spiral, the chords, (and even the $2/\pi$ radius circle at the $n \rightarrow \infty$ limit) <u>need not remain within the same plane as represented through the exponential function</u>. The wave could apply in spacetime of any curvature, on any surface synthesizable using planes of the concatenated triangles. Figure 49 for example, displays what might be called a "**canonical form**" description that does not utilize a cone. In all synchronized-mode embodiments, periodicity derives from the natural numbers, articulated at respective odd-perfect-square locations of $(2n+1)$ from the origin. The wave can be non-localized in representing smallest quanta. It can depict a possibility and probability wave. Irrational values can be avoided. Rather than the harmonic exhibiting time reversal symmetry, s_n (and chord-length) variation with n, gives an arrow to time. Peaks and zero-crossings at quadrant positions associate with non-prime locations for n. The wave begins at the origin event, not at minus infinity in time as with sinusoids. Reference to the origin exist throughout the entire wave for n, prior cycles P, quadrants, tangential magnitude s_n, and radial extent between quadrants δ_n. Several irrational, imaginary, and infinite constants of mathematics like, π, e, $\sqrt{-1}$, ∞, become superfluous to the model. <u>Two more arbitrary constants of physics representing a Planck time and a Planck length become unnecessary, both being</u>

the direct effect of one Planck angle $\Delta\theta = h$ having chord-lengths representing those constants on a unit sphere.

The unified wave limit at $n_{max} = 2/\Delta\theta^2$, establishing the finest resolution polygon within the conic unit circle, **corroborates that it is impossible to compound something at an infinite rate**, as occurs along a perfect circle in the exponential. In physical reality, a Planck time (**not zero time**) represents the smallest possible interval for each compounding advance. **The exponential function thereby conflicts with objective reality** to the extent it permits reiteration via compound growth (or angle) to occur at markedly smaller than Planck-time intervals. Time progressing along discrete Planck time chord-lengths of the conic-unit-circle or unit sphere provides the appropriate compounding-limit-function for exponential-type harmonics. The shortest compounding interval becomes a Planck time, or more explicitly a **Planck angle $\Delta\theta$**, since it characterizes a chord-length on the unit-circle. That same excess **Planck angle of space will also automatically denote a minimum Planck length** on the momentum-position unit circle. An exemplary $\arcsin(2/\pi)$ apex-angle cone converts the unit circle's total perimeter extent into 4 rather than 2π angular units so that each quarter cycle equals a radian and one circular quadrant. The four radians per cycle between perfect squares remains valid **for any possible configuration of triangle planes that comprise the spiral**. Each quarter-cycle period-of-time avoids an irrational relationship involving π and in fact identifies with a non-prime integer n. These express fundamental properties inherent to all harmonic waves.

185

In the spatial real world, the smallest possible chord-length or compounding extent cannot reach zero but limits at a Planck length. It demonstrates that as defined, conventional exponential harmonics could only apply for circumstances wherein Planck's constant equaled precise zero. Moreover, as a function, this unified wave can mathematically accommodate any desired degree of spacetime curvature by allowing the minimum angle or chord-length to limit at whatever desired magnitude. Analytic representation includes a higher degree of freedom permitting all possible levels of inter-integer indeterminism $\Delta\theta$ and all possible spacetime curvatures. A "polygon" having Planck length sides $\Delta\theta$, (rather than the perfect circle) describes the real world basis for a harmonic figure. **How could the known-smallest-Planck-length extent be compatible with a polygon that requires smaller length chords, as needed in the limit construction of a unit circle?** The infinite trail of decimal digits possible in the mathematical definition of π becomes physically meaningless. Empirically experienced π' will have its least significant digits "truncated" by the inherent uncertainty associated with $\Delta\theta$ in $\pi' = \pi(1+\Delta\theta/2)$. The system of natural integers designates harmonics following an origin event via "quantized numbers" n, rather from assumed Newtonian continuity of infinitesimal advance around a perfect circle.

This discussion began with the variable energy-time progressing along a Euclidean axis in analogue of how the independent variable time t conventionally advances along such an axis. An increment of traversed energy-time was shown as equivalent to crossing a minute excess-angle in a cyclic wave. For minimal-size constituents, the orthogonality axiom was explained and invoked. This granted numerous solution modes, several of which express as a "Pythagorean Spiral" on a plane, and on a cone of apex angle $\beta = \sin^{-1}(2/\pi)$, [or on a unit-sphere.] The logic for integer advance along the axis stemmed from traversals of the excess-angle in spacetime. That produces a population of discrete energy-time (or excess-angle) units of minimal size. The spiral plotted in a cone resulted in time's advance along chords around the conic-unit-circle representative of the excess-angle. **A "transformation" thereby resulted from energy-time progression along a Euclidean axis abscissa into advance along minute-size polygon-like chords around the conic-unit-circle or unit-sphere surface**. That transformation from succession along a linear axis to a route circulating the unit circle mimics the exponential harmonic definition wherein linear-independent-variable t (normally along a conventional axis) becomes arbitrarily re-expressed to progress around the unit circle in $\exp(i\omega t) = \lim n \rightarrow \infty$ of $(1+i\omega t/n)^n$. However, no clue or explanation is provided for how or why this exponential function transfers from a conventional Euclidean axis linear independent-variable to suddenly circulate as time around the unit circle. **The exponential function is simply defined ad hoc utilizing an imaginary operator $i \equiv \sqrt{-1}$, without a logical conduit describing the rational by which it comes about. Employing that operator in exponential evaluation unknowingly constitutes an invoked form of the orthogonality axiom, but without background justification or knowledge of why.** By contrast, the unified wave derivation explicitly illustrates how and why circular progression originates

187

through the orthogonality axiom. Since formation of the unified wave can conceptually allow n to approach infinity as $n_{max} \rightarrow \infty$, it encompasses the harmonic exponential definition.

Indeed, the presented derivation articulates the reason, method, and relationship by which the Euclidean axis linear-independent-variable converts into cyclic traversals around the unit circle, via physics of the orthogonality axiom. It analytically and graphically depicts how, for smallest indivisible units, harmonic waves emerge from a Euclidean axis linear-independent-variable. No comparable justification occurs furnishing a foundation for the harmonic exponential, which arbitrarily enacts an operator from nowhere to artificially synthesize the harmonics. That functional procedure mathematically requires chord lengths smaller than a Planck time and Planck length thus representing physical impossibility. In its formation, the unified wave by contrast exposes elegant symmetry synchronizing four intrinsic radians within natural numbers with four quadrants of the wave. The four radians demonstrate as mathematically absolute necessitating dimensional-unit consistency for radians and cycles. The determination includes analytic ability to accommodate any minimal chord-length $\Delta\theta$ in approach to the unit circle, (satisfying any excess angle/cycle).

The solution allows many other possible modes for the wave as well as De broglie and probability waves of quantum mechanics. It produces the indeterminism of the uncertainty principle logically commensurate with the shroud of each smallest constituent population member $\Delta\theta$. Irrational numerical values are excluded from physical realization. Orthogonal progression becomes a logical consequence of linear-independent variable advance whenever fractional values for indivisible entities constitute a "disallowed" condition. That produces a "reason" for waves. The resultant harmonics can extend to include components for rest mass and motional energy. Resultant waves retain balanced synchronization with perfect squares of paired consecutive numbers from n = 0 to n = ∞. The mathematics of natural numbers thereby inextricably couples to physical reality. Peaks and zero-crossings of the wave remain everywhere synchronous with non-prime values. Solutions mathematically-at-least accommodate the gamut of space curvatures as a continuous variable, allowing any excess-angle/cycle $\Delta\theta$. Sensible analysis "explains" what Planck's constant is. The simple physical description of an additive increment to mathematical $\pi \equiv$ (perimeter/diameter) can justify its logical "cause", which circumstance is **more general** than requiring such increment to be **exclusively zero**. This thesis unveils mathematical and physical characterization of spacetime under those more general conditions. The derivations may help clarify various perplexities of physics like wave/particle dualism, times arrow, the uncertainty principle, the red shift, the enigma of irrational numbers, and what is specifically being conserved in (and the coalescing of) the conservation laws. The inquiry may provide a single "classical" foundation for quantum theory and string theory. It can plausible characterize resonances of fundamental particles and interactions. Modes of the unified wave are conjectured to depict all elementary particles including those particles in the sea of quantum. The mechanism producing

189

the wave is so basic it will be shown to infer the phenomenological foundation for elementary processes like the DNA helix ladder (See Appendix A). These attributes juxtapose to those of counterpart exponential harmonics with ad hoc derivation and containing known inconsistencies plus an inherent singularity.

Most scientific disciplines are highly compartmentalized because end use or utilitarian function dominates studies in those respective fields. Biologists examine living organisms, often through chemistry that in turn is influenced by physics. Few biologists analyze their subject through the physics of atomic nuclei, although those nuclei comprise all living things. Minimal continuity-of-analysis links end behavior with the truly underlying processes in each discipline. Relationships of mathematics within the number system are at least as scientifically basic as sub-nuclear phenomena. Such fragmentation of continuity has isolated practices of science into specialties that might otherwise benefit from the serendipity of one unifying foundation. Waves provide a concept with potential to coalesce those various disciplines. They can characterize particles and matter as well as motion and forces. A generalized wave embedded within the natural numbers has potential to couple many branches of quantitative science. The DNA helix ladder exemplified as a wave within natural numbers implies a unifying continuity from sub-nuclear through biological to cosmological behavior. All forces can interpret as the result of interacting waves. That is certainly true of "action-at-a-distance" forces like gravity and electromagnetism. Atom-to-atom push and pull forces also transmit through nuclear or lattice structure waves. The study and understanding of forces can equally well be addresses as wave phenomena. This theory of waves may plausibly provide a variant approach to a unified theory of force fields. Wave methods can also yield a basis for possibility waves, probability waves, matter waves, particle waves, acoustic waves, water waves, etc. This study suggests they all intrinsically relate to the natural numbers and unavoidably infiltrate the geometry and topology of physical and biological things.

Besides the presumed basic quantities of physics that describe phenomena, i. e., time, length, mass, and charge, it would seem there is **angle**. When spacetime is curved, angle is not exclusively defined over large extents by Euclidean distances (as in a plane figure). An additional parameter is required to specify non-Euclidean magnitude. Parameter n_{max} characterizes the state of possible curvature. Except at $n = \infty$ where the four quadrants defined through A, B, C, D become exact mathematical radians (and conditions are Euclidean, with zero curvature), **rotation-amount constitutes a "possible degree of freedom"** like length or time. Moreover, one could not employ the notion of a wave in physics without the concepts of angle, distance, and time. The radian is a fundamental absolute unit of rotation, being precisely one quadrant of a unified wave as $n \to \infty$. Few other physical variables, if any, have such absolute basis within the natural numbers sequence. Such rotation amount is not defined through relative comparison of perimeter-to-diameter for a perfect circle on a perfect plane. That perimeter/diameter = 4 circumstance "coincidentally" happens to equal the angle of one unified wave quadrant on the cone as $n \to \infty$. The 2π-radian description on a circle requires **perfect planarity and exact equi-distance from center**, or zero curvature in space, a mathematically expressible but not physically achievable condition. By contrast, the definition of an absolute radian as a quadrant of the unified wave **comprises the sum of a very large number of small angles** θ **whether they are planar or not**. That definition places constraints only on the planarity of each relatively small triangle in the spiral. **That qualification is vastly less stringent than the perfect large-area plane required to even define the** 2π **radians of the circle. Also, planes of those triangles extend only to the extremely small angle** $\Delta\theta$ **as** $n \rightarrow n_{max}$. As $n \rightarrow \infty$ the angle subtended in each plane approaches zero so each triangle formed by chords of the conic unit circle has miniscule planar area.

16) REFLECTIONS ABOUT DIMENSIONAL UNITS

The following considerations may clarify issues about dimensional units. Stating Planck's constant is a dimensionless "angle per cycle" means h is dimensionless in $E = h\nu$ and energy proportions to frequency. Planck's constant can then be expressed as energy-time/cycle since Energy = cycles/second, making (Energy) x (time) equal to cycles. Therefore, energy-time/cycle = cycle/cycle = dimensionless, consistently verifying that energy proportions to frequency. If the unit spacing between integers n of figure 14 depicts the least quantum of action h of quantum mechanics, it would actually signify energy-time/quadrant or energy-time/radian on a cone rather than energy-time. A quadrant on the conic-unit-circle however designates one of 4 per cycle and has a perimeter length of unity and one radian of angle. Per-unity normalization may minimize distinction between energy-time and energy-time per quadrant. If for example, Planck's constant were the dimensionless number 0.01, on a circle it signifies an excess angle of 3.6°. If it were 10^{-6}, the excess angle would be 3.6×10^{-4} degrees. That would characterize the highest resolution "almost-regular" polygon around the unit-circle as having $\approx 10^6$ sides, each distinguishing an angle ranging close to 3.6×10^{-4} degrees. The question then emerges: What is the numerical value Planck's constant should have and how does that value accord with the differing numerical amounts it assumes under say MKS or CGS units? Both those systems are arbitrary but provide self consistent though different numerical values for the identical Planck's constant magnitude h. Therefore other analogous systems of dimensional units must exist that would provide the same constant magnitude for Planck's constant but result in a different numerical description. Assume for example h' = 10^{-30} was the de facto excess angle per cycle of spacetime. To satisfy that condition (if dimensional units of seconds remained unchanged in a new dimensional unit system), units for energy need then be roughly midway between the Joule and the Erg. Under those new dimensional units, the same **magnitude** energy-time/cycle of Planck's constant could then numerically equal 10^{-30} ≡ h'. In pre-selecting the dimensional-unit system to result in energy-time/cycle = 10^{-30} = h', that unit system would no longer be arbitrary.

Now the question remains to establish the precise numerical value for the excess angle/cycle prevailing in spacetime and adapting a system of non-arbitrary units to provide that numerical value. To do that we need to know what the de facto excess angle/cycle really is. If it derives from isotropic homogeneous expansion, (which seems plausible) it will depend upon that component of expansion. It is however noteworthy that various inhomogeneous expansion mechanisms might be in effect. Farthest galaxies may be receding faster for other reasons so simply attributing the excess angle to a given cosmological region's excess expansion may be unsuitable. Without assuredly knowing the degree of homogeneous expansion at this time, it would seem prudent to simply interpret its effect as an additive increment to π whose exact numerical value, Δθ/2 remains to be determined in π' = π(1+Δθ/2). We leave the existence of the additive increment to π as a prediction

194

of the theory with its energy-time/cycle magnitude known and its precise numerical value remaining to be determined, thereby establishing the necessary dimensional units. For energy and momentum calculations, conventional measures of h = $\Delta\theta$ used in either MKS or CGS units remain appropriate. A non-zero excess-angle $\Delta\theta$ would help explain certain anomalous conclusions of the October 1994 COBE experiments. They predicted a Hubble constant of 80, suggesting the universe to be about 8 billion years old if the excess-angle $\Delta\theta$ (and cosmological constant Λ), were zero. Any non-zero excess angle invariably serves to increase apparent age since cosmic expansion would compensate for gravitational contraction after the big bang. The universe would then sustain longer in an "inflationary-type" mode from its origin to the present. Evolutionary models of suns suggest some to be 12 to 18 billion years old however. The existence of excess-angle $\Delta\theta$ might help ameliorate the time discrepancy beyond 8 billion years while remaining consistent with Einstein's cosmological constant Λ.

Non-uniform concatenated triangles compounded to mathematically form harmonic waves, the arrow of time, irreversible processes, violation of time reversal symmetry, increasing entropy, the expanding universe, a cosmological constant, <u>may all be manifestations of a non-zero excess angle</u> <u>$\Delta\theta$</u>. Exponential-type harmonics exp(iωt), derived from e, π, i, and ∞, comprise from **uniformly compounded** triangles thereby characterizing Euclidean spacetime, symmetrical time reversal, a static universe, non-causality plus the presumed uniform absolute "master clock" partitioning of time from -∞ to +∞ through any sinusoidal frequency solution. That representation is obviously deficient toward portraying the sequence of empirical events for physical reality. Not only does the existence of $\Delta\theta$ satisfy microzone phenomena, but its influence also extends throughout the entire cosmological realm.

One distinction between a so called classical and a quantum system is the lack of an absolute dimensional reference scale in the former. Unrestrained scalability upward or downward in classical systems results from neglected absence of an unequivocal reference that specifies a definite smallest-magnitude property of spacetime. A comparative standard for the intrinsic structure of radiation and matter has been overlooked in classical Newtonian theory. The omitted absolute property allows scalability and the use of arbitrary dimensional units, which degree of freedom has been adapted into contemporary quantum theories. However, **a definite-magnitude absoluteness of spacetime exists as Planck's constant**. That ultimate smallest property is often represented as a Planck length by contingent normalization to the unit circle <u>thereby converting the excess-angle property into a smallest perimeter chord-length</u>. A minute angle at radius unity appears identical to a chord-length on the unit circle. That absolute-magnitude property of spacetime voids the classical degree of freedom and unrestrained scalability otherwise manifested in the form of arbitrary units. To appropriately model actual reality, the dimensional-unit scale <u>need incorporate the prevailing absolute numerical value into analysis. Physical objects materialize in terms of that unmitigated numerical value, the real excess-angle per cycle of spacetime</u>. It is therefore not so surprising that arbitrary units should be relinquished by embedding the de facto numerical property of the environment within the dimensional units employed. Things "measure" in terms of how many of those intrinsic smallest numerical items apply in empirical experiments. That numeric unit becomes the standard to which physical observables compare. Physically realizable structures should express as quantized integers referencing that absolute unit, e.g., as an integer number of Planck lengths or Planck times. <u>In retrospect, models of reality should replicate the prevailing "absolute-size" smallest-entity using non-arbitrary units</u>.

Contemplate the choice of dimensional units for the physical variables of time or distance. Such units should originate from the smallest-possible encountered-entity using integers of that minimum and counting those units **upwards**. This is in contrast to **dividing down** from a larger-valued encountered-entity. Consider time for example, where three larger-valued parameters set the ambient conditions. These are the year, or rotation about the sun; the month, or cycles of the moon; and the day derived from earth rotation on its axis. These three non-synchronous intervals formed our historical dimensional-unit basis for time as the year, month and day. Our ancestors used those consonant periodicities to characterize event locations. They provided countable interludes consistent with cultural and accuracy needs. The day was then integer-fraction sub-divided into hours, minutes, and seconds to accommodate a smaller functional interim. Empirical encounters in technology then necessitated smaller time increments of the second, as milli, micro, nano, pico, fempto-seconds. No conceptual limiting process existed for this division operation with multipliers of 10 to some minus exponent always seemingly inclusive in a multiplicative manner. At about 10^{-43} seconds, physical phenomena emerged suggesting a smallest experienced interval at a Planck time. It became somewhat irrelevant to subdivide further.

The process of invariant division as an operator was presumed acceptable down to that smallest interval. Mathematically the process of division itself as an operator does not depend upon scale making it seem suitable to continually divide down to and past this minimum encountered entity. This train of division philosophically obscures a need to invoke the orthogonality axiom in approach to that minimal interval. It also obscures the engagement of an enormous density of irrational values in the presumed continuity down to a Planck time. On the other hand, had one started with the dimensional unit of a Planck time and counted upwards, one could describe all possible intervals as rational integer numbers of Planck times. Moreover, by invoking the orthogonality axiom in association with this minimal limit, the influence of the space-between-integers could be taken into account to accommodate such missing information. That reasoning constitutes the blueprint of this theory. Conventional familiarity originating by dividing down from large dimensional units concealed the improved dimensional-unit-models that result by counting-up from smallest non-divisible units. The latter can accommodate inter-integer gaps between smallest units. Exclusion of "inhabitable values" within those gaps is a mandatory part of any sound model and has been the oversight of Newtonian mechanics.

There are other ways to examine dimensional units relevant to $E = h\nu = \Delta\theta\ \nu =$ (excess-angle/cycle)(cycles/second) = (excess-angle traversals/second). Under any interpretation for the meaning of h or the dimensional units for h, at one cycle per second **E would be numerically the same as h for whatever units h were in**. At 2 cycles/second E would be 2 units of whatever units are utilized for h, etc. When dimensional units for h alter, the applicable unit of energy must correspondingly differ to **represent the same energy** at whatever number of cycles/second prevail. That is an inherent property in the linear-system relationship $E = h\nu$. Now, for the 1 cycle/second case we can pick one specific set of units for h that allows it to numerically equal the de facto excess angle/cycle. That choice would correspondingly change the numerical, (but need not change the magnitude) value of energy since h = (energy)(time/cycle). Reiterating, consider only $\nu = 1$ cycle/second for example where E = h. Observe the effect of changing dimensional units for energy while retaining its absolute magnitude constant. If the dimensional units of energy are chosen small, many of them are necessary to achieve a given magnitude of energy. Then h will be a large number. That large numeric would however represent the same equivalent energy if dimensional units used doubled in size so h numerically halved, and so forth. Therefore, h can be made to have any numeric value by merely changing the size of the dimensional units used for energy.

That is always verifiable at a frequency of 1 cycle/second wherein E = h @ 1-cycle/second. If h is numerically large because energy is dimensionally designated in minute dimensional units, energy and h must still always equate at 1 cycle/second.

Some other factor must constrain what dimensional units to use because the equation works for all possible dimensional units of energy. Yet, we know the quantum of action h represents some physical non-commutative unit of the de facto environment. Energy should best be described in units of that smallest physical entity prevailing within the system. Existence of a smallest quantum of action in the real world prohibits the use of arbitrary dimensional units, which more appropriately should associate with multiples of that actual smallest unit. Since ν is a cyclic phenomena delineating rate of change of angle, the relationship effectively shows that with each cycle, energy portrays angular rate of advance through a smallest geometric angle. Such normalized numerical value depicting that actual angle/cycle can be assigned to h and it simultaneously stipulates the appropriate dimensional units of energy. That must be the case for ν = 1 cycle/second and wherein E = h. In those appropriate dimensional units, energy is simply the rate (in cycles/second) of passing through that excess angle. The cycles/second ν multiplier on h articulates that in the most explicit way. All that remains is to establish what the normalized excess-angle fraction physically is, and that will define the appropriate non-arbitrary units to be utilized in characterizing energy. Parameter h, being an intrinsic property of spacetime, must numerically delineate the appropriate units to employ for energy, just as the intrinsic property c must specify the appropriate normalized units for velocity v, as articulated by the radicand $\sqrt{(1-v^2/c^2)}$. The angular elaboration for h = $\Delta\theta$ simply satisfies that criterion and all the other relationship conclusions mathematically expressed in this treatise.

Astronomical measurements and the cosmological constant Λ of general relativity, tell us the universe is expanding. An inherent curvature of spacetime is consistent with a prevailing excess-angle as compared to Euclidean description. An excess angle is one way to annotate that curvature. Human expectation that Euclidean geometry without curvature applies to our physical world has obscured correct elucidation of the simplest possible relationship E = hν.

Reiterating relevant dimensional units for the equation E = hν is in order recognizing that the smallest chord comprising the conic-unit-circle or the unit-sphere has length $\Delta\theta$. Time is expressed as either time or seconds. The following equivalent statements can be made.

E = hν = [energy-time/cycle]x[cycles/second] =   (16*1)

= [energy-time/quadrant]x[quadrants/second] =

= [energy-time/radian]x[radians/second] =

= [excess-angle traversed/cycle]x[cycles/second]. From a different vantage, canceling cycles and using radial extent unity such that the conic unit circle becomes the normalized base-figure-of-relevance, the equations become:

E = [number of smallest-perimeter-extents of the unit circle traversed]/[second] =

= [number of minimal chord-lengths traversed along the conic-unit circle]/[second]

= [the rate with respect to time of traversing the excess-angle]=

= [the previously mentioned definition for energy {in terms of smallest energy units} as the time rate of traversing the excess-angle.]   (16*2)

Consistency under the orthogonality axiom thus sustains, with Planck's constant representing an (excess-angle-traversed)/(complete cycle). Analogously, linear momentum depicts the rate with respect to distance of traversing the same excess-angular increment ($\Delta\theta/2$ within π'). The rotational rate of traversing that additive increment to π' portrays angular momentum. The three motional conservation laws for energy, momentum and angular momentum are accordingly seen[1] coalesced into a single law conserving progression rate through the excess angle with respect to time, distance, and rotation.

All forces can interpret as the result of interacting waves. That is certainly true of action-at-a-distance forces like electro-magnetism and gravity. Atom-to-atom push-and-pull type forces also transmit through nuclear or lattice structure waves. Therefore, the study and understanding of forces can equally well be addressed in terms of wave phenomena. A unified theory of waves provides a variant approach to a unified theory of force fields. The wave approach also includes a basis for probability waves, matter waves, acoustic waves, water waves, particles, photons, etc. Conclusions from this study indicate all waves intrinsically relate to the natural numbers and thus to the formation, geometry and topology of physical and biological things. With DNA being shown as exemplary of an intricate form of wave, most phenomena, inclusive of life, can interpret as a manifestation of unified wave generality. These consequences of this treatise stem from simply interpreting the ordinal number sequence under the altered paradigm of the orthogonality axiom.

Natural numbers as <u>integers</u> only tell "**<u>how many</u>**", not "how much". It may be philosophically difficult to accept that the space between 0 and 1, or 1 and 2, etc. may not specify along a Euclidean line. Such intervening space may occur orthogonal to the prevailing direction from the origin. That would appear as a basic rule if each unit of the natural integers really means one smallest constituent (one monad). When enumerating divisible things comprised of say 10^{20} sub-entities, then 10^{20} occupyable points could presumably apply between integers. Negligible orthogonal traverse would occur progressing from say unity along the axis, to an arc from the origin one part in 10^{20} greater than unity along that axis. Right-angle progression concurrent with population advance would be negligibly small and unobservable. It is likely the general circumstance should allow some non-zero curvature to space necessitating a non-zero $\Delta\theta$. Unfamiliar as the concept might be, orthogonal advance between integers is seemingly the norm rather than the exception, and that may be why de Broglie waves are unavoidable. From n = 0 to n = ∞ a repetitious cycle occurs **<u>within the natural numbers</u>** for **<u>exactly</u>** every increase in s_n by 2. Waves that result should substantiate that a radian does not define by where perimeter equals radius because that credo only applies for a perfect circle on a perfect plane. The limit as n $\rightarrow$ ∞ of unified wave analysis herein provides the absolute definition of a radian as a unit of **cumulative angular rotation (summed θ_n's irrespective of their independent directions**). **The radian is seen as a property of the ordinal number sequence**, which numbers intrinsically associate with angles. Such definition for radians is not restricted to a plane, which however stylish is a mathematical fantasy of Euclid having only academic association with the real world. The absolute radian defines from the mathematical harmonic periodicity within natural numbers. It constitutes one "reason" why dimensional units of cycles cannot be dropped in the denominator of

h = E/ν = energy-time/cycle.

For large n, equation 14*8 approaches $\sqrt{[2/(n+1)]}$, which associates angles in figure 15 with the consecutive-unit-sequence between integers. It relates angular properties of the natural numbers to the step increase by unity (in n) between all adjacent integers. That natural integers possess angular properties has not been commonly known making the concept unfamiliar. However the orthogonality axiom applies when numbers are utilized to "count" smallest things. Angular properties of the numbers then loom into the enumeration process via inability to traverse fractions. This is particularly true of spacetime variables following an origin event because time appears positive and monotonic thereafter. When counting macroscopic objects, each unit-increment comprising "one-thing" is actually sub-dividable representing a large divisible number. Angular phenomena characterizing those conditions grow less pronounced.

As n → ∞, segments of a process tend to appear linear and continuous as with the exponential. When s_a = 1, figure 12 allows comparison of inter-integer transitions under macroscopic continuity contrasting the case of monadic indivisibility for s_a = 0. Presumed continuity allowing infinite partitioning has overshadowed angular progression properties for monads. Angular changes manifest cyclic repetition in the course of variable advance (with population growth). Population accrual and cycles intermesh hand-in-hand. When unity-increase occurs for monads, the extent a number is removed from zero impacts angular consequences differently than just the algebraic linearity of adding +1 implies. Adding unity to 2-removed-from-the-origin has different angular impact than adding unity to 2000-removed, though that difference shows up as an advance of one algebraic unit for both cases. Angular attributes of integer n accruals are influenced by the unified wave spiral, namely by θ_n = arcos[n/(n+1)].

The unified wave spiral is comprised exclusively of concatenated triangles, each having three defined sides without[*] invoking π, e, i, or ∞. <u>Two sides are always integer, (n and n+1) with the third side √(integer) = √(2n+1)</u>. With all vectors of the construction so designated, [rather than specifying the spiral in terms of right angles between n and √(2n+1)], **<u>no need exists to introduce orthogonality or imaginary</u>**

---

[*] The symbol π employed in these analyses need not be the transcendental-constant-π of infinite digits, but a finite-digit parameter derived from this investigation. The sum of chord-lengths at n_{max} around the base of the synchronous spiral forming on a cone when ξ = 1, divided by the diameter of that base, could pattern the derived parameter symbolized by the π used in equations herein. It would be very close to a truncated version of traditional-π though not the same and not an arbitrary constant. It would contain some uncertainty and be a ratio more closely representing arbitrary physical reality in curved spacetime than transcendental π.

operator √-1 = i. The entire spiral function is defined only in terms of concatenated triangle-side-lengths defined by integers. The expression is solely in terms of integer n without need to stipulate any triangle angles, orthogonal operators, or π. The structural algorithm can lack any discussion about angles. The four radian limit toward large n and the Planck length "polygon" formed at n_{max} circumvent the n = ∞ unit circle and the need for π and ∞ within the wave's definition. As angular radian amount, the integer 4 replaces 2π and **it derives from a model based solely upon integers**.

Infinity is an impossible constant for many reasons, among them is its **simultaneous existence within mutually exclusive sets as both odd and even, and also as both rational and irrational**. The unified wave avoids the infinite limit process engrained within exponential harmonics by keeping "constants of mathematics", π, e, i, and ∞, superfluous to the physics. Those arbitrary constants are avoided in the Occams razor sense of a simpler model that can provide ostensibly the same result as unit-circle-traversed exponentials. The fact that symbolic nomenclature for the operator exp(iωt) appears uncomplicated compared to a graphic spiral based exclusively upon integers says nothing about the relative complexities or comparative generality of those depictions. The unified wave foundation is a lot simpler than the infinite limit requirement of the harmonic exponential because of anomalies associated with infinity.

Moreover Planck's constant **traditionally interpreted in terms of three separate-embodiment constants** of: <u>action</u>, a <u>Planck length</u>, and a <u>Planck time</u>, constitute **three different-dimensional-constants for action, length, and time,** (plus indeterminacy of the uncertainty principle). **<u>Each is an arbitrary constant</u>**. In the unified wave format <u>a single excess-angle $\Delta\theta$</u>, as Planck's constant **<u>jointly specifies</u>** <u>Planck length</u> and <u>Planck time</u> in terms of chords of a single **Planck angle**, which is also a Planck's constant unit-of-action (and includes uncertainty of the space between integers n). Planck lengths (and Planck times) derive from chords-at-a-Planck-angle along the unit-sphere. The three traditional dimensional units for <u>action</u>, <u>time</u>, and <u>length</u>, umbrella under <u>one unifying source in the form of a **Planck Angle**, an excess angle in spacetime</u>. For the unified wave, one angle $\Delta\theta$ = h replaces the **seven traditionally invoked constants, <u>e, π, i, ∞, Planck time, Planck length, and action h.</u>**

Figure 18 additionally portrays how the commonplace photoelectric-effect-experiment (in "appropriate" absolute dimensional units), describes a **<u>Planck Angle</u>**. It is simply the angular constant of proportionality between energy and frequency, <u>the slope of the straight line in the figure</u>. **<u>Those seven arbitrary, untidy constants of conventional models become supplanted by the single physical-construct of excess-angle $\Delta\theta$ = h.</u>** That constitutes a measure of relative intricacy. However utilitarian those constants have been mathematically they perpetuate the intellectual straightjacket of Euclidean spacetime. Simplistic geometric construction from concatenated triangle sides based only on integers

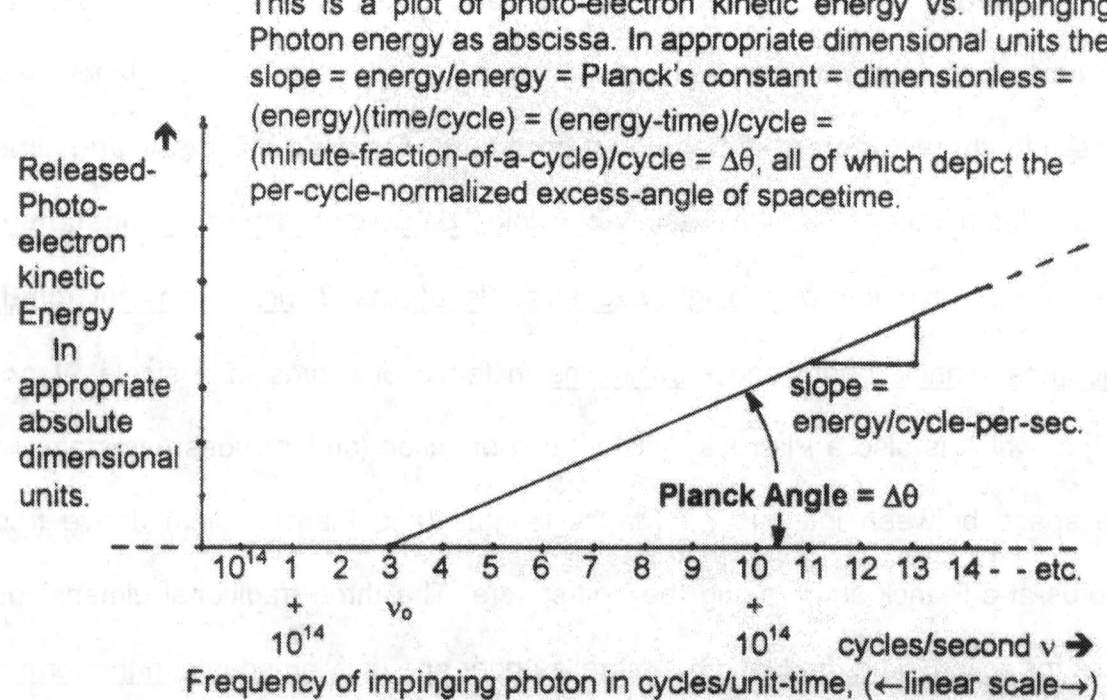

This is a plot of photo-electron kinetic energy vs. impinging Photon energy as abscissa. In appropriate dimensional units the slope = energy/energy = Planck's constant = dimensionless = (energy)(time/cycle) = (energy-time)/cycle = (minute-fraction-of-a-cycle)/cycle = $\Delta\theta$, all of which depict the per-cycle-normalized excess-angle of spacetime.

Released-Photo-electron kinetic Energy In appropriate absolute dimensional units.

slope = energy/cycle-per-sec.

Planck Angle = $\Delta\theta$

10^{14} 1 2 3 4 5 6 7 8 9 10 11 12 13 14 - - etc.
+ v_o
10^{14} 10^{14} cycles/second v →

Frequency of impinging photon in cycles/unit-time, (←linear scale→)

Figure 18. This shows a typical plot for the physical experiment when measuring the photoelectric effect. Photons of different frequency v are made to impinge on a photocathode. The kinetic energy of photo-released electrons is plotted as ordinate vs. cycles/second v as abscissa. Only enlarged segments of both scales are shown. Both are linear and greatly expanded to the extent that each cycle-per-unit-time and each smallest energy-unit denote as a separate division. For impinging photons of frequency less than a specific v_o, no photo-electrons are emitted. The resultant curve will be a straight line of positive slope. When "appropriate" absolute dimensional units are chosen for energy, [and similarly with time shown here in seconds], the **angle** between that straight line and the abscissa constitutes the **Planck Angle**. That represents the actual excess angle prevailing in spacetime. **It provides an "experimental measurement" test basis for this theory**. In the proper dimensional units, the Planck angle is the rate at which photon energy increases per cycle per second. Were that angle to somehow get larger, the energy per rate of traversing it (i.e., per cycle/second) would increase. The greater the time rate of traversing the existing excess angle, the greater the corresponding energy must be. Planck's constant depicts a wave's energy increase per each additive cycle/second of that wave. It is the slope of the relationship between energy and frequency of a non-rest-mass particle. It represents an excess angle per cycle that such wave must pass through. Energy equals the time rate of passing through that angle. **A smallest unit of energy-time delineates the fraction of a cycle portrayed by traversal of that smallest angle. Energy-time/cycle is that fraction of a cycle per full cycle and that is Planck's constant = $\Delta\theta$ = h = dimensionless.**

and whose arbitrary planes also grant non-localization, provide a far more general structure than one dependent upon a host of extraneous man-made constants. The unfamiliar and thus seemingly elaborate multi-sided spiral figure has infinitely fewer sides than the unit circle of the harmonic exponential.

The exponential harmonic solution-element, which as a "frequency" begins at minus infinity in time, exemplifies the inherent tenure of time reversal symmetry and non-causality in that exponential. By contrast, **the unified wave begins at the origin causing the event and proceeds thereafter**. Illusory negative time relative to the cause is not created by the mathematics. Even if n were permitted to approach infinity, the unified wave can still retain an arrow of time emanating from an origin. With asymmetry of the spiral retained in approach to ∞, the arrow of time sustains. The mathematical result can become time symmetrical but only at the n = ∞ point. Only when n vanishes as a mathematical parameter in the final expression (as occurs in the exponential's definition) does the change in chord-lengths-from-which-the-wave-formed also vanish. That point is not reached in the unified wave definition, which makes headway through Planck-time chords of the conic unit circle. The limit advances as far as n_{max}, not to ∞. It exemplifies how the exponential harmonic effectively utilizes only the n = ∞ endpoint, irrespective of the limit procedure containing n employed in its mathematical definition. **Parameter n vanishes in the final expression**, which then contains a singularity harbored through cyclic continuity around an idealized unit circle. That suggests why the exponential affiliates with "phantom" mathematical constants and operators like e, π, i, ∞. An idealized Euclidean-geometry universe is not the real stuff of this world. The real entity possesses an excess-angle Euclid overlooked, **which angle also serves to limit all actual parameters to finite rational mathematical values <u>related to that angle</u>**. To make all oscillators comprising a Black Body Radiator finite, Planck discovered effects of that angle at the beginning of the 20th century; but he did not recognize the constant he used for correction was an excess angle.

A plausible critique of certain derivations within this theory might be that mathematical identities, equations or limit relationships are often utilized in preference to mathematical proofs. The "reason" for this practice is that it is impossible to rigorously prove anything within an assured consistent set of logic. Laws can only be unproven, generally by finding a condition where they are knowingly invalid. A so called "proof" in mathematics or physics can never provide guarantee that all the known or implied postulates involved in that proof absolutely apply[13]. Outwardly consistent logic used in the course of the proof may represent a special case of an unknown more general condition, under which general case the proof would be invalid. Unaware of the more general condition, the proof would seem true and absolute with the underlying assumption being overlooked. It is thus preferable to search for a condition that disproves a law or tenet presumed true, than to attempt a proof that may inadvertently contain unknown assumptions. If subsequently found, the disproof condition will often lie outside the boundary of the law's validity and therefore provide a clue regarding an overlooked or erroneous hypothesis used in expressing the law. An example situation would be Newtonian mechanics through the 19'th century where the assumption velocity could increase indefinitely was unknowingly invalid. Another example treated herein is the presumption that zero excess angle exists in spacetime.

Recognition of that excess angle raises the question of what constitutes a valid proof. Should the proof be derived within the more general domain inclusive of the excess angle, or is it acceptable to use Euclidean logic for the proof? That question made it seem reasonable to express the subject relationship, equation or identity in a form that allows explicit evaluation for any given case. If or when a condition of invalidity can be determined, it should elude toward whatever unaccounted-for assumption makes the tenet invalid. Conditions of disproof generally infer the applicable boundary of validity and the unrecognized premise that was inadvertently implied. To elaborate the above with specifics, should a legitimate proof be derived using an excess angle of $\Delta\theta$, or zero? Are 4 d-radians at $n_{max} = 2/\Delta\theta^2$ what constitutes an actual radian in spacetime, or should it be a mathematical radian at $n = \infty$? More generally, should the proof be generic, allowing any numerical value for $\Delta\theta$, to accommodate every feasible curvature of spacetime? Should inequalities like $\sqrt{(2n+3)} - \sqrt{(2n+1)} \approx 1/\sqrt{(2n+2)}$ be considered "equal" at $n_{max} = 2/\Delta\theta^2$, or only at $n = \infty$? Such questions veil the issue of what constitutes a valid proof.

17) POSSIBLE IMPLICATIONS IN BIOLOGY AND CHAOS THEORY

Unified wave solutions should occur in disciplines where sinusoids now appear in continuous analysis, including physics, chemistry, engineering, mathematics, and biology. To the extent variables actually embody populations rather than being continuous, classes of problems within those broad disciplines should yield these unified wave solutions. The new waves likely contain a recipe for the dynamics between chaos and self-organization, for the objective existence of rationality, time's arrow, and entropy increase. One seemingly fruitful scientific area concerns the dynamics of chaos. Analysis there might reveal how the unified wave may shift modes from indeterminate randomness to overlapping periodic vectors in directions A, B, C, D, or vice versa. The process is perhaps easiest to consider by visualizing waves in the space surrounding a momentum-position spatial rather than temporal axis. The issue then decomposes into what factors govern subsequent directions of consecutive tangential vectors, while they still maintain orthogonality to their respective radial vectors. A solution mode forming a standing wave having perfectly periodic harmonics, [as near the conic angle $\beta = \sin^{-1}(2/\pi)$] can at least conceptually be extremely fragile. The wave comprises from the accrual of an enormous number n of different-individual-angles θ_n whose sum total must achieve exact periodic overlap to infinity. Any successful interaction to interrogate such a mode of the wave may destroy the precarious overlapping of sequential A, B, C, D directions. Yet, even with that standing wave overlap broken, the period associated

213

with accumulated θ_n's, (or the wave's energy) would remain unaltered. Upon successful interrogation, (interaction) the wave would plausibly undergo some perturbation in angular accrual. If angular summation resulting in exact A,B,C,D, overlap does not sustain, the interrogating entity could cause the mode to loose its interference properties and revert to randomness. Interrogation, measurement, or observation of the state of the system thereby has ability to alter the system's state, the result appearing as though human intervention through measurement "collapsed the wave function". Mathematical precariousness regarding A,B,C,D, overlap symmetry provides a vehicle for standing waves to possibly terminate. Precise conditions that cause bifurcation between harmonic and non-harmonic solution modes are highly relevant for these systems but may not yield easily to analysis. However, interpreting uncertainty due to $\Delta\theta$ as governed under a law conserving minimum indeterminism provides a more general "philosophical reason" why interrogation cannot penetrate that intrinsic uncertainty imposed under that conservation law. The precise state of the system remains "hidden" to the extent of the information shroud of the smallest monad and further information than that limit grants can never be acquired.

Planar spirals of figures 1, 2, and 3 portray a dynamics wherein the plane encompassing both radial and tangential vectors never changes, where no such incentive to change exists. Alternatively for example, a proximal force field or potential well might well exert some influence on the direction of each next s_n or position n. The locus of vector locations might constrain in response to the field's influence. Prevailing fields could conceivably affect the dispersal pattern of vectors. For specific circumstances, periodic standing wave solutions might be possible and these would represent allowed stationary states within that field distribution. It is suggested that quantum mechanical solutions may constitute calculations of these circumstances. It is further speculated that if represented in an appropriate "phase space", unified waves might perhaps portray the bifurcation mechanism between chaos and periodicity. These might alternatively associate with attractors of equilibrium theory. The threshold between stationary state harmonic solutions at $2\beta/N = (2/N)$ arcsine$(2/\pi)$, and non-localized potential randomness of the unified wave might conceivably model the "butterfly effect" between states of order and chaos. This interplay between order and disorder can rest within the same generic function, which framework retains all possibilities along the sequence following an origin event. The unified wave might thus provide a format wherein a stimulus could produce bifurcation between cyclic harmonics and discontinuous effects.

These analysis suggest "self organization" within the natural numbers could conceivably be feasible for spiral axes close to overlap near conic-apex-angles $(2/N)\sin^{-1}(2/\pi)$, with N integer. Though presented here with great supposition, unified waves may grant periodic phenomena in time and space as solutions to basic oscillatory equations. Such solutions might plausibly characterize how an energy driving force could "self organize" into a temporal oscillatory mode, or a chemical reaction source could generate recurrent spatial patterns. The combination of these raw materials, energy and reacting matter which are the same, in unified wave form may predispose not only forces and electromagnetic waves, but also the impetus of life itself. As displayed in figure 17, when the conic angle differs slightly from $\sin^{-1}(2/\pi)$, polarization in the 4-quadrant spiral-points-of-symmetry "twist". This allows the wave to exhibit properties that bear analog to the DNA helix ladder. The spiral can then partition into two intertwined helices (articulated in Appendix A, or by n_1 and n_5 in Table 1, or by directions A and E). These helices exhibit repetitive four-quadrant serial-sequences (or half-cycle "ladder rungs") between them. This self-organization condition that might emerge from randomness, characterizes a pair of discrete helices sequentially exhibiting two of four possible ladder rungs (quadrants) between them. That structure emanating from natural numbers might well relate to the evolutionary formation of living entities. **Equation 4*6 indicates that such a helical ladder generated from the spiral retains the unique feature of absolute successive cycles P_n referenced relative to the origin for each position n of the sequence**. All other features of the unified wave also apply. For an entire procession however long the chain n, the spiral emerging from ordinal numbers constitutes an ideal template for genetic encoding, self-organization and self-replication systems. Except for the zeroth cycle having n = 1, 2, 3, the four quadrant partitions that subdivide all further cycles to infinity always satisfy divisible (non-prime) integers n. It is hard to believe that such a formidable array of features in so fundamental a matter as the natural number system could be coincidental.

18) SUBSEQUENT VIEWPOINTS REGARDING MONADIC

VARIABLES

Transitions interpreted between integers along the Euclidean axis are **unidirectional outward from the origin**. Spaces between axial integers seem to possess a "degree-of-occupyability" dependent upon divisibility of the entity being enumerated. Such axes portray change for large macroscopic objects or highly ordered objects. By contrast, the orthogonality axiom applies for growing populations of non-divisible entities, or entities without sub-attributes. Alternative ways of stating the orthogonality axiom also exist, without stipulating angles, for example as the concatenated sequence of triangles of sides n, (n+1), $\sqrt{(2n+1)}$. However vector directionality or triangle side-lengths need not be the exclusive methods to portray transitions. For traditionally-understood transitions between natural-number-integers along the Euclidean axis, two restrictions for non-monadic population change are notable. (1) Uni-directionality; the transition between any integer n and (n+1) **always occurs in the identical axial direction**, and (2) each transition from n accomplishes a net magnitude change of unity to reach value (n+1). That describes conventional integers along Euclidean axis.

For monadic populations, two restrictions also apply. (1) Each transition magnitude from n to (n+1) is the "**geometric mean" of their sum**, or $\sqrt{(2n+1)}$. **Nothing need be said about direction**, (which statement may be less restrictive

217

than condition (1) above for non-monadic entities. Restriction (2), is the same as above and would be: Each transition accomplishes a net magnitude change of unity to reach value (n+1). Both sets of criteria end up with identical **magnitudes from the origin** for each state, being the sequence of natural numbers, 0, 1, 2, 3, - - - n, (n+1), - - -. For the geometric mean description however, **a cluster of possible angles** affiliate with each integer, while for the Euclidean axis case, only one angle (zero) associates with each integer. The Euclidean axis case will later be seen as a limit case involving one specific angle of the generic circumstance. The general case allows many possible angles, only one of which might be specifically zero. Other equivalent inter-integer transition descriptions exist besides the **orthogonality axiom** and the "**geometric mean**" designation. A particular elucidating picture stems from a root mean square sum interpretation to be subsequently explained.

The spiral as rendered on a plane illustrates one "**geometric mean**" **type transition process** for the natural number sequence. **Every transition from integer n to (n+1) has magnitude $\sqrt{(2n+1)}$ and that sole characterization can describe a monadic progression procedure even when the rendering figure does not entail a plane**. Description in terms of angles is unnecessary. Rendering on a plane merely enhances visualization. Each angle θ_n [or triangle side $\sqrt{(2n+1)}$] could more inclusively progress in any direction from n to reach (n+1). However unfamiliar, the natural numbers inherently allow (or do not prohibit) at least a **component of orthogonal progression between each pair of numbers**. Such component is intrinsic to enumerating discrete non-divisible objects. If counted things were divisible, integers would be assigned to each smaller part, reverting to an integer description for those sub non-divisible parts. Rendering vectors geometrically, as a spiral shown in figures 2 or 4 is **totally unnecessary in order to recognize the array of possible angles each natural number links with.** Those possible angles affiliated with each integer are engrained in the respective natural numbers, **the angle zero being only one special case for the Euclidean axis,** however routine that case has been. **Correspondingly, the four-radian limit at n → ∞ is not an artifact of the geometric spiral configuration. Those radians are *inherent* to the natural numbers n as defined through perfect squares of (2n+1).** (18*1)

The graphic spirals merely enable conceptual visualization by picturing an "axis or format for monadic modes" presented on a plane, cone, sphere, or as the modes of figures 7 to 10, or figure 49. Those various "snapshots" show only a few of **an enormous number of additional possible modes for orienting the natural-number-sequence into standing-wave or non-standing-wave type solutions. In all such modes, each respective angle θ_n remains unchanged, with certain or all of those angles occurring in different directions to create the different modes!** No matter how the angles orient they constitutes a possible mode, **although whenever directions A, B, C, D do not overlap, a standing wave solution will not be feasible.** The standing wave solutions "**are in the numbers**", as the Pythagoreans would say, not in the graphics. Such solutions would be very difficult to describe in a "plot" were a plane or conic surface not utilized.

The **geometric mean criterion** for monads, that **transitions between each n and (n+1) have magnitude √(2n+1), is adequate to describe waves from the natural numbers**. That statement defines how progression occurs. Statements of orthogonality, right-angleness, the imaginary √-1, directionality and graphical representations are then seen as **superfluous, immaterial to the unified wave harmonics which permeate the number system itself**. Parameters e, π, i, and ∞ are extraneous to the fundamental wave generation process.   (18*2)

Under that definition of transition between integers, f**our replaces the familiar 2π harmonic divisor.** [A considered title for a mathematical version of this theory was **2π = 4**.] Equally compelling to that mathematical reorientation is recognition that the roots of quantum, and biological phenomena elegantly emerge directly from the natural numbers. For monadic dynamical variables, √(2n+1) represents each **gap** between n and (n+1), with n and (n+1) being only unity apart in **static** population number. It's as simple as that in terms of how monadic populations must grow. **They are not forced to progress along a single axial direction, which freedom grants more generality than being obliged to do so.**

A "tangential" (or annular) region-of-abstraction exists for a monadic variable dynamically progressing through each gap between integers denoting fixed population values. Population members possess a shroud of obscurity conjoined

220

with the intangible interval between adjacent numbers. Traversal to the next population value above n is not analogous to idealized-point-type population members passing through an idealized Euclidean-line threshold to enter the ensemble designated as (n+1). If zero-size points depict the enumerated objects, Newtonian continuity would prevail. Ambiguity associates with traversing that region of abstraction between n and (n+1). Moreover, in general, a degree of correlation can associate with that transition. That correlation will subsequently be conveyed as parameter s_a in figure 12. For monads, outgrowth of that ambiguity contributes **intangible influence $\sqrt{(2n+1)}$**. The transition's **intangible** influence may exceed unity although algebraic magnitude increase relative to subsequent or prior radial vectors will always be precise unity. Percentage-change-induced (and effective angle θ_n,) diminish monotonically however with ever increasing population, as is reasonable to anticipate on a statistical basis.

The criterion of inter-integer transitions proportioning to the geometric mean of the sum of the involved integers materializes as an origin-angle $\theta_n = \tan^{-1}(s_n/n) = \tan^{-1}[\sqrt{(2n+1)}/n]$ associated with that transition. Such transitions are described by **vectors**, not **scalars**. The term $[\sqrt{(2n+1)}/n]$ depicts the **fractional influence** of the transition, whose decrease with increased n is plausible. The angle whose tangent is that fractional influence, portrays angular consequence of the transition. Thus, at least four separate, but alternative or equivalent "criteria" statements can be made regarding monadic transitions between integer n and (n+1).

1) The **orthogonality axiom** stipulates the transition from n to (n+1) is always **orthonormal to the direction of vector n emanating from the origin**. Direction of that right angle goes unspecified. **Transitions can have a vector component in any (or all) possible directions orthogonal** to the-origin-to-vector-n-direction.   (18*3)

2) The **geometric mean** criterion indicates that transitions from n to (n+1) must have **magnitude** $s_n = \sqrt{(2n+1)}$. No direction need be specified. **Transitions can occur** in any and all possible directions, **having magnitude** dependent **only** upon the "**geometric mean**" of n+(n+1), as $s_n = \sqrt{[(n+(n+1)]}$.   (18*4)

[Also, the square of each transition exceeds the square of its predecessor by 2 since $s_{n+1}^2 - s_n^2 = 2$. From n = 0 to ∞, constancy in that incremental change of 2 prevails in transition magnitude squared and that could be listed as another criterion.]

3) The origin-angular requisite for transition from n to (n+1) entails associated **angular shift from the origin** $\theta_n = \tan^{-1}(s_n/n) = \tan^{-1}(\sqrt{[(2n+1)/n]}$.   (18*5)

Transitions induce that range of angular **shift in all possible directions about the prior vector from origin-to-n.** Magnitude or direction of the transition between n and (n+1) can go unspecified with only origin-angle θ_n being specified.

4) A statement based on the root mean squared sum of all transitions following n = 0 will be forthcoming.

5) For the generic case describing partially monadic and **partially macroscopic** entities, using figure 12, the **transition-angle emerging from vector n toward vector (n+1),** would be: $\gamma_{n\&sa} = \tan^{-1}[\sqrt{\{(n+1)^2 - (n+s_a)^2\}}/s_a]$ (18*6)

as will be explained at equation 18*18.

Any of these 5 stipulations can delineate the description of how populations advance between integers, #5 being the most general. Choosing any one of the first four criteria automatically includes the other three-of-the-four as also applying. They provide four alternative descriptions saying the same thing for monadic accrual. They demonstrate criteria of uniformity and invariance in transition relationships from n = 0 to n = ∞. The all-inclusive case #5 extends from monads to macroscopic objects. **For monads, a new origin angle at (n+1) differs from n "<u>by amount θ_n nutated to all possibilities about the prior vector direction of n</u>". Such added angular possibilities <u>associate with each successive advance from n to (n+1).</u> <u>Each such "small conic bundle" of new possible directions emerges with every unit increase in population.</u>** Between consecutive perfect squares of (2n+1), the cumulative sum of those θ_n small conic-bundle-angles, (<u>**regardless of their individual directions**</u>) asymptote to exactly 4 radians as n → ∞. That constitutes a complete cycle so that **angular properties** and the 4-radian limit vest within advance through the natural numbers, <u>**irrespective of graphic presentations**</u>. Periodicity relates to and derives from the perfect squares within the number system. The four radians express the angular extent traversed between perfect squares as the origin-angular-change per transition approaches zero, i. e., when $\theta_n \to 0$ as n → ∞.

This <u>**angular**</u> alteration characterizing transitions between populations may be <u>totally unfamiliar</u>. For integer-only populations portrayed in terms of natural numbers, **change must interpret as a vector process**. Since the 3-dimensional

direction of each θ_n around prior vector n remains unspecified, the

concatenated vector sequence becomes directionally non-localized.

Newtonian continuity would allow infinite incremental change along the axis,

proceeding from n to n+ε, where $\varepsilon \to 0$. Thus, "transition magnitude" in that case

would be: $s_n = \varepsilon \to 0$ so that angle $\theta_n = \tan^{-1}(s_n/n)$ became so small as to be

negligible by definition. Then, transition magnitude to (n+ε) and angle approach

zero, which can be neglected whatever their directions. We therefore typically

interpret Newtonian continuity along an Euclidean axis as a scalar sequence.

When the entity being described is an ensemble of non-zero-size items (a

population contrasting Newtonian limits of nothingness), vector depiction becomes

mandatory. Then any-and-all of condition (1), (2), (3), and (4) above jointly apply

and conclusions of this treatise become generically legitimate.

Transit between the natural numbers acquire a geometric-mean transition-

magnitude s_n and origin angle contribution θ_n associated with every value n.

Symmetry, periodicity, and wave generating properties of those natural numbers

emerge intrinsically, as elucidated by the spiral with all its depicted consonant

modes. For monads, the natural angular rectilinear divisor of geometric space (as

well as partition of numerical "space") becomes 4, not 2π. The well-known

algebraic natural number sequence for Newtonian processes distinguish from an

alternate set, the unfamiliar **geometric mean natural number sequence** for

monadic processes. **Both result in identical integer magnitudes sequentially**

one unit apart. The latter possesses a bundle of affiliated possible angles with each number n and accommodates quantum phenomena. Freely speaking, the geometric mean evaluates as "**geometrically midway**" between the sum of starting and ending values of the transition. **In that sense, each monadic transition employs a tangential extent "midway" between starting and ending values**.

The #4 defining criterion for monadic progression can be expressed as follows. Every integer (n+1) is **the root mean squared (RMS) sum** of the **prior integer** n and the **transition magnitude** $\sqrt{(2n+1)}$ to reach (n+1) from n. Thus, in figure 12, $\sqrt{[n^2+(\sqrt{(2n+1)})^2]} = (n+1)$ (18*7)

describes this case of the natural numbers with s_a = 0. This RMS approach is another perhaps more general way to construe the <u>orthogonality axiom</u>, or the <u>geometric mean</u> interpretation, or the <u>changing origin angle</u> θ_n. Two modes of this algebraic RMS-sum representation with s_a = 0 render graphically on a plane or a cone as in figures 1 to 4. They characterize the case of population increase with total lack of correlation between **assemblage** n and the **transition** to (n+1). The RMS stipulation bases on: $\sqrt{[(\text{prior state})^2 + (\text{transition } \sqrt{\{2n+1\}})^2]} = (n+1)$

This restates as: $\sqrt{[(\text{prior state 1})^2 + (\text{transition 1 to 2})^2]} = (\text{new state 2}).$

For the next transition from state 2 to 3

$\sqrt{[(\text{new state 2})^2 + (\text{transition 2 to 3})^2]} = (\text{new state 3}).$ (18*8)

Upon substitution this accrues as:

$\sqrt{[(\text{prior state 1})^2 + (\text{transition 1 to 2})^2 + (\text{transition 2 to 3})^2]} = (\text{new state 3})$

Therefore, **every state is the RMS sum of all transitions from the initial state upwards,** i. e. $\sqrt{[}$ (**initial state**)^2 + (**first transition**)^2 + (**2'nd transition**)^2 + (**3'rd transition**)^2 + ---- + (**every prior transition**)$^2]$ = (**any state**).   (18*9)

The initial state is unity (or the transition from n = 0) and each consecutive transition s_n has magnitude equaling the square root of an odd number, i. e. they are, $\sqrt{1}$, $\sqrt{3}$, $\sqrt{5}$, $\sqrt{7}$, $\sqrt{9}$, $\sqrt{11}$, $\sqrt{13}$, - - - etc.   (18*10)

Then squares of those transition magnitudes are 1, 3, 5, 7, 9, 11, 13, - - - etc. The sum of those squares would be:

1 = 1, 1+3 = 4, 1+3+5 = 9, 9+7 = 16, 16+9 = 25, 25+11 = 36, 36+13 = 49- - - etc. Each of these are themselves consecutive perfect squares so their square roots, representing each consecutive state, (or the respective RMS sums) become:

1, $\sqrt{4}$ = 2, $\sqrt{9}$ = 3, $\sqrt{16}$ = 4, $\sqrt{25}$ = 5, $\sqrt{36}$ = 6, $\sqrt{49}$ = 7, - - - etc. to n.   (18*11)

Those are none-other than the natural number magnitudes all the way until n → ∞. It is therefore perfectly reasonably that **each new population is the RMS sum of all prior transitions**.

The orthogonality axiom in fact represents one way to educe the natural number magnitudes. Every natural number is the RMS sum of all orthogonal transition magnitudes up to that number! That is apparent from the three spirals of figures 1 to 3. Those represent the sum of the odd integers since transitions $s_n^2 = [\sqrt{(\text{odd integer})}]^2$ = odd integer, which sum from unity upwards is always a perfect square having the consecutive integers for square roots. *Every n* *of the monadic spiral is the RMS sum of all prior sector magnitudes (all prior*

transitions √(2n+1) regardless of direction). (18*12)

That describes a **fundamental criterion-of-growth, relating perimeter (tangential) extent to radial extent from n = 0 to ∞**. The radial path from origin to any vector node equals the RMS sum of tangential-path-sectors from origin to that node. The statement puts no constraints on any existing angles or directions. Being equal means, **(the RMS sum)/n = 1, for all n.** (18*13)

This RMS equivalence between all chord-length perimeter-extents (transitions) and "radius" (n) is more foundational than radians being the ratio of perimeter to radius. That canon only applies for a fairy-castle mathematical figure, the circle, hinged on everywhere having exact equi-distance from a central point on a perfect plane. Being the apropos rule, the RMS summation elaborates maximum randomness and lack-of-correlation between inter-integer transitions. **Accrued direction angles θ_n remain irrelevant in the RMS equivalence interpretation, as compared to requiring a perfect circle on a perfect plane**. Quantum processes for monads grow by that method. Exponential progression only satisfies the limit case of that criterion, when every sector or chord of the unit circle becomes of zero extent and n = ∞.

By contrast, the RMS growth decree applies to all numbers n at every scale of the number system including n → ∞. Plotting the geometric figure as a spiral in a plane merely allows observance of how all angles θ_n must accrue between perfect squares, no matter which way those angles would otherwise be headed. It grants insight as to how periodicity arises **within the natural numbers**. The geometric mean description portrays what <u>natural numbers actually signify</u>, **<u>the assumable values for a population of discrete non-divisible things <u>without implying conventional continuity between the integers</u></u>**. The RMS summation criterion yields the same integers 1, 2, 3, 4, 5, - - - etc., to $n \rightarrow \infty$ as achieved by linearly adding unity to each prior population membership n. All algebraic relationships between those integers remain the same as for conventional interpretation. The geometric mean criterion relates exclusively **upon the number in <u>relative proximity</u></u>**. **The RMS criterion** relates solely to **all transitions** upward from $n = 0$ (or any n_0), which is what counting numbers are about. The orthogonality criterion says, "Always go to the next integer at a right angle" and it maximizes linear independence. The angular shift requisite specifies only change in origin angle. The square of each transition magnitude exceeding its predecessor by 2 references a specific continuity in incremental transition advance. These 5 criteria achieving the same goal as the sixth, (i.e., unity separation between integers) is indicative of their greater generality than the sixth taken alone.) **Customary interpretation of a continuum existing between the integers narrows that meaning to the sixth case taken alone as a special case.**

<u>It is the interval between integers embodying the missing information shroud of each quanta that designates how they amass to population n</u>. Stated in RMS form for monads, <u>population n depends only on the transitions</u>

leading to n. <u>The number of transitions since zero is the same as population n.</u> <u>The RMS sum of those transition magnitudes is the same as population n.</u> <u>The natural numbers tell how many non-correlated gaps have been gone through from zero to get to any value n.</u> In a sense then, <u>natural numbers are a measure of, or delineate the quantity representing, transitions since zero.</u> **For monads, they are both the RMS sum of transition magnitudes since zero, and the algebraic sum of their number**. While almost all of the transition magnitudes are irrational, their <u>RMS sums are not only rational, **but all are integers out to infinity**</u>. Between any n and (n+k) one can write:

$$\sqrt{[n^2+2kn+(1+3+5+7+9+11+13+\text{- - - etc., but for only k terms)}]} = (n+k) \qquad (18*14)$$

That depicts the aggregate for k transitions beyond value n. It applies for all possible rational values of n, with k integer and ≥ 0. It also provides a useful identity for evaluating an integer number of added transitions. The sum of any n plus k transitions evaluates via this integer-transition manner just as conveniently as the more-commonplace additive-interpretation along a Euclidean axis infers fractional-transitions as possible. In a similar manner whereby (RMS) summations accommodate random-waveform correlations, RMS sums for this circumstance likewise entail such correlations. These correlations also intermesh with the harmonics since cycles within the natural numbers occur at odd-integer transition-magnitudes, irrespective of graphical presentation.

Figure 12 typifies the all-inclusive case, the bridge connecting "**geometric**" or monadic-interpreted **natural numbers** wherein $s_a = 0$, and the opposite-limit-case of "**algebraic**" interpreted **natural numbers,** where s_a = unity.

Since $(1-s_a) \equiv \xi$, sector magnitude $s_n = \sqrt{[2n\xi+1]}$ (18*15)

parametrically accommodates that continuum range, with each origin-angle for all n and s_a becoming:

$\theta_n = \tan^{-1}\{\sqrt{[2n\xi+1-s_a^2]}/(n+s_a)\} = \tan^{-1}\{\sqrt{[(2n+1)-s_a(s_a+2n)]}/(n+s_a)\} =$ (18*16)

$\tan^{-1}\{\sqrt{[\xi(2n+2-\xi)]}/(n+1-\xi)\}$

When $s_a = 0$, angle $\theta_{n,sa=0} = \tan^{-1}\{\sqrt{[(2n+1)]}/n\}$ for monadic entities. Continuous variation (dependent upon s_a) extends to the other limit case when $s_a = 1$, wherein,

$\theta_{n,sa=1} = \tan^{-1}[0/(n+1)] = 0$ for Newtonian variables (18*17)

along Euclidean axes. There, all vectors lie co-linearly along the axis. Figure 49 will subsequently portray a harmonic mode categorically exhibiting synchronous overlapped ABCD directions covering the entire range where $1 \geq \xi = (1-s_a) \geq 0$.

The full compass of natural number enumeration cases thus generalize in terms of fraction s_a, or ξ. As extrapolated from figure 12, in all cases, the plane normal to vector n at s_a intersects the spherical surface comprised of all possible (n+1) vector-tips, thereby forming a circular locus. For descriptive purposes, the bundle of possible (n+1) vectors around that circle, iterated with its understood predecessors on respective circular loci at all preceding radial n values might be called a *"vect-array"*. **Radial magnitudes for those cases comprise the**

conventional natural numbers n, while a <u>vect-array cluster of possible angles</u> can <u>annex to each number n</u>. Whenever $(1-s_a) = \xi$ exceeds zero, transitions between adjacent natural numbers assuredly **behave as vectors**. Each natural number n would then affiliate with a vect-array cluster of possible vectors having differing angles. The "**case**" parameter ξ, granting the range of <u>possible</u> angles that accompany each natural numbers may be unfamiliar, but it spans the expanse from <u>Newton's **Continuity** at $\xi = 0$ ($s_a = 1$), through Leibniz' **Monadality** having $\xi = 1$ ($s_a = 0$)</u>. For all cases of $1 \geq s_a \geq 0$, transitions between numbers involve progression from integer n to (n+1) at some <u>nondescript **angle**</u> relative to the prior concatenated vector. From figure 12, that angle at each n is:

$$\gamma_{n\&sa} = \tan^{-1}[\sqrt{\{(n+1)^2 - (n+s_a)^2\}}/s_a] \qquad\qquad (18*18)$$

where, $0 \leq s_a \leq 1$. Here $\xi \equiv (1- s_a)$ is called the "<u>case</u>", "<u>scatter</u>", or "<u>monadicity</u>" of the natural numbers and is believed a rational value. When $s_a = 0$, then $\gamma_n = 90°$ representing the extreme condition of orthogonality applicable for monads. When $s_a = 1$, then $\gamma_n = 0°$ describing the other endmost condition, the Euclidean axis circumstance for macroscopic, highly divisible, or highly organized processes. Parameter s_a is treated as a rational fraction delineating natural number cases $0 \leq \xi \leq 1$. Each case ξ **<u>necessitates vector transitions at that ξ between all consecutive integers</u>** and engenders an angle γ_n dependent upon n and ξ.

Figure 12 shows graphical construction accommodating **graduated degrees of orthogonality and divisibility**. It elucidates a spiral formation method where transition s_n has both an **a**xial and **o**rthogonal component designated as s_a and s_o. Axial component s_a will always comprise a fraction of the unit-distance between n and (n+1) and the remainder of that distance defines as ξ = 1-s_a. Each transition advance can be envisioned as progressing "diagonally" from n along s_n to (n+1). Alternatively, by way of components of s_n, it can progress from n to (n+s_a) along the axis and then in an orthogonally plane along s_o to (n+1). Here, component s_a of the transition is being allowed in the axial direction attendant with diminished orthogonal component s_o. On the far left, the concatenated triangle exemplifies how progression from n = 1 to n = 2 would situate in forming the resultant spiral. It represents the same triangle as the one shown along the axis having origin angle θ_1. The spiral forms by concatenation on the left from the axis-depicted consecutive triangles, (as when going from n = 2 to n = 3 for example). The continued remainder for 3, 4, etc. would graphically construct similar to what has been previously described. The spiral rotates "angularly" on the left by concatenated linking of triangles shown along the axis having origin-angles θ_1, θ_2, θ_3, etc. Those triangles have respective vectors s_n labeled s_1, s_2, s_3, etc. Every axial-component s_a at n has respective $s_o = \sqrt{[(n+1)^2 - (n+s_a)^2]}$. Each triangle's diagonal s_n must therefore be

$$s_n = \sqrt{[s_o^2 + s_a^2]} =$$

$$\sqrt{[(n+1)^2 - (n+s_a)^2 + s_a^2]} = \sqrt{[2n+1-2ns_a]} = \sqrt{[2n(1-s_a)+1]} = \sqrt{[2n\xi+1]} \qquad (18*19)$$

This result constitutes a most unique unification of all wave processes covering the range from monads to Newtonian continuity. In accommodating *degree-of-divisibility*, the spiral incorporates an ξ less than unity and ends up with diagonal transition lengths of $\sqrt{[2n\xi+1]}$, instead of $\sqrt{[2n+1]}$ for the case when $\xi = 1$. **That constitutes an elegant simplicity engrained within the natural numbers to encompass the enumeration of "partially divisible" entities distinguished by** ξ. The case of $\xi = 0$ represents zero non-divisibility (or total divisibility) while $\xi = 1$ (where transition orthogonality maximizes), depicts maximum non-divisibility. The entire range $0 \leq \xi \leq 1$ covers the gamut of intermediary possibilities. <u>**This "modifier"** ξ **on the efficacy of the information shroud surrounding each population member constitutes an unfamiliar variable inextricable from natural numbers and enumeration procedures**</u>. It will later be shown that <u>symmetries intrinsic to the $\xi = 1$ case preserve for all values of ξ. **Values of n where quadrants of the wave occur also remain invariant with** ξ. Axis directions ABCD of figure 3, shown to never associate with a prime value of n, correspondingly retain that feature throughout all ξ. Arbitrary values of ξ do not alter the intrinsic 4 radians/cycle and 1 radian/quadrant synchronization generated by natural numbers</u>. Though the range of ξ displays important properties of numbers and waves, discussions will first focus on the $\xi = 1$ case before analyzing waves of divisible entities.

Figure 12 considers a population n that delineates a growing spacetime dynamical variable. It comprises of smallest monadic sub-units, whatever they may be. Those minimal sub-units embody certain attributes when entering a set or category associated with the variable. Those attributes can be represented by transition-correlation-parameter s_a, which can be thought to express degree-of-divisibility extending from indivisible at $s_a = 0$ to totally divisible at $s_a = 1$. It can also signify degree-of-correlation. Referring to figure 12, the general case describing natural numbers n **regards all of them as having an affiliated cluster of possible vectors.** Degree-of-divisibility parameter s_a influences that bundle of vector angles. Because figure 12 is presented on a plane for illustration, it can only render comparison of **consecutive** angles θ_n, not their resultant absolute angular direction in "spherical coordinates". When n increases from the origin with $s_a = 0$, vectors can diverge rapidly to soon be in any possible spherical direction thereby exhibiting non-locality. However, the relative directions of those vectors (the associated vect-arrays) possess a probability distribution that becomes more and more constrained as $s_a \rightarrow 1$. Then each possible increment contributes an additional "phase-space-type-angle" θ_n, which diminishes with increased n. When $s_a = 1$ maximum angular correlation prevails between all consecutive-integer states. **This case for Newtonian continuity with zero non-locality has vectors for all integers in-line along the Euclidean axis.** Each vect-array bundle would then coalesce into a single vector direction along the Euclidean axis. The gamut of natural number cases thus elaborates through figure 12, and the above discussion. The situation **effectively grants another degree-of-freedom ξ (= 1-s_a), or additional "correlation",** to the sequence of natural numbers. The commonplace scalar designation of consecutive numbers merely represents the case of $\xi = 0$. For this wider interpretation with $0 \leq \xi \leq 1$, consecutive, integers remain one unit apart algebraically (the same as usual) and vect-arrays affiliate with each n. The simple RMS growth criterion of equation 18*12 no longer applies when $\xi < 1$.

234

19) CORRELATION EFFECTS AMONGST POPULATION MEMBERS

It is observed that as s_a increases from zero, each origin angle θ_n affiliated with a respective n diminishes to reduce the tendency of vectors to "randomize" away from the axis. Each consecutive vector at angle θ_n would remain **closer in direction from the origin to its predecessor** due to smaller θ_n angles. Alternatively stated, the orthogonal transition component at each n decreases with increased s_a, reducing the likelihood of vectors to diverge from the Euclidean axis, or from where they were at a lesser n. The consequence of parameter s_a thus manifests as determinism in vector direction affiliated with each respective radial vector n. It can be thought of as a directional correlation parameter as well as signifying degree-of-divisibility. Cases other than $s_a = 1$ represent feasible though unfamiliar instances of the natural numbers, one for each possible rational value of s_a. **Parameter s_a might also interpret to delineate the degree to which a "newly added" population member correlates with existent members n**. That speculative conjecture could help parse further meaning to the different ξ cases. For cases other than $s_a = 1$, a respective angular possibility-bundle affiliates with each integer of the conventional natural numbers. For $s_a = 1$ an angle similarly exists, **but it is zero so respective vectors for all integers end up as traditionally interpreted in-line along the Euclidean axis**. Transition angles intrinsically associated with each inter-integer advance have thus been unfamiliar.

Each natural number invariably references the zero origin and is therefore a vector from that origin to the spherical surface at radius n from the origin. For the case when s_a is slightly less than unity, consecutive integers are no longer confined to lie along co-linear lines but begin accruing a component of angular randomness relative to prior numbers as describable by a vect-array. For small n that bundle diverges only moderately from the axis, but takes on increased possibility to deviate from the axis-line as n continues to increase. The effect simulates a "random walk" diverging from the axis at a rate related to $\xi = (1-s_a)$ and n. For $\xi \neq 0$, whenever n reaches any value of significance, an absolute reference direction for each vector n no longer exists. [Actually, since nothing defined the direction of the initial unity-magnitude vector at n = 0, vector-tip locations on all spherical surfaces, (even those of small n) are indeterminate.] Only relational angles-θ_n to proximal vectors remain meaningful. While on a plane, Euclidean-axis graphics in figure 12 inter-compare angles generated between respective consecutive integers, in an "actual concatenated embodiment" those same angles might end up heading in most any spherical direction about the origin. The spiral shown in formation to the left of the axis depicts one incarnation of those angles dispensed on a plane. Presentation of the triangles along the axis shows the construction method and semblance amongst them, wherever their absolute locations might become in spherical coordinates. For monads, it turns out that any single complete cycle will thoroughly randomize the possible direction of vectors even if known at the start of the cycle. Interpretative significance amongst all the

236

vector elements brought out in discussion remains valid however, no matter what the absolute orientation of those elements.

Whereas s_a can portray the degree of correlation that inheres in any specific case of the natural numbers, $\xi = (1-s_a)$ denotes the directional **"scatter"** characterized by that case. **Scatter** ξ complements correlation s_a **indicating the broadness of the vect-array cluster of possible angles. When $\xi = 1$ and $s_a = 0$, scatter maximizes**. For that case, angles affiliated with an enumeration process have maximum randomness. When $s_a = 1$, scatter ξ would be zero since total correlation and zero transition-direction-randomness then prevail for population growth. The range of scatter modes aggregate with correlation s_a to a fixed sum. For every case of the natural numbers, it is always true that

[correlation s_a]+[scatter ξ] = unity = $s_a+\xi$ = $s_a+(1-s_a)$. (19*1)

[Examples of negative correlation $(-s_a)$, or scatter greater than unity are later presented in figure 46, but no attempt is made here to interpret significance in that region.] **Clusters of <u>possible</u> vector directions always affiliate with each integer of the natural numbers.** Allegedly that suggests why the <u>**probability**</u> of locating an actual result emerges as a product $\Psi\Psi^*$ in quantum mechanics. Whatever the specifically appropriate (though unknown) probable angle within the scatter cluster, conjugate multiplication would return the result onto the real axis. A similar consideration applies to the cluster of angles associated with each sector length. Accordingly, the natural numbers actually possess an additional degree-of-freedom than integer-magnitudes. It is expressed through scatter ξ or transition correlation s_a.

Certain relationships in figure 12 simplify when expressed in terms of scatter $\xi = (1-s_a)$, thereby providing additional reason to give that parameter a name. It should be noted that the descriptive interpretations of divisibility, correlation, case, monadicity, or scatter, assigned these variables herein represents a "best effort" to connote the phenomenological role they play. Accuracy of such interpretation bears no interdependence with the mathematical exactitude vested in those variables. Those are simply words that might aid visualization with far more subtle meanings possible. Conjoined with every integer magnitude-n exists a vect-array cluster of possible angles. Processes of conventional algebra have not typically accounted for the monadic constituency underlying spacetime variables and the **attendant bundle of concomitant indeterminate angles anchored to every natural number n**. When $\xi = 0$, all angles θ_n are zero, which has historically been the interpretative case of the natural numbers. It is deficient a degree of freedom representative of monadic scatter.

There exists perhaps another way to construe these analyses. It is quite speculative but worthy of mention since this entire hypothesis addresses the "smallest" entities of physical reality and this conjecture may further elucidate them. Angle γ_n in figure 12 is somewhat like a "phase shift in phase space" between extant population n and the transition of adding a new member. When two "signal components" are significantly unrelated in phase (an analogue for large γ_n at any given n), less correlation prevails than if they were closer in phase for smaller γ_n. The RMS sum of those "signal components" tends to reflect that phase difference, much akin to how RMS sums treat relationships between two uncorrelated frequency signals. Comparative γ_n angles might similarly interpret as a relative phase angle in phase space related to the transition.

Figure 14 might graphically clarify why the RMS sum of sectors extending from n = 0 to any n will equal that n. It provides a possible "meaning" to the RMS sum criterion when $\xi = 1$. Beginning at n = 1, the (transition-vector-length-s_n) squared for n = 2 is, $[1^2+s_1^2] = [1^2+\sqrt{(2^2 -1^2)^2}] = 2^2$. Since $s_0 = 1$, this is actually $2^2 = (s_0^2+s_1^2)$. To get vector-length-squared for n = 3, s_2 must be added, resulting in $3^2 = (s_0^2+s_1^2+s_2^2)$. Adding s_3 to proceed to vector-length-squared for n = 4 yields $4^2 = (s_0^2+s_1^2+s_2^2+s_3^2)$. This process continues indefinitely for any n in the form

$$n^2 = s_0^2+s_1^2+s_2^2+s_3^2+s_4^2+ - - -s_{n-1}^2. \tag{19*2}$$

Therefore $n = \sqrt{[s_0^2+s_1^2+s_2^2+s_3^2+s_4^2+ - - -s_{n-1}^2]}$, which is the RMS sum. A possible way to conceptually characterize each s_n, would replace the words "transition magnitude" with the phrase: "***tangential time from t_o to reorient the entire population into the new configuration as required upon acquiring one more member***", called the "***reorient time***", $\equiv R_n$. **Here, time would be measured from t_0 in figure 14 so each orthogonal s_n is time-from-the-origin to the respective next transition. Then equation 19*2 can be written as:**

$$n = \sqrt{[R_0{}^2 + R_1{}^2 + R_2{}^2 + R_3{}^2 + R_4{}^2 + - - -R_{n-1}{}^2]} \qquad\qquad (19*3)$$

Figure 14 then illustrates how every magnitude n would equal the RMS sum of all prior times the ensemble population took to "collate" each new population member. Interval R_n **starts from the origin** proportioning to something **related to the entire population**, not only the last member. Although $\sqrt{(2n+1)}$ describes the transition between the last two members, containing n it also signifies the entire population. Thus, s_n seems to convey an interval related to **the-entire-prior-time for the currently-prevailing population to reconcile each additional quanta**. As population grows, reorientation time grows, and so do s_n and R_n since those scales denote the time utilized. More quanta added require more time to reshuffle.

For $\xi = 1$, each tangential sector circulating the spiral perimeter can represent a reorientation interval for the assemblage to equilibrate with the new arrival. It takes time to re-sort the increasing ensemble. Magnitude vector n then comprises from (or is commensurate with) the RMS sum of all prior restructure intervals. Waves of figure 14 emerge from inherent periodicity within natural numbers upon accumulating the population. Under those conditions, cycles-within-the-numbers occur automatically upon continued rearranging n things as n traverses perfect squares of (2n+1). Such reconfigurations liken to "retardation delays" from continually reoriented smallest constituents as new members join the group. Time-to-assimilate grows with population, making s_n longer by a periodic increment of 2 for each four quadrants traversed. That's just a mathematical truism of the numbers elucidated by the spiral. The RMS sum depends upon square roots and squares, which follow inherent periodicity of the natural numbers. Vectors of the drawn spirals merely provide a metaphor for processes and quantities involved. As analysis herein shows, this intrinsically periodic case describes maximum redistribution with minimum correlation between new and existing membership. Figure 14 delineates that periodicity in terms of four-quadrant-cycles asymptotic to 4 radians/cycle.

It is appropriate to seek deeper possible interpretation for this, however speculative that inquiry may be. One conjectured scenario is that transitions comprising smallest constituents (say Higgs particles of the vacuum) produce a "retardation delay" in reorienting, as their number increases with continued wave progression, or whenever the initial disturbance "fan-out" affects additional quanta. The disturbance in propagating through the quantum sea (vacuum), must reshuffle an ever increasing number of states n. Their reorientation entails transition time s_n proportional to total time transpired and injects a drag or delay in the process. Since we have assigned the monads being addressed as the smallest singletons of existence those "vacuum quanta" might characterize the population. For a propagating wave, that delay could plausibly affiliate with the nanosecond per foot retardation associated with electromagnetic radiation passing through that sea of quantum. That "drag" could produce the limiting rate of advance in vacuum. Because it occurs in the microzone, behavior could be independent of macrozone relative velocities. Whatever smallest entities comprise the false vacuum, their "reshuffle time" as annotated by RMS summation would establish propagation rate of the disturbance through that quantum sea. The conjecture has some plausibility because akin to propagating waves, the resultant time-per-cycle unavoidably remains constant throughout the process. Inherent periodicity within the number system creates that constancy. For $\xi = 1$, figures 12 and 14 show that **the RMS sum of independently-separate prior transitions <u>always equals</u> the RMS sum of the prior prevailing population and the last transition**. <u>**The same retardation delay**</u> can thus attribute to **either <u>many quanta</u> entering the population "almost simultaneously"**, (simulating the collective condition of all prior transitions), **or <u>one at a time</u>** premising single additions to the prevailing population. Any and every combination of accrual situations thereof remains satisfied. <u>**The RMS circumstance tolerates all combinations of the sequence by which population members accrete under minimum correlation**</u>. Each quantum state entails retardation, adjusting for the new membership environment, which procedure engenders cyclical repetition as figure 14 illustrates. For an object with rest mass when $s_a \neq 0$, this RMS tolerant circumstance for population advance becomes violated in proportion to s_a.

Execution of these RMS summations may be an alternative way to interpret propagating electromagnetic radiation based on the interactive "drag" of quantum-sea monads. This type model might satisfy relativistic phenomena as portrayed by figure 12, contrasting Maxwell's Equations, which do not. Under this understanding, the false vacuum bears some similarity to an ether medium. This conjecture perhaps remains consistent with one unit of energy-time/quadrant being everywhere indeterminate. That implies any monad having energy-time/quadrant equaling one action unit can emerge and vanish anywhere in vacuum, so long as its product of prevailing energy and existent time does not exceed Planck's constant. This would allow the possibility of virtual photons prevailing everywhere in false vacuum. It also infers that each integer n represents an "allowable" position or condition in time [and space] with non-occupyable gaps between those integer positions. Time and space as perceived from the macrozone <u>would be discrete</u>. **It further concludes that all natural numbers can be both the algebraic sum of the sequence of unity-separation transitions, as well as the RMS sum of those same transitions**. Another feasible interpretation for the RMS sum might be that uncorrelated possibilities increase in relation to total time transpired, which manifests as an RMS sum of those times. Above and beyond these far out speculations, as a population of n monads, a cyclical recurrence participates within natural numbers linked to integer transitions of magnitude $\sqrt{(2n+1)}$. **<u>A transformation takes place from monotonic progression in n, to a cyclic process in s_n</u>**. For monadic populations under the orthogonality axiom, that transformation is indigenous to the natural numbers and to absolute radians of rotation. Even the most skeptical need admit that such waves constitute a mathematical actuality. The only question could be; "How does this mathematical process explain observed phenomena?" In conjunction with all other evidence the question is certainly a candidate justifying further evaluation.

20) DISCUSSING TRANSITIONS BETWEEN NATURAL NUMBERS

Symbolisms mathematicians use to represent factual things sometimes create non-obvious problems in arithmetic and math. While symbols greatly simplify the presentation and handling of information, they increase abstraction and possibility for error. To an adult the distinction between **numerals** like 3 and the **factual thing** like [* * *] are ostensibly equivalent. They may not be equivalent to an infant or child who has not yet learned to convert factual things into symbolized quantities. If for whatever reason, what the symbol connotes is not totally explicit, even adults can readily make inadvertent mistakes. Concerning the natural numbers, we have historically learned a set of rules for manipulating **the numerals** with seemingly logical consistency. Conception of the processes those rules implement does not necessarily provide visualization for exactly what would happen to the represented number of **factual things**. After continual usage, rules for handling numerals often get applied (and extrapolated) by rote, rather than by detailed logical analysis of the factual things the numerals represent. Fundamental oversights can then arise. Commonly representing divisible objects by natural numbers when they should more generally also encompass monads cites an example. It may be difficult to accept that natural numbers describe monads because they so often portray partitioned things. The practice of putting large numbers into scientific notation tends to obliterate that the foundation natural numbers address monadic integers. The Euclidean axis with infinite points designated between integers has become the customary model for the number

245

scale. Natural numbers can also portray divisible things with the proviso that to be comprehensive, **such portrayal must also encompass monads.** The divisible case has traditionally allowed fractional values between integers without including a transverse component to account for any monadic trait in what is being enumerated. As such, that treatment only covers the tallying of infinitely divisible objects. **Allowing an orthogonal transition <u>component</u> between integers is simply <u>more general</u> than when advance occurs either along one radial Euclidean axis line or is totally orthogonal to that line**. <u>**The circumstance where an orthogonal component is allowed is all embracing and does not violate Peano's Postulates.**</u>

The stipulation that for monads, natural numbers constitute exclusively integer-values is as indispensable as allowing fractional values when counting divisible entities. Including the gamut of values $0 \le \xi \le 1$ allows encompassing monads plus divisible entities, and the range between. The orthogonality axiom, applied to a proportional degree through parameter ξ becomes a **mathematical method of covering the span from monads to macroscopic objects.** Marker ξ in the interval between integers is necessary to accommodate the entire domain of enumeration possibilities. The inter-integer comparative parameter portrayed through ξ **<u>constitutes an unrecognized fundamental variable</u>**. As consequence of the orthogonality axiom and ξ, natural numbers delineate nested-spherical-surfaces having consecutive integer radii. **Numbers express vectors not scalars**

with ξ elaborating <u>how much of the axial direction gets "bypassed" by</u> <u>orthogonal traverse</u>.

For the ξ = 1 spiral depicted in three dimensions, vector s_n always emanates normal to the origin-to-n direction and invariably terminates somewhere on a circular locus generated on the (n+1) sphere. For ξ = 1 the plane in which all possible vectors s_n must lie inevitably forms a circular-locus-intersection having radius $s_n = \sqrt{(2n+1)}$ on the sphere-of-radius-(n+1). Dependent on the value of $0 \leq$ ξ ≤ 1, precisely where the s_n vector-tip might lie on the side of the plane distal from the origin constitutes an ambiguous portion of the (n+1) spherical-surface. Cumulative ambiguity of absolute location increases with n even as the indistinct contribution from each consecutive transition gets smaller. Thus by the time the spiral has made one complete revolution (where P $\geq$ 1), vector-tip-location on any integer-radius spherical-surface will have become totally indeterminate. Yet, those concentric integer-radius-spherical surfaces specify the only allowable states for monads, with radial-direction-gaps between spherical-surfaces prohibited. The functional space between integer-radius spherical-surfaces can be dynamically traversed, but not statically engaged. Concentric-nested-spherical surfaces therefore provide a more elementary depiction of the natural numbers than the intersection of those surfaces with one radial line, (than with integers marked along a Euclidean axis). Those spherical surfaces encompass monadic conditions. The concentric-spheres can degenerate into being points along a straight Euclidean line only when $\xi \to 0$ wherein angular progression of the planar spiral approaches zero rotation. Then natural numbers would then not be describing monads but very divisible macroscopic objects. For monads, the location of a radial vector-tip for n will be indeterminate on any sphere of radius n, except to the degree the location of a prior vector-tip might be known. For monads the angular difference between consecutive vectors always varies as $\theta_n = 2[\sqrt{(2n+3)}-\sqrt{(2n+1)}]$ radians. If a vector direction were conceivably known at n, than the direction at (n+1) will be somewhere at any origin-angle θ_n removed from the direction for n. For monads of zero divisibility, that represents the maximum angle between consecutive integers. It is feasible that transitions may be possible with $\xi > 1$, but no interpretation is attempted for that condition.

The essence of this theory rests on an exposition that natural numbers decipher differently than conventionally construed. Integers themselves only tell how many, not how to get from one to the next. Relative to zero things, there can be 1 thing, 2 things, 3, things, 4 things, etc. Those numbers mandate nothing about the transition process between one integer and its successor. **That transition process is totally free to occur all possible ways not regulated by some other constraint**. The number 1 specifies unity away from zero for example, and is characterized by a spherical surface of unit radius surrounding the origin. Anything described by the integer 1 can be somewhere on that spherical surface. That provides a general interpretation of what the natural number 1 means, one removed from zero. The number 2 relative-to-zero designates that a population of 2 could be anywhere on the spherical surface of radius 2 from the origin, two away from zero. The number 2 does not necessarily relate to the number 1, it relates to the **origin** as 2 away from zero. Analogously, each number n depicts a possible concentric spherical surface about the origin having integer radius n. That presents a generic description of what natural numbers state in terms of possible locations relative to an origin starting value. Integers themselves furnish no restraint concerning how advance from say 3-from-the-origin, gets to the value 4-from-the-origin. Consequence of that transition must end up on the sphere of radius 4 from the origin, but how such progression reaches there is undefined. Spacetime variables (energy-time) and (momentum-position) monotonically advance through the discrete natural numbers. Following an event, time or position increase positively from the origin and in a granular manner. Those variables thus accrue monotonically; in simplest phenomena no negative values of relevance follow the precipitating event at the origin.

Various ways exist to advance from a sphere of radius n to a location on an outer concentric sphere of radius (n+1). A generalized way for advance from n to (n+1) to occur allows the transition be totally (or at least partially) independent of n. That most inclusive case happens when the value of n or location on sphere n, has minimal bearing on the process of getting to (n+1). Any stipulation based on a governing relationship or correlation mechanism involving location at n to establish how advance should transpire will always be less general than when transitions take place independent of such relationship. The most encompassing situation prevails when advance between integers has least kinship to n. Then, various constraining relationships might be added, each such case being more specific and less universal than when no relationship exists. For that most general condition, the advance will require maximum directional independence of the origin-to-n direction. The criterion stipulating how such advance will occur should be identical for all values of n. Conditions of greatest generality are satisfied by ξ being anywhere between zero and unity. For the ξ = unity case progression is totally orthogonal to the prevailing direction from origin-to-n. **Then components of the transition minimally correlate with, (add to, or subtract from, or have dependence upon) the location of vector n prior to the advance**. Such transitions could then emanate anywhere outward from the plane tangent at vector n to the sphere-of-radius-n. That plane would be normal to the direction from origin-to-n and have zero directional-component along the path of origin-to-n. For ξ = 1, transition vector s_n within that plane would reach spherical surface (n+1) anywhere at the intersecting circular locus between the tangent plane and the spherical surface at radius (n+1). With the entire range $0 \leq \xi \leq 1$ allowed, all outward directions into a hemisphere from vector-tip n become possible. That logic provides the most general description of how each transition between natural number n and the next higher integer (n+1) can take place. The other extreme for ξ =0 grants long-established advance between all integers along the same co-axial line, namely along the unidirectional Euclidean axis.

If for example ξ were unspecified allowing the full range of all possible transition routes, than it will be shown a synchronous-solution-mode manifesting at a fixed ξ_1 implies a harmonic solution is feasible on a conic surface of apex angle $\beta_s = \sin^{-1}[(2\sqrt{\xi_1})/\pi]$. The existence of such a synchronous mode would have involved integer-transitions for all n values to have occurred with axial component $s_a = \xi_1 - 1$. In that manner ξ would constitute an open-degree-of-freedom that specified prevailing transition location ξ_1, apex angle $\beta_s = \sin^{-1}[(2\sqrt{\xi_1})/\pi]$, and cumulative quadrant extent $\sqrt{\xi_1}$ as later depicted in figures 37 and 49.

Symmetry, simplicity, and invariance are important attributes in physics and science. The symmetry of basic numbers generating cyclic repetition with four equal-radian quadrants is apparent under the orthogonality axiom. Simplicity of perfect squares of the Pythagorean Theorem influencing harmonic solutions is similarly evident. Eliminating six unnecessary arbitrary constants, e, π, i, ∞, (a Planck length), (a Planck time), provides a further example of austerity. The greater the number of independently describable invariant-criteria, the greater typically will be generality. Everyone is familiar with the fixed angular-direction-requirement associated with conventional unit-change between consecutive integers. That is a known universal invariant criterion for the description of sequential numbers. It is unrecognized that under Euclidean interpretation such unit-change is being repeatedly compelled to occur within the same direction (axially) as applied for the smaller number, **a restriction invoking a very concise property of the prior number**. Such restriction limits invariance and universality.

Under the orthogonality axiom, unit-change between consecutive integers also occurs, so that <u>any invariance due to unity algebraic separation would apply to both cases</u>. Such "unit-algebraic-spacing" invariance criteria would be consistent under both interpretations. However, for monadic transitions, less dependence exists upon the origin-to-smaller-n-direction. The transition can occur in a range of possible directions, or within a range of different angles, rather than being co-linear with all prior directions (or all prior transitions), as occurs under traditional understanding. Several additional invariant criteria thus associate with monadic (or partially divisible) transitions that do not apply under conventional forward movement along the Euclidean axis. Each criterion would be independently sufficient to characterize progression properties. For $\xi = 1$ they are listed below. <u>[Parentheses describe the case for the range $0 \le \xi \le 1$].</u>

1.   **Each transition of magnitude $\sqrt{(2n+1)}$ can always occur in a multiplicity of possible orthogonal directions for each n and such stipulation is universal to all n**. [Transition magnitudes become $\sqrt{(2n\xi+1)}$ and a multiplicity of possible axial components would be allowed for $0 \le \xi \le 1$].

2.   Each transition always maintains a resultant magnitude that equals the "geometric mean" of the summed two integers between which the transition occurs. [For $0 \le \xi \le 1$, the transition is the mean of two summed numbers, wherein n was being multiplied by ξ, as $\sqrt{\{n\xi+(n\xi+1)\}} = s_n$.] **Both yardsticks include criterion universal to all n as an invariant of the transition.**

3.   **The root mean squared sum of all prior transition-magnitudes from n = 0 upwards will always equal the new population number. <u>Population n_1</u>**

plus the RMS sum of however many transition magnitudes from n_1 to n_2 will equal n_2. This is a universal criterion for reaching any number from any smaller number independent of the final or starting number. It is a further invariant that characterizes natural numbers under monadic accrual.

4.   **The difference between adjacent transition magnitudes squared separating any two consecutive integers will always be 2, irrespective of the integers**. That is an invariant criterion for all transition-magnitudes squared. [The difference in s_n^2 for $0 \leq \xi \leq 1$ will be 2ξ.]

5.   The nature of stacked concentric spherical surfaces that each integer value represents with respect to the origin is another invariant criterion under orthogonality axiom interpretation. Natural numbers themselves specify no preferred direction for transitions between numbers. They articulate only the magnitude of each possible population value relative to the origin. **Elucidating integer values as spherical surfaces about the origin demonstrate an invariance criterion for direction, absent in conventional rendering**.

6.   **Consecutive numbers reference zero and are unity apart, which condition might be elaborated in terms of Peano's Postulates for $\xi = 1$ and $0 \leq \xi < 1$**.

7.   **Allowing transitions to be vectors with a range of possible angles is more general than demanding than that they be specific vectors co-linear with that of the prior integer, (i.e., stipulating that they be scalars).** Accordingly, six invariant criteria characterize orthogonality-axiom transitions, only one of those yardsticks being that consecutive integers be unity apart. Those additional invariant

criteria for the intractable way in which change can occur demonstrate the increased degree of generality manifested in the new exposition. To the extent these criteria are universal to all n, they remain invariants of the counting process. The more such criteria there are, the more general that process can be.

An excess spacetime angle $\Delta\theta$ emerges mathematically as a compounded monadic entity whose enumeration will typically encounter a "maximum number" less than infinity. That number n_{max} relates to the curvature of space and delineates as $n_{max} = 2/\Delta\theta^2$. It also represents the maximum possible assemblage of smallest indivisible things indicating the quantitative range-limit of a physical process. It infers a limit of $1/\Delta\theta$ discernable cycles in a world where each cycle can only be resolved into $2/\Delta\theta$ parts. The maximum total number of partitions of spacetime becomes $2/\Delta\theta^2 = n_{max}$. Undeterminable higher numbers become impossible as **causal consequence** of the "missing information shroud" apropos the monad's indiscernible innards, in this case $\Delta\theta$. Prevailing indeterminism of the system will equal the inaccessible gap between integers, which also equals monad "size". Fractional inter-integer intervals of the compounded excess angle will be unattainable. The point being made is that system **uncertainty does not extraneously introduce, but causally stems from and everywhere equals inability to further subdivide a monad** into sub-constituent attributes. The missing information about that smallest element permeates the system in various forms, never being recoverable or its effects avoidable. That is why an excess angle $\Delta\theta$ can associate with natural numbers and it also articulates intrinsic indeterminism within that system. It is alleged that smallest indeterminate angle depicting space curvature is Planck's constant in physical reality.

Irwin Wunderman

21) A GENERAL DESCRIPTION FOR INTER-INTEGER TRANSITIONS.

Parameter s_n analytically portrays "**transition extent**" in progression from one spherical surface of integer value n removed from the origin, to the outwardly concentric spherical surface (n+1) removed. For ξ almost zero and $s_n \approx 1$, repeated transitions maintain in almost a straight co-axial line from the origin close to a Euclidean type axis. Each origin-angle θ_n would then change very slightly, accruing with each advance to the next higher spherical surface. For $\xi \neq 0$ though however close, some off-axis cyclic advance would still occur. As ξ gets larger, s_n increases from unity and each origin-angle $\theta_n = \arccos\{(n+1-\xi)/(n+1)\}$ between adjacent-concentric-sphere-locations increases also. Those increases reveal how transitions at any ξ alter relative directional-locations on each next higher spherical-surface. Each subsequent (n+1) location becomes **<u>potentially further and further away in direction</u>** from an earlier direction relative to the origin. Parameter ξ thus specifies "**<u>monadicity</u>**", or "**<u>angular scatter</u>**", or **<u>lack of directional correlation</u>** with population advance.

Mathematically, ξ can increase beyond unity without altering the functional expression for s_n. Though transition relationship $s_n = \sqrt{[2n\xi+1]}$ can apply on either side of the orthogonality condition where $\xi = 1$, the region $\xi < 1$ conveys **<u>reduced correlation</u>** amongst assimilated population members as their numbers accrue.

Non-correlation minimizes at $\xi = 0$ and maximizes at $\xi = 1$. Were ξ to exceed unity, an altered form of correlation (or "over-correlation") would seemingly occur, but with vague meaning. Scatter parameter ξ is thus seen to cite **comparative-correlation in the direction-of-consecutive-transitions** between population integers. It is unfamiliar, as are the concentric spherical surfaces about the origin that natural-number-integers represent. Applicable for whatever ξ, $s_n = \sqrt{(2n\xi+1)}$ indicates **transition magnitude** during monotonic growth from each integer n to the next higher integer (n+1). For $\xi = 1$, population n equals the RMS sum of all prior transition magnitudes s_n. It is also the algebraic sum of the number of transitions. For ξ however smaller than unity, s_n (and the RMS sum of prior s_n magnitudes) diminishes, and it no longer equals to n. At the limit $\xi \to 0$, the RMS criterion summing transitions s_n would reduce to $\sqrt{(1^2+1^2+1^2+1^2+- - -)} = \sqrt{n}$, instead of n. Parameter ξ therefore characterizes the scope between $\xi = 0$ where population n produces unit-transition magnitudes along the Euclidean axis ranging to where it is both the RMS and the algebraic sum of those magnitudes at $\xi = 1$. Over the range $0 \geq \xi \geq 1$, the RMS sum varies from $\sqrt{n}$ to n, encompassing a domain between directional-correlation to maximum-independence. Gradations in the universal mechanisms by which enumeration occurs thus vests in the value of ξ, concomitant with all its associated interpretative meanings as outlined herein.

Parameter ξ has historically been a priori presumed equal to zero forcing each shift between integers to have the same direction and magnitude

defined by $s_n = \sqrt{(0+1)} = 1$. That enhances the viewpoint that the integer sequence possesses continuous intervening fractional values with Newtonian continuity along a Euclidean axis, rather than contain **disallowed inter-integer** positions for monadic accrual conditions. Mathematical inclusion of non-zero (as well as zero) ξ in the passage to each higher integer, as transition magnitude

$$s_n = \sqrt{(2n\xi+1)} \text{ [instead of } s_n = 1], \qquad\qquad (21^*1)$$

will always be more general than setting parameter ξ to zero at the onset. The gamut of trajectories for accrued populations can then be examined, permitting associated harmonic and indeterminacy processes to manifest in analysis. Mechanisms for de Broglie waves and stable state periodicity can correspondingly be explored and understood. For population growth, **parameter ξ represents the fractional amount of Euclidean axis bypassed by orthogonal transition, another qualification suggesting its rationality.** Fixed values of ξ parametrically depict spirals from the generic family of population growth with an orthogonal component. Besides $\xi \leq 1$, nothing mathematically precludes values of ξ as high as 2, 3, or 4 for example; though the physical meaning of that region is unknown.

It will be shown that **the symmetrical ABCD overlap conditions exhibited by the unified wave for $\xi = 1$ is virtually identical to synchronization that occurs for all ξ values**. Explaining why synchronization tracks for all ξ can get more involved graphically than by algebraically noting $\theta_n = \cos^{-1}[(n+1-\xi)/(n+1)]$ is everywhere only slightly less than $(\sqrt{\xi})\cos^{-1}[n/(n+1)]$. **At large n** they converge to equate. For $\xi = 1$ those angles sum to synchronize every four radians on a cone of apex angle $\beta = \sin^{-1} 2/\pi$. For $\xi < 1$, the origin angle of each triangle forming a synchronous spiral would diminish by amount $\sqrt{\xi}$. Therefore spiral modes having chords along the surface of a narrower cone of apex angle reduced by $\sqrt{\xi}$ to $\beta' = \arcsin (2\sqrt{\xi}/\pi)$ as in figure 37 can remain synchronous for all ξ. Due to second order effects, even more precise synchronous tracking can result than given by the expression

$$\{\cos^{-1}[(n+1-\xi)/(n+1)]\}/\{\cos^{-1}[n/(n+1)]\} = \sqrt{\xi} \qquad (21*2)$$

Whatever deviations from perfect synchronization at very small n might exist on a perfect idealized cone is also made non-critical by synchronization being possible on a slightly modified conic surface like bullet shaped. This permits $\check{R}_s$, the radians-at-synchronization to ubiquitously equal $4\sqrt{\xi}$ radians/cycle. It permits ξ to emerge as a fundamental parameter in enumerating spacetime variables following an origin event. Besides describing angular scatter, it relates to apex-angle "narrowness" of the cone on which harmonic modes explicate in counting procedures. It also has other relevancy. Figure 32 shows the analytically derived and measured radians-at-synchronization ($\equiv \check{R}_s$) as a function of ξ.

Figure 32 plots the radians at which synchronization occurs for the unified wave spiral as a function of ξ, when graphed on a cone. Radians are measured as the sum of perimeter chords around the conic cross-section per repeat cycle, divided by the value of n at that cycle. Radians-at-synchronization diminish from 4√ξ as √ξ diminishes from unity because chord-lengths reduce as $s_n = \sqrt{(2n\xi+1)}$ filling the circular base of a smaller cone. The four-quadrant rectilinear symmetry as established at the cone's cross sectional circle for ξ = 1 remains ostensibly unchanged for all ξ. However, the "effective perimeter" around the cone's circular cross-section decreases for the same extent n from the apex. "Radians-at-Synchronization" ≡ R_s as calculated from equation 21*6 would be 4√ξ. Measurements taken from the graphics are seen here to match that angle within experimental error. The parabolic relationship $R_s = 4\sqrt{\xi}$ makes each harmonic quadrant have the cumulative extent of its chords equal to √ξ, which extent appears graphically as a straight-line in figure 12 commensurate with the leg of each inscribed triangle.

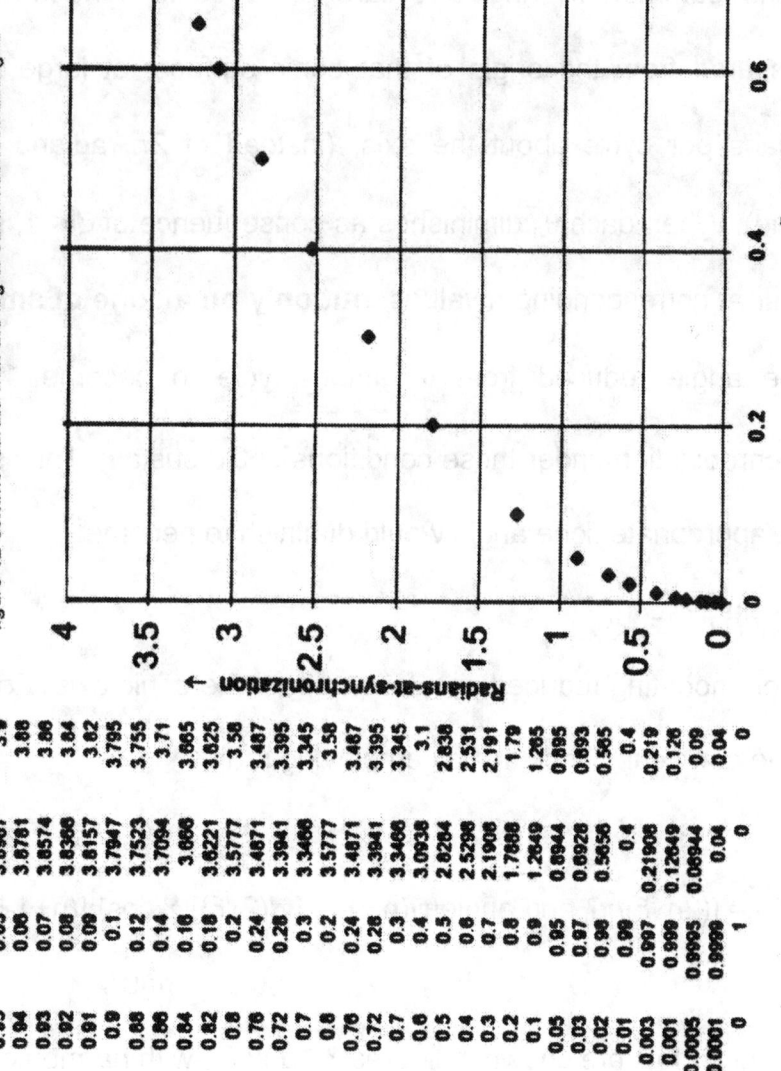

COMPARISON OF MEASURED RADIANS AT SYNCHRONIZATION AND 4√ξ

ξ	s_n	4√ξ	Measured Radians at Sync.
1	0	4	4
0.995	0.005	3.989	3.99
0.99	0.01	3.9799	3.98
0.98	0.02	3.9597	3.96
0.97	0.03	3.9395	3.94
0.96	0.04	3.9191	3.92
0.95	0.05	3.8987	3.9
0.94	0.06	3.8781	3.88
0.93	0.07	3.8574	3.86
0.92	0.08	3.8366	3.84
0.91	0.09	3.8157	3.82
0.9	0.1	3.7947	3.795
0.88	0.12	3.7523	3.755
0.86	0.14	3.7094	3.71
0.84	0.16	3.666	3.665
0.82	0.18	3.6221	3.625
0.8	0.2	3.5777	3.58
0.76	0.24	3.4871	3.487
0.72	0.28	3.3941	3.395
0.7	0.3	3.3466	3.345
0.8	0.2	3.5777	3.58
0.76	0.24	3.4871	3.487
0.72	0.28	3.3941	3.395
0.7	0.3	3.3466	3.345
0.6	0.4	3.0938	3.1
0.5	0.5	2.8284	2.838
0.4	0.6	2.5298	2.531
0.3	0.7	2.1908	2.191
0.2	0.8	1.7888	1.79
0.1	0.9	1.2649	1.265
0.05	0.95	0.8944	0.895
0.03	0.97	0.6928	0.693
0.02	0.98	0.5656	0.565
0.01	0.99	0.4	0.4
0.003	0.997	0.21906	0.219
0.001	0.999	0.12649	0.126
0.0005	0.9995	0.08944	0.09
0.0001	0.9999	0.04	0.04
0	1	0	0

To characterize the mechanism by which synchronization maintains for $0 \leq \xi \leq 1$, note that as consequence of ξ, the origin angle of each spiral triangle diminishes as $\theta_n = \cos^{-1}[(n+1-\xi)/(n+1)]$, {compared with $\theta_n = \cos^{-1}[n/(n+1)]$ for $\xi = 1$}. In the latter case where the numerator contains only n, synchronization with respect to the summed θ_n angles occurs on a conic surface of apex angle $\sin^{-1}2/\pi$. Measured from the origin of that conic surface, at large n, there would be four radians per cycle about the axis, (instead of 2π radians per cycle plotted on a plane). When each θ_n diminishes as consequence of $\xi < 1$, synchronization can still occur at corresponding n values, **but only on a cone of smaller apex angle**. If the cone angle reduced from 4 radians/cycle to become $4\sqrt{\xi}$ radians/cycle $\equiv \check{R}_s$, synchronization under those conditions could sustain. This is displayed in figure 32. The appropriate cone angle would diminish to become

$$\beta_s = \sin^{-1}[2\sqrt{\xi}/\pi], \tag{21*3}$$

accommodating reduced circulation about the conic axis. Polar plotting an end view of the cone can check this by employing radius

$$r(n,\xi) = [(2n\sqrt{\xi})/\pi], \text{ and angle } \psi(n,\xi) = [\pi/(2\sqrt{\xi})] \sum_{n=0}^{n} \cos^{-1}[(n+1-\xi)/(n+1)]. \tag{21*4}$$

Such plots are shown in figures 33 and 34 with harmonic repetition apparent.

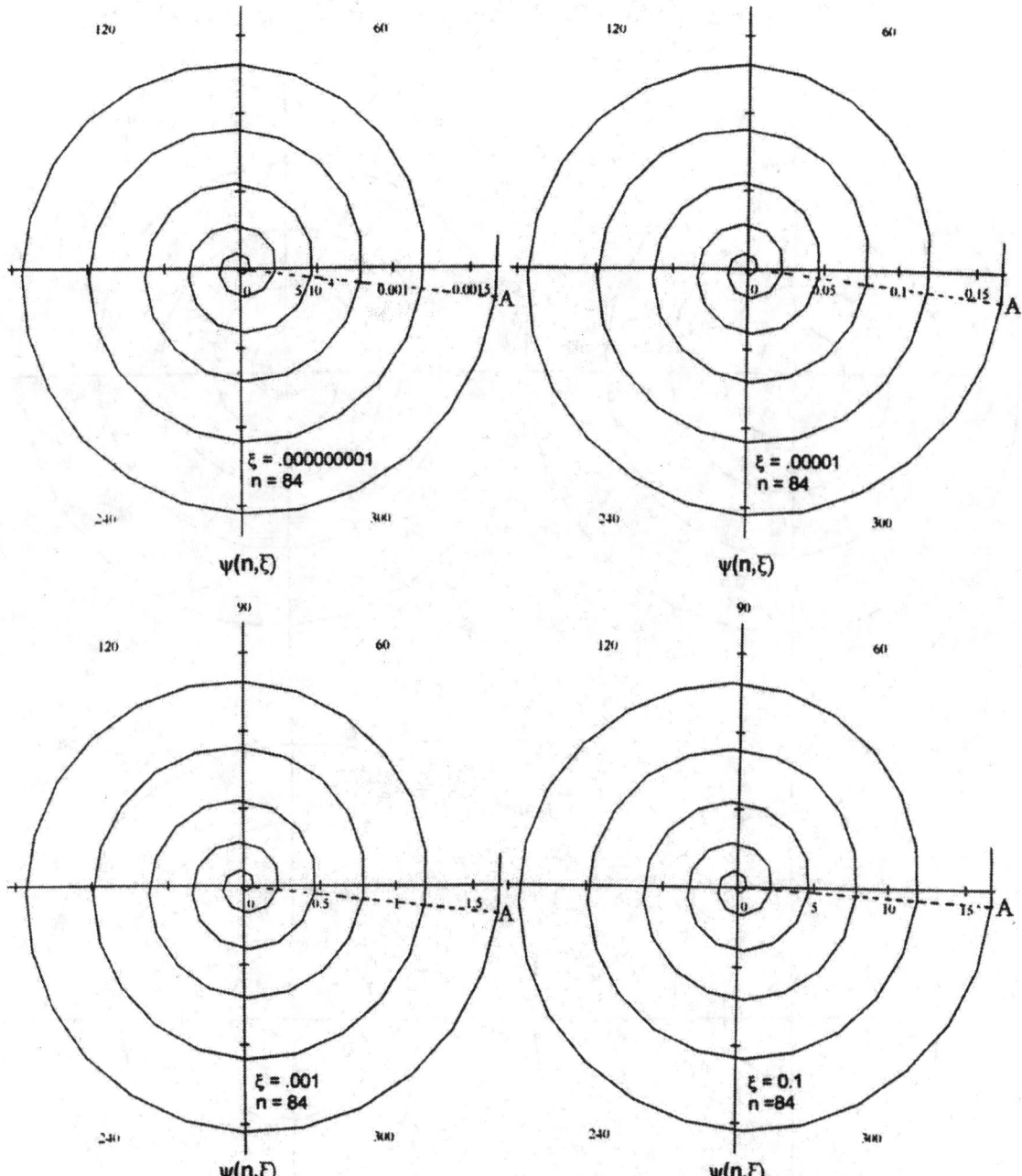

$\psi(n,\xi)$

$\psi(n,\xi)$

$\psi(n,\xi)$

$\psi(n,\xi)$

Figures 33 and 34 show synchronous spiral modes with different ξ values depicted as an axial view looking into the apex of a conic surface characterizing those modes. Tangential sectors are chords within the conic surface for n up to 84. For P = 6 cycles shown, synchronization registers along the dotted lines whenever n = 4, 12, 24, 40, 60, and 84. In all these graphs radius r(n,ξ) = (cone-perimeter-per-unit-n)/2π = $\check{R}_n/2\pi$ = $4\sqrt{\xi}/2\pi$ = $(2\sqrt{\xi})/\pi$, is plotted at every cumulative angle given

by $\quad \psi(n,\xi) = \dfrac{\pi}{2\sqrt{\xi}} \displaystyle\sum_{n=0}^{n} a\cos\left(\dfrac{n+1-\xi}{n+1}\right)$

Wave "polarization" is seen to vary from about -6° at ξ = 0 to +24° at ξ = 1. It increases roughly another +80° as ξ extends from 1 to 2.

263

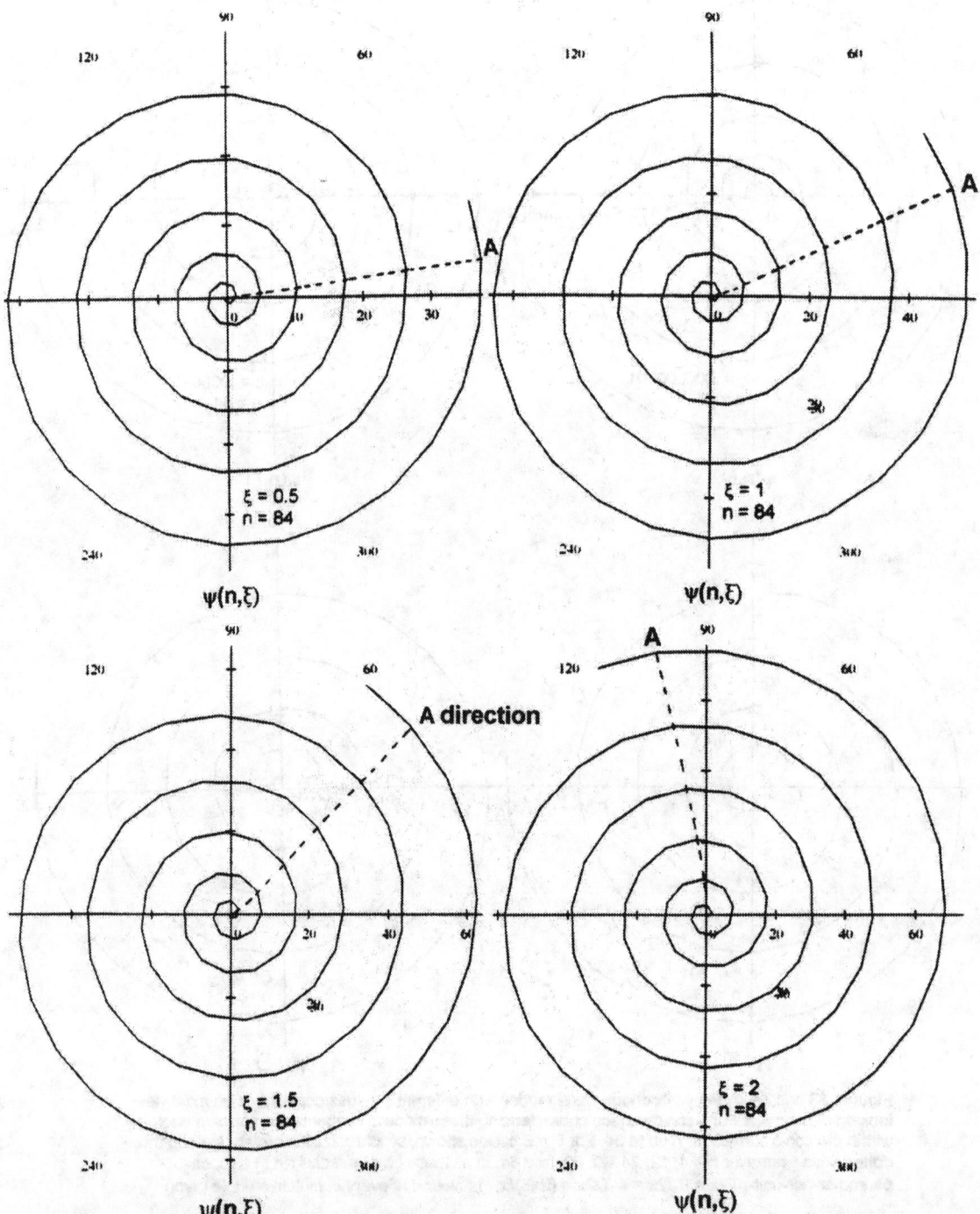

Figure 34 continues the characterization of figure 33. The prevailing apex angle for synchronization at each ξ is given by $\beta_s = \arcsin [(2\sqrt{\xi})/\pi]$ so the appropriate cone becomes very narrow and pointed as ξ → 0: The radius scale noted along each abscissa signifies how pointed the cone is since extent from the origin is the same in each graph and equals 84 at the outermost node.

264

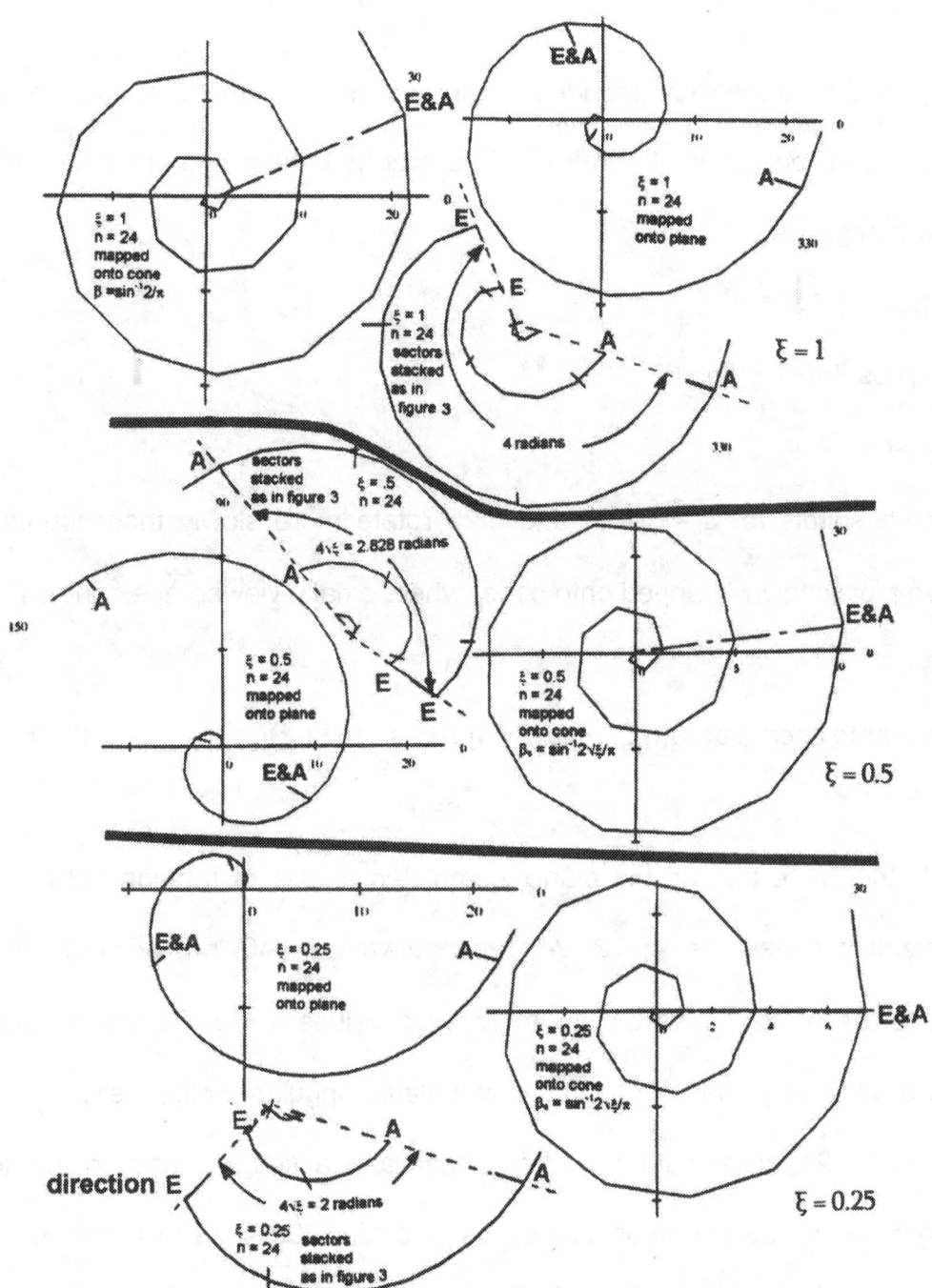

Figure 35 helps illustrate a construction process to delineate synchronous modes of the unified wave. The diagrams display how the "fan" of figure 3 would be formed for $\xi = 1$ in the top three graphs, for $\xi = 0.5$ in the middle graphs and $\xi = 0.25$ in the bottom three. The spiral mapped within each appropriate cone constitutes an axial view of that cone's apex showing synchronicity and polarization of the wave. Here radius is $2n\sqrt{\xi}/\pi$ with respective angle at each n given by $\Psi(n,\xi)$ in figures 33 and 34. The spiral-mapped-onto-a-plane plots show radius n vs. the angle $\cos^{-1}[(n+1-\xi)/(n+1)]$. A radial-inward mark is included for reference at nodes n = 4, 12, 24 where synchronicity occurred on the cone mapping. The third three plots of stacked sectors were literally cut out of the prior mapping-onto-planes at those radial inward marks. Each such fan-shaped section of sectors was then concentrically stacked about the origin as was done in figure 3 leaving axis-crossing fragment-lines for visual allusion. Opposing dotted lines of each fan can then be joined to form each cone within which that spiral mode can be synchronous portrayed as in figure 36.

Figure 35 alternatively provides a narration that reverts to the transparency construction method used with figure 3. The graphs entitled "mapped onto plane" accumulate angle as

$$\psi(n,\xi) = \sum_{n=0}^{n} \cos^{-1}[(n+1-\xi)/(n+1)]. \tag{21*5}$$

Those three spirals for ξ =1, 0.5, and 0.25 rotate more slowly than counterpart respective plots shown "mapped onto cone" where axially-viewed apex angles

would appear to accrue as $\psi(n,\xi) = [\pi/(2\sqrt{\xi})] \sum_{n=0}^{n} \cos^{-1}[(n+1-\xi)/(n+1)]. \tag{21*6}$

We then note that on the plane, cumulated angles of rotation between the synchronization nodes n = 4, 12, 24, (and outward as 40, 60, 84, etc.) remain constant for each fixed ξ. In the figure those n values 4, 12, 24 are marked by radial-inward line segments. If each such cumulated angular-section between 0 & 4, 4 &12, 12 & 24, were "cutout" and re-plotted as stacked sections, (or rotated a fixed angle making sections concentric), the resulting "fans" as in figure 3, would comprise $4\sqrt{\xi}$ radians/cycle $\equiv \check{R}_s$. (21*7)

Connecting the dotted sides of those fans together forms a conic surface within which direction A chords of the spiral could be synchronous. Of course that statement illustrating this construction process entails the slight approximation that

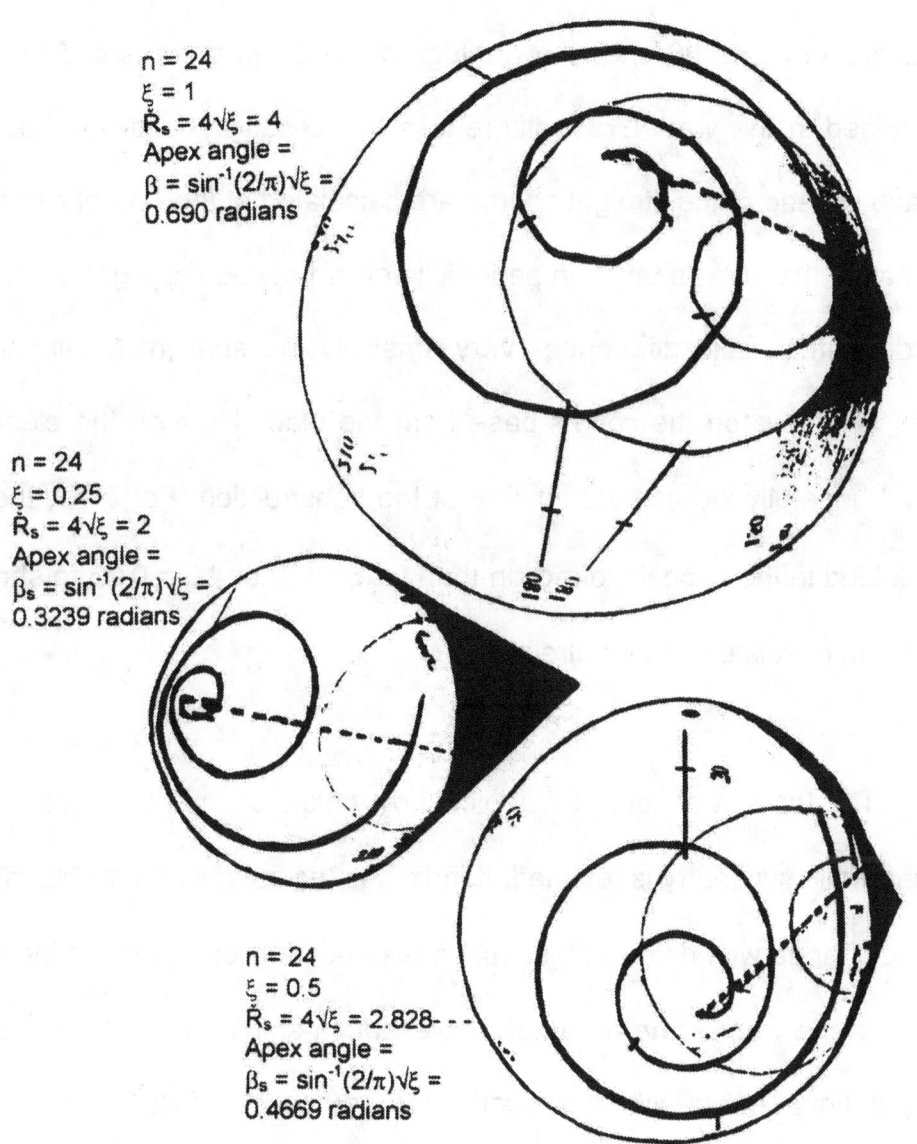

n = 24
ξ = 1
$\check{R}_s = 4\sqrt{\xi} = 4$
Apex angle =
$\beta = \sin^{-1}(2/\pi)\sqrt{\xi} =$
0.690 radians

n = 24
ξ = 0.25
$\check{R}_s = 4\sqrt{\xi} = 2$
Apex angle =
$\beta_s = \sin^{-1}(2/\pi)\sqrt{\xi} =$
0.3239 radians

n = 24
ξ = 0.5
$\check{R}_s = 4\sqrt{\xi} = 2.828$- - -
Apex angle =
$\beta_s = \sin^{-1}(2/\pi)\sqrt{\xi} =$
0.4669 radians

Figure 36 is a photograph of transparencies of the three stacked-sector "fans" in figure 35 after they had been folded into cones by connecting their dotted lines. The photo was taken with sunlight from the left leaving shadows of the black spiral lines on clear acetate falling onto a white sheet at the base of each cone. It was hoped, (with only slight success), that would increase the illusion of three dimensionality while depicting almost axial views. Each dotted line indicates where both surfaces of the fan joined. For ξ = 1 it was folded in the opposite direction than for the other two cones so that the spiral rotates the other way. This provides a physical description for how the synchronous condition illustrates on cones of apex angle $\beta_s = \sin^{-1}[(2/\pi)\sqrt{\xi}]$.

the curved chord-lengths around the conic surface fill the same angular extent as straight chords. With θ_n being small the approximation is good, and to aid visualization figure 36 shows a photograph of 3 spiral transparencies literally cut and folded in this way. Exact differences that result by following the curved conic surface instead of the straight chords are calculated at the end of Appendix B. Even at small n, the $1/P^2$ variation in periodicity error between straight chords and curved chord-lengths make differences very small. Using sunlight for illumination in the photo, shadows on the cone's base from the black lines on the clear acetate film should hopefully aid comprehension of the construction. For ξ =1, the acetate film was folded in the opposite direction than for ξ = 0.5 and ξ = 0.25 to show the unified wave can circulate in either direction.

The top plot of figure 37 depicts how polarization angle shifts between two ξ values while symmetry is retained. It indicates the values of n that identify quadrants do not change with different ξ so all peaks and zero-crossings of the **waves occur at the same non-prime n-values.** Polarization shifts associate with straight-chord computational errors. When the cords are permitted to follow the conic surface (or unit–sphere's surface), rather than traverse within the surface, no such errors result. Discussions of Equation 21*8 and Table 4 treat the perfect periodicity that results without any shift in polarization. Polarizations affiliated with different ξ's constitute only different "offset angles" for the entire wave. Non-prime values characterize divisibility. While parameter $s_a \equiv 1-\xi$ can signify divisibility of the wave's enumerated

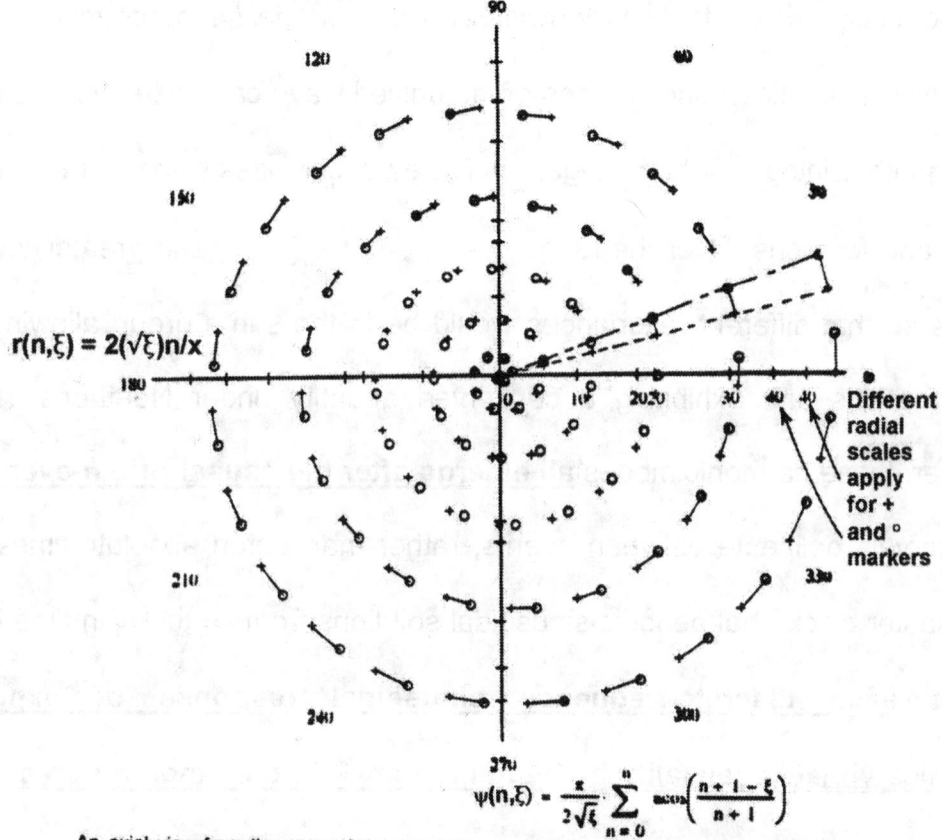

$$r(n,\xi) = 2(\sqrt{\xi})n/x$$

Different radial scales apply for + and ° markers

$$\psi(n,\xi) = \frac{x}{2\sqrt{\xi}} \sum_{n=0}^{n} \cos\left(\frac{n+1-\xi}{n+1}\right)$$

An axial view from the apex of the spiral shown coalescing two radial scales for comparing radius-normalized node positions for ξ = 0.7 and ξ= 0.9.

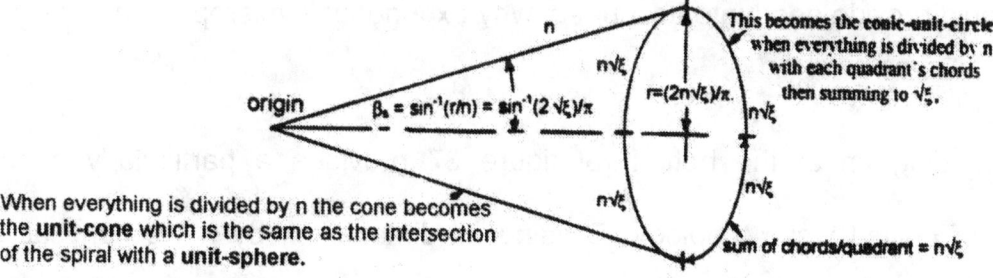

Figure 37. The upper plot superimposes the angles of two spirals having n up to 84. The + symbols delineate ξ = 0.7and the σ symbols depict ξ = 0.9. A straight line connects respective data points for values of n above 40, illustrating accumulation of angular shift with ξ. Two different radial scales were respectively employed allowing exclusive contrast of the angles. The apex-angle and conic surfaces would differ between the two ξ cases so this comparison only exemplifies how the polarization angle advances with ξ. The lower diagram shows the mathematical and graphical simplicity of all harmonics that follow an origin event. Each quadrant of every cycle comprises from summed chords s_n whose cumulative transverse component in the non-axial direction amounts to n√ξ. For s_a close to unity the chords are close to unit-long, primarily in the axial direction. Although they then affect the RMS sum, that component direction has almost no effect on the radians-at sync $Ř_s$, which is measured from the origin and remains 4√ξ. Radians/quadrant = $Ř_s$/4 = (n√ξ)/n = √ξ for all ξ. At large-n and ξ >>0 the unit-length axial-component of the transition hardly effects the RMS sum of transitions s_n, which sum then also approximately equals (n√ξ)/n = √ξ and the radians/quadrant. Origin apex angle of the cone is $β_s$ = $\sin^{-1}$(r/n) = $\sin^{-1}$(2√ξ)/π. That picture characterizes how sinusoidal harmonics are generated in contrast to the exponential function.

269

constituents, ξ relates to lack-of-divisibility or *monadality*, approaching zero for totally divisible entities. Since modes of the unified wave can arise over the range of ξ having either integer or half integer spin, they encompass **supersymmetry** for all bosons and fermions. Over the range $0 \leq \xi \leq 1$ (and greater) symmetry sustains so that different resonances would be in the same group allowing gauge transformations and exhibiting a conserved quantity under Noether's Theorem. Moreover, these harmonic modes **all emerge after the causal-origin-event** so that time intervals delineate between events, rather than unfurl absolute-time-derived-by-a-"master-clock" that periodic sinusoidal solutions from -∞ to +∞ in time infer. By contrast **each discrete-frequency sinusoidal response of exponential harmonics violates causality by beginning at -∞ in time thereby preceding its causal event and portraying a master clock extending to +∞.** That characteristic alone should provide adequate account of why exponentials misrepresent reality.

The diagram at the bottom of figure 37 provides a particularly concise synopsis of unified wave topology. Dividing all dimensions by n we can call the result the "**unit-cone**" having a "**unit-conic-circle**" cross-section of radius 2/π at its truncated base. The unit-conic-circle would be **unity-extent from the apex origin along the cone's side-length** (not from the center of the circular base) and comprise four quadrants each of accrued extent √ξ. That same circular locus would exist at the intersection of the cone with a unit-sphere, so the analyzed circle could equally well be on the unit-sphere. Equation 14*17 showed that each broadened-sector extent in figure 14, (having magnitude $[\sqrt{(2n+3)} - \sqrt{2n+1}] = \Delta s_n \approx 1/\sqrt{(2n+2)}$) was equivalent at large n to half the length of its respective unit-conic-circle chord between **J** and **K** in figure 15. With each cycle those cords can respectively progress around the conic-unit-circle, or a similar circle on the unit-sphere. Thus, those broadened-line accrued-chord-lengths in figure 14, summing in extent to each quadrant and cycle, **signify discrete-increment progressions of time around the conic-unit-circle of figure 37 or the unit-sphere of figure 49**. **It analogues in distinct increments, how time advances around the unit circle in exponential harmonics**.

Equivalence is presented between broadened-line portions of each s_n in figure 14 and unit-sphere chord-lengths **J** - **K** of figure 15. That equality is valid at large n. However a more precise relationship can be articulated for all n. Each broadened-line segment in figure 14 represents an incremental increase in s_n over the prior s_{n-1}. Every cycle synchronizes with consecutive odd integers of s_n so **summed incremental increases per cycle must equal exactly 2**. Since as n approaches infinity, the conic unit circle comprises precisely 4 radians, **twice those broadened-line-lengths (accruing to 2x2=4 in length) would then always fit exactly around the** perimeter **of the 4-radian circle**, (which has a perimeter of 4). Even when n is very small, exactly twice the broadened-line summed extents in figure 14, (e.g., $2(\sqrt{3}-\sqrt{1})$, $2(\sqrt{5}-\sqrt{3})$, $2(\sqrt{7}-\sqrt{5})$, etc.) will each

Demonstrating that the SUM(root of(odd#)-root of(odd#-2)) between perfect squares of (2n+1) always sums to exactly 2.

Radians traversed around the 4-total radians of a $\beta = \sin^{-1}(2/\pi)$ cone exactly equals twice the sum of s_n increments between perfect squares of $s_n=\sqrt{2n+1}$

Table 4

A	B	C	D	E
2n+1	sq.root(odd#-2)	sq.root(odd#)	rootSn–rootSn-2	SUM of D
	S_{n-2}=root(2n-1)	S_n=root(2n+1)	$S_n{}^{.5}-S_{n-2}{}^{.5}$	between
			=C-D	perfect squares
			incremental S_n	of (2n+1)
1		1		
3	1	1.73205081	0.732050808	Sum = 2
5	1.732050808	2.23606798	0.50401717	For n=1
7	2.236067977	2.64575131	0.409683334	to 4
9	2.645751311	3	0.354248689	
11	3	3.31662479	0.31662479	
13	3.31662479	3.60555128	0.288926485	
15	3.605551275	3.87298335	0.267432071	
17	3.872983346	4.12310563	0.250122279	Sum = 2
19	4.123105626	4.35889894	0.235793318	For n=5
21	4.358898944	4.58257569	0.223676751	to 12
23	4.582575695	4.79583152	0.213255828	
25	4.795831523	5	0.204168477	
27	5	5.19615242	0.196152423	
29	5.196152423	5.38516481	0.189012384	
31	5.385164807	5.56776436	0.182599556	
33	5.567764363	5.74456265	0.176798284	
35	5.744562647	5.91607978	0.171517137	Sum = 2
57	5.916079783	7.54983444	1.633754652	For n=13
39	7.549834435	6.244998	-1.304836437	to 24
41	6.244997998	6.40312424	0.158126239	
43	6.403124237	6.55743852	0.154314287	
45	6.557438524	6.70820393	0.150765408	
47	6.708203932	6.8556546	0.147450668	
49	6.8556546	7	0.1443454	
51	7	7.14142843	0.141428429	
53	7.141428429	7.28010989	0.138681461	
55	7.280109889	7.41619849	0.136088598	
57	7.416198487	7.54983444	0.133635948	
59	7.549834435	7.68114575	0.131311313	
61	7.681145748	7.81024968	0.129103928	
63	7.810249676	7.93725393	0.127004257	Sum = 2
65	7.937253933	8.06225775	0.125003815	For n=25
67	8.062257748	8.18535277	0.123095024	to 40
69	8.185352772	8.30662386	0.121271091	
71	8.306623863	8.42614977	0.11952591	
73	8.426149773	8.54400375	0.117853972	The radians of each
75	8.544003745	8.66025404	0.116250293	transition equals
77	8.660254038	8.77496439	0.11471035	$2[\sqrt{2n+3} - \sqrt{2n+1}]$
79	8.774964387	8.88819442	0.11323003	exactly, from n =0 to ∞
81	8.888194417	9	0.111805583	
83	9	9.11043358	0.110433579	
85	9.110433579	9.21954446	0.109110878	
87	9.219544457	9.32737905	0.107834596	
89	9.327379053	9.43398113	0.106602079	
91	9.433981132	9.53939201	0.105410882	
93	9.539392014	9.64365076	0.104258747	
95	9.643650761	9.74679434	0.103143584	
97	9.746794345	9.8488578	0.102063457	
99	9.848857802	9.94987437	0.101016569	
101	9.949874371	10.0498756	0.10000125	
103	10.04987562	10.1488916	0.099015944	Sum = 2
105	10.14889157	10.2469508	0.098059201	For n=41
107	10.24695077	10.3440804	0.097129667	to 60
109	10.34408043	10.4403065	0.096226076	
111	10.44030651	10.5356538	0.095347244	
113	10.53565375	10.6301458	0.09449206	
115	10.63014581	10.7238053	0.093659482	
117	10.72380529	10.8166538	0.092848532	
119	10.81665383	10.9087121	0.092058288	
121	10.90871211	11	0.091287885	Etc. to
123	11	11.0905365	0.090536506	n = ∞

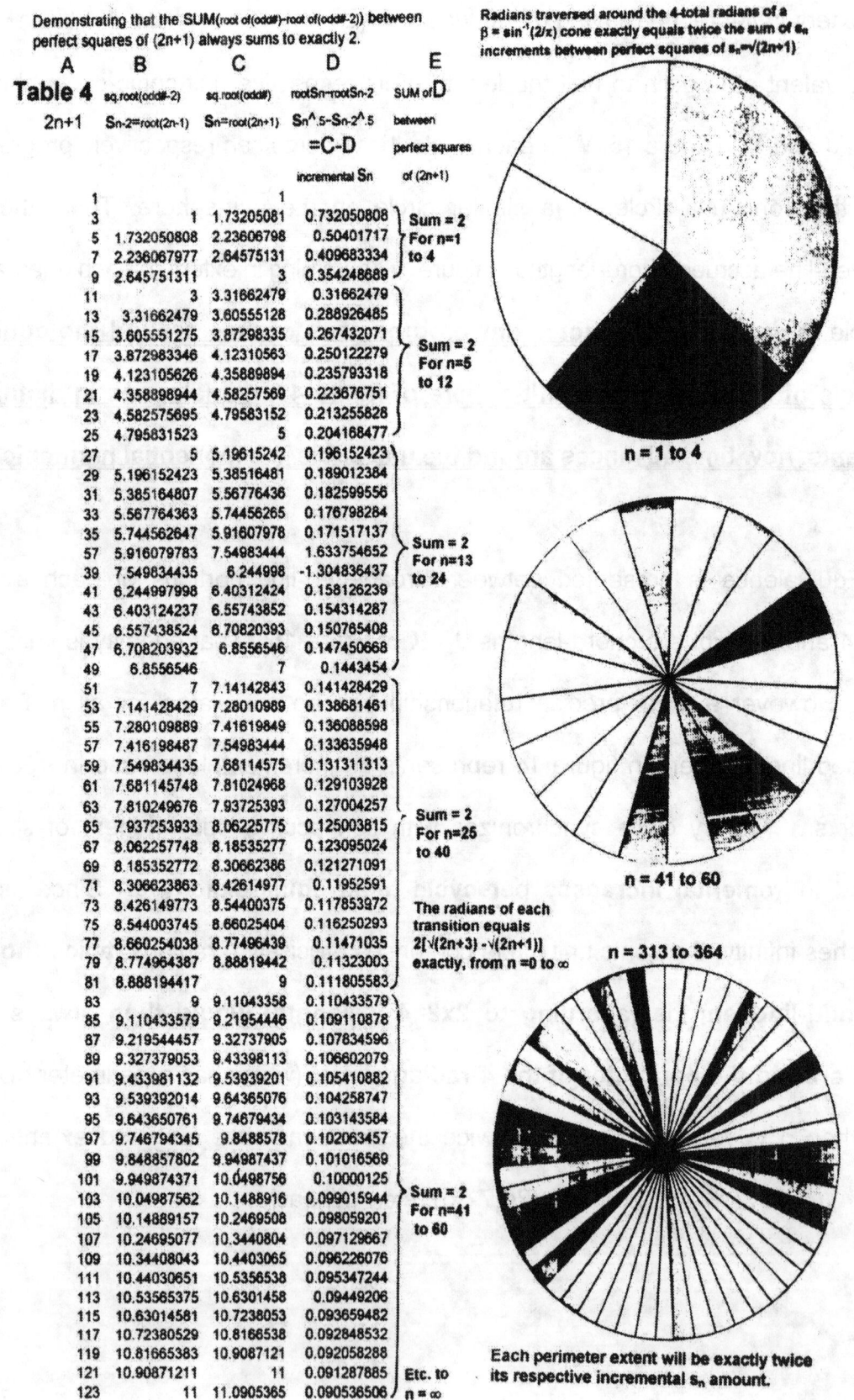

n = 1 to 4

n = 41 to 60

n = 313 to 364

Each perimeter extent will be exactly twice its respective incremental s_n amount.

sum to precisely 4 radians traversed around the $n \rightarrow \infty$ derived 4-radian circle. Since those broadened-line extents represent time increments, their equivalence on the absolute 4-radian circle also depicts time increments, as progressing around the perimeter of that circle **in analogue of the harmonic exponential**. As $n \rightarrow \infty$, the absolute 4-radians are perfectly synchronous, <u>**so for all n, time (and rotation) progress in discrete radian increments of**</u>

$2(\sqrt{3}-\sqrt{1})$, $2(\sqrt{5}-\sqrt{3})$, $2(\sqrt{7}-\sqrt{5})$, - - - $2[\sqrt{(2n+3)}-\sqrt{(2n+1)}]$, etc. to $n \rightarrow \infty$.　　(21*8)

No matter how far removed from $n = 0$, all angular sub-partitions of each cycle **[as $2\{\sqrt{(2n+3)}-\sqrt{(2n+1)}\}$ radian increments],** can always be examined between consecutive odd integers of $s_n = \sqrt{(2n+1)}$. Based on integer n, this provides a discrete operator permitting **exact** dissection of any cycle of the wave. <u>The resulting sum of those chord-segments invariably yield perfect periodicity for all cycles over all n, even though sub-partitions that comprise those cycles can each differ and be mostly irrational.</u> **As delineated in Table 4**, that sum between perfect squares of (2n+1) suffers no imperfection in periodicity. <u>**Periodicity stems directly from the perfectly equal spacing between odd integers in magnitude s_n and will thus always be exact**</u>. This "origin" of periodicity within the natural numbers is worth reiterating. Exponential harmonics progress around a unit-circle of 2π-radians. Unified waves can be envisioned to progress around a conic-unit-circle or unit-sphere that comprises 4-radians with a perimeter extent of 4. For all n, broadened-line incremental-segments of s_n in figure 14, (increments of time) must algebraically sum to have exact-extent of 2 per-cycle-of-perfect-squares in (2n+1). When twice those incremental s_n chord-segments follow around the perimeter of a 4-radian circle in discrete analogue of continuous exponentials, they will **always experience perfect repeat periodicity as equation 21*8 indicates**. Each cycle's four quadrants will not precisely equal at small n, but toward n $\rightarrow \infty$, every quadrant approaches one absolute radian.

Figure 37 clarifies how and why before division by n, no amplitude associates with the wave. The spiral continually grows. The apex angle of the cone changes with $\sqrt{\xi}$, so the "perimeter" of the conic-unit-circle, or on the unit-sphere changes correspondingly. Being the intersection of n-vectors on the unit-sphere, when the spiral figure is divided by n, the circular locus remains **unity from the origin along the conic or spherical surface. Diameter ($4\sqrt{\xi}/\pi$) of the circular base changes, while unity side-lengths of the cone remain unchanged**. The magnitude of each summed quadrant grows or diminishes along the conic-unit-circle base or unit-sphere in accord with the $\sqrt{\xi}$ expressed value. Radius $r = (2n\sqrt{\xi}/\pi)$ of the circular locus provides a manifestation of amplitude within the wave, but from the vantage of fixed unity-side-length on the cone it remains "hidden". It effectively "occurs on an expandable circle". Figure 49 shows another way to represent the figure 37 mode on 4 planes forming a "square base" on the unit-sphere's surface. It avoids π and maintains each quadrant's triangles within the same plane. The separate planes also demark wave-energy interchange intervals.

Figures 38 through 42 enhance the display of periodicity inherent to the natural numbers by graphing superimposed radial rather than tangential vectors of the spiral's apex view. Effects over the range $0 \leq \xi \leq 1$ show up in the representation. The expression for cumulative angle $\psi(n,\xi)$ accommodates $\xi \neq 1$ conditions. For these mappings on a plane, $\sqrt{\xi}$ in the denominator of the angle expression "compensates" for the axial view of radians-at-synchronization being $4\sqrt{\xi}$ for all ξ. Radius employed in all these presentations was kept constant to exclusively emphasize the angular harmonic symmetry so radial scale is not consequential here. Four-quadrant-symmetry is seen to remain preserved over the range of ξ. The primary effect of ξ varying between unity and zero is a nominal shift in polarization orientation of about $30°$ counter-clockwise with increasing ξ. This shift amount under straight-chord analysis is plotted in figure 43.

It should be recognized that periodicity and four-quadrant symmetry brought out through these depictions <u>occurs within every wave exhibiting ABCD overlap</u>. Each represents a possible harmonic-resonance solution-state subsequent to an origin event. <u>As shown for ξ = .000000001, even as ξ converges *arbitrarily close to zero in approach to continuity*, harmonic qualities of the natural number system sustain!</u> However no harmonics would appear along the Euclidean axis for ξ = 0 though they exist for any value minutely greater than zero. They exist mathematically from n = 0 to n = ∞ although the physical reality of excess-angle $\Delta\theta$ (Planck's constant) limits that range to n_{max}. The orthogonality axiom has simply been a missing part of the repository of scientific interpretation in the process of counting.

Figure 38 effectively shows a "magnified-radius" axial-view of a cone of apex angle β = arsine $2/\pi$. Radii n progress outward from the origin to each respective node point X on the conic surface. Tangential sectors are not depicted to elucidate the node-point symmetry of the quadrants, which are delineated in-line near 120° and 300°. Cumulative angle $\psi(n,\xi)$ of the spiral rendered does not correct for the oblique appearing radii of extent shown as r(n) = n. Here, ξ = 1 and r(n) = n is the radius of a polar plot of angular coordinates $\psi(n,\xi) = \sum\limits_{n=0}^{n} \text{acos}\left(\frac{n+1-\xi}{n+1}\right)$

A total of 114 radial vectors are shown with the radial scale being set by that maximum n. An actual end view of the spiral would appear the same except every radius would be reduced by $2/\pi$.

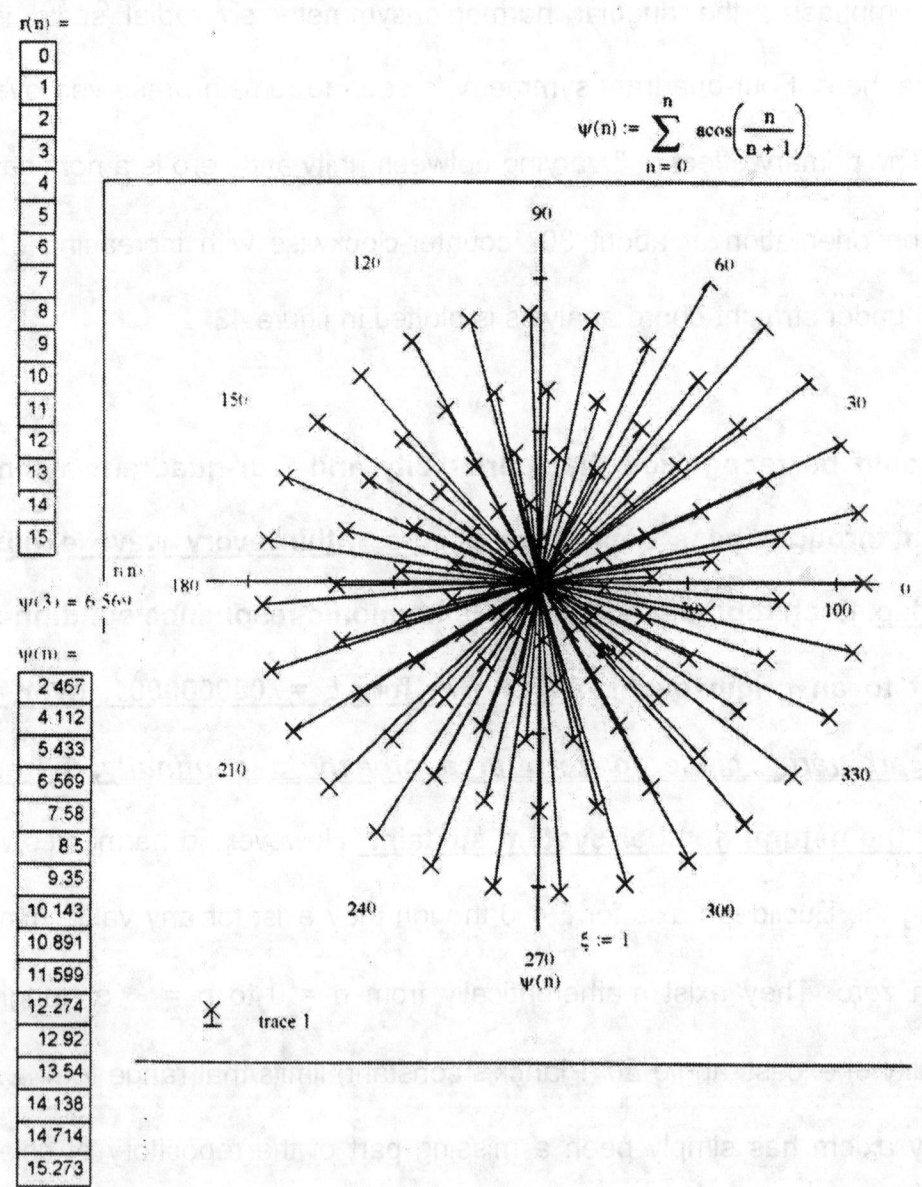

276

Figures 39 and 40 indicate the effect of changing ξ over the range from 1 to .0000001 on a plot of the spiral with n having 200 radii. To elucidate synchronicity, radial lines to each node are drawn rather than tangential sectors. For ξ = 1, node directions overlap in a rectilinear fashion near **direction A** and at right angles thereto. These polar plots demonstrate the variation in radians-at-synchronization $Ř_s = 4\sqrt{\xi}$ by plotting the expression for cumulative angle as $\psi(n,\xi) = \frac{\pi}{2\sqrt{\xi}} \sum_{n=0}^{n} \text{acos}\left(\frac{n+1-\xi}{n+1}\right)$

vs. radius = n, (rather than radius $(2n\sqrt{\xi})/\pi$ as an actual end-view of the cone would appear).

As the value of ξ becomes less than unity, the four quadrants rotate clockwise but stay synchronous as depicted in these plots.

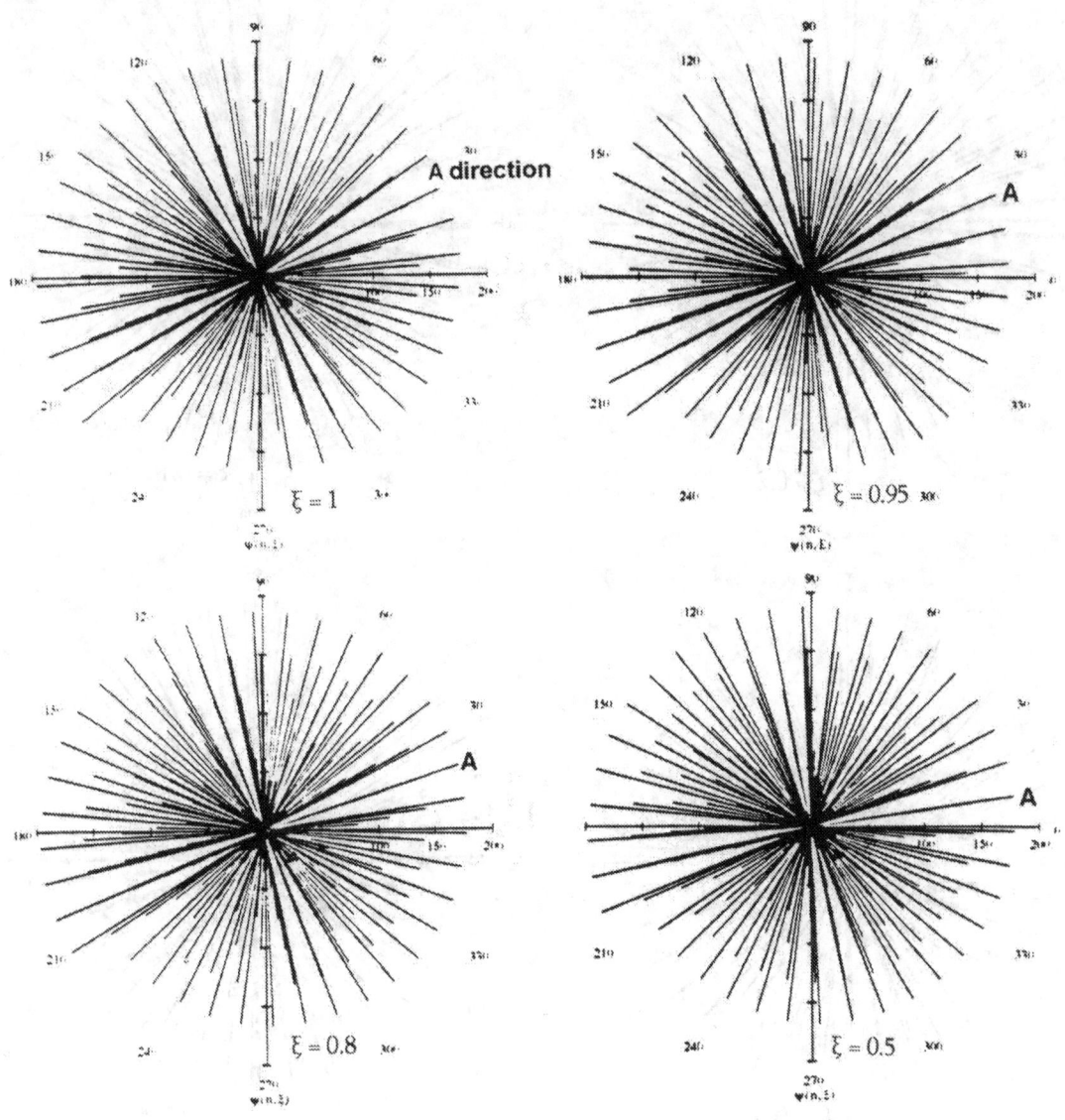

277

Figure 40 continues the figure 39 sequence of decreasing ξ while quadrant node patterns retain relative synchronization. The absolute polarization angle of the four synchronous radii directions have little significance in terms of ability for modes of the wave to self-interfere or behave as a standing wave. These plots were arranged to demonstrate the wave's intrinsic synchronicity for n = 200 and any fixed value of 1 ≥ ξ > 0. They also show how radians-at-synchronization remain at 4√ξ by allowing (π/2√ξ) to multiply the cumulative angle in the expression employed in the plot, i.e.

Radius is displayed as n (rather than (2n√ξ)/π as each end-view of the respective cone would appear)

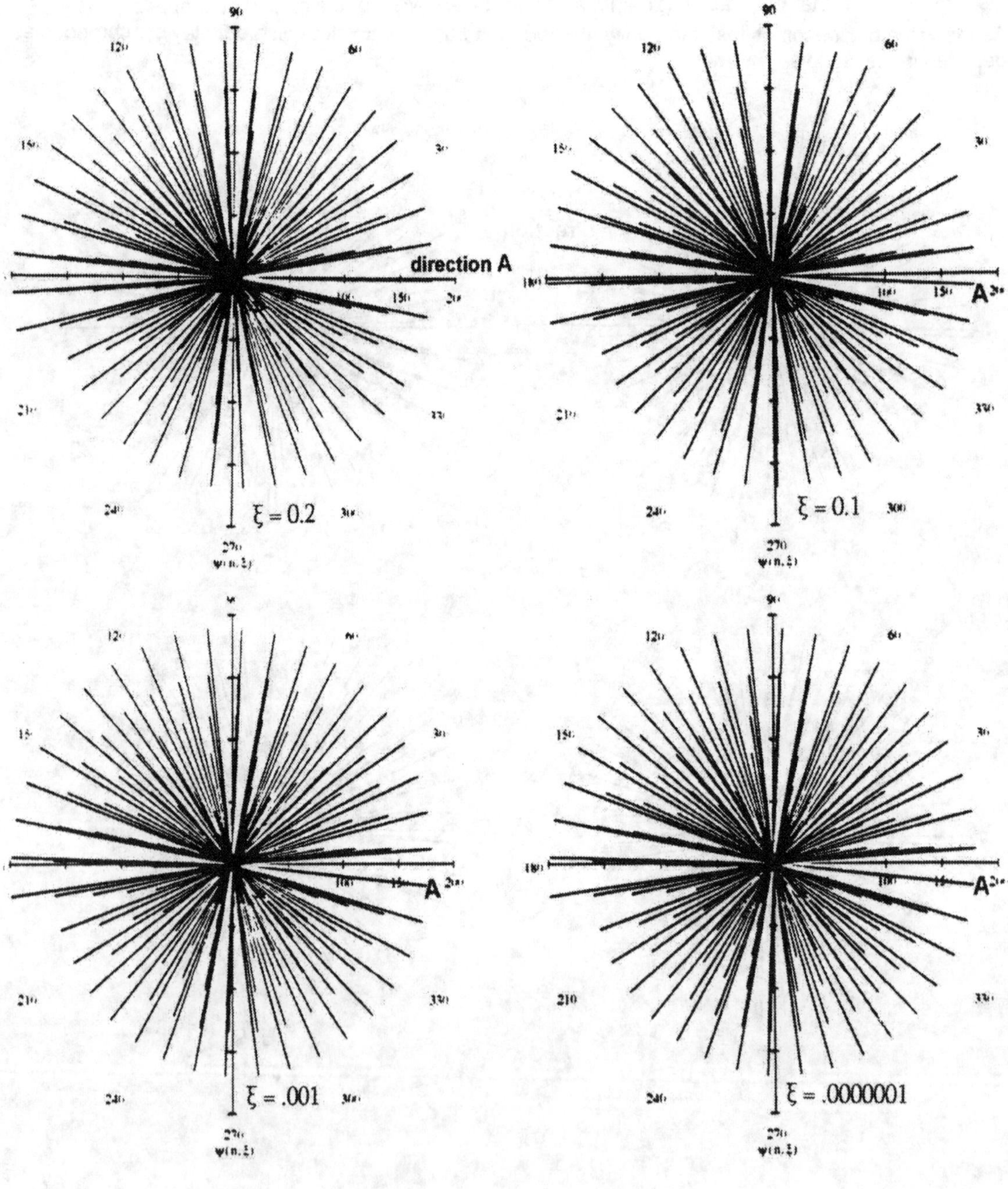

$$\psi(n,\xi) := \frac{\pi}{2\sqrt{\xi}} \sum_{n=0}^{n} a\cos\left(\frac{n-1-\xi}{n-1}\right)$$

direction A

ξ = 0.2

ξ = 0.1

ξ = .001

ξ = .0000001

Figure 41 uses the same format as figures 38 through 40 except maximum n is here increased to 2000, for ξ = 1, 0.5, and 0.01. The fourth graph increases n to 6000 different radii with ξ being unity. All plots display the inherent four-quadrant symmetry and periodicity intrinsic to natural numbers. That symmetry continues out to n_{max} (and n→∞). The graphs demonstrate the enormous number of potential unified wave natural modes, which possibilities compound even further by the radial vectors not being constrained to a single surface as portrayed here. It is noteworthy that when taking the sum of 6000 consecutive angles a mean deviation of even 1/1000 of a degree would scatter the symmetry. When n reaches 10^7 = $2P^2$+2P (and P ≈ 2235.568 cycles, or for example about 1.5mm away from a 671nm red photon emission source), a mean alteration of a millionth of an angular degree per spiral triangle would scatter the wave's symmetry. Such precariousness intrinsic to the wave's **nonetheless-periodic-mode** expressly would allow loss of that mode's interference properties upon any interrogation of the wave, (e.g., upon inserting a sensor to determine which slit the wave goes through, or trying to establish the internal state of the system). At one meter from the emission source that same photon would have undergone $1/(671 \times 10^{-9})$ = 1.49- - -$\times 10^6$ cycles = P. There, n = $2P^2$+2P = 4.44- - -$\times 10^{12}$. A mean alteration of one pico-degree in each origin angle of the spiral would scatter the symmetry and terminate interference. **Such phenomenon of loosing interference upon interrogating the wave disturbance is otherwise known as** <u>**collapse of the wave function**</u>. Analysis here shows the mathematical process of how and why it comes about.

$$\psi(n, \xi) = \frac{\pi}{2\sqrt{2}} \sum_{n=0}^{n} \cos\left(\frac{n+1-\xi}{n+1}\right)$$

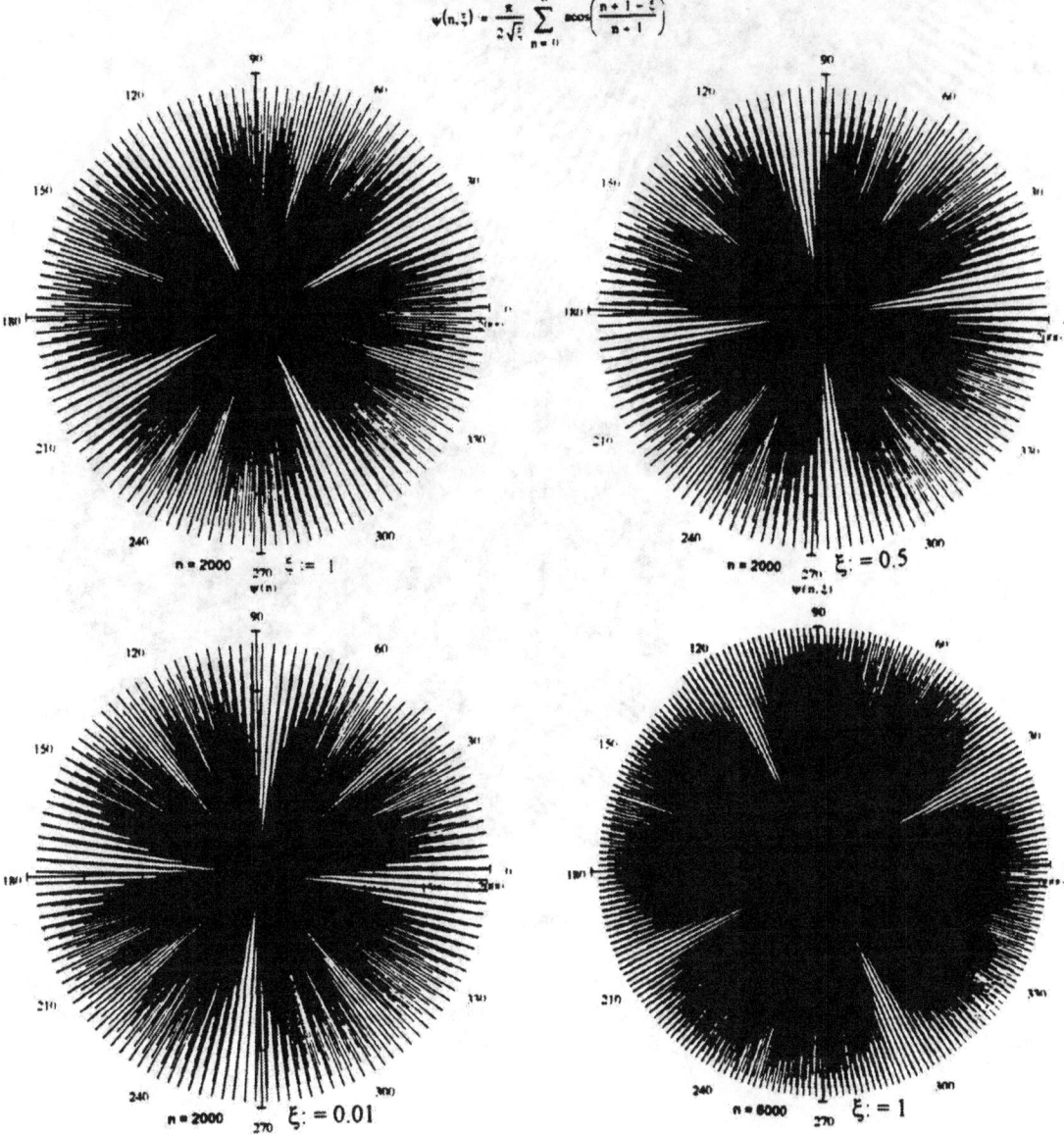

Figure 42. In the upper plot a reversed transparency of the figure 41 graph having ξ = 1 and n = 2000 superimposes onto itself. The lower graph similarly superimposes the ξ = 1, n = 6000 pattern onto its own mirror image. Retention of four-quadrant symmetry is obvious signifying the intrinsic quadrature nature of the natural numbers under angular progression through those quadrants.

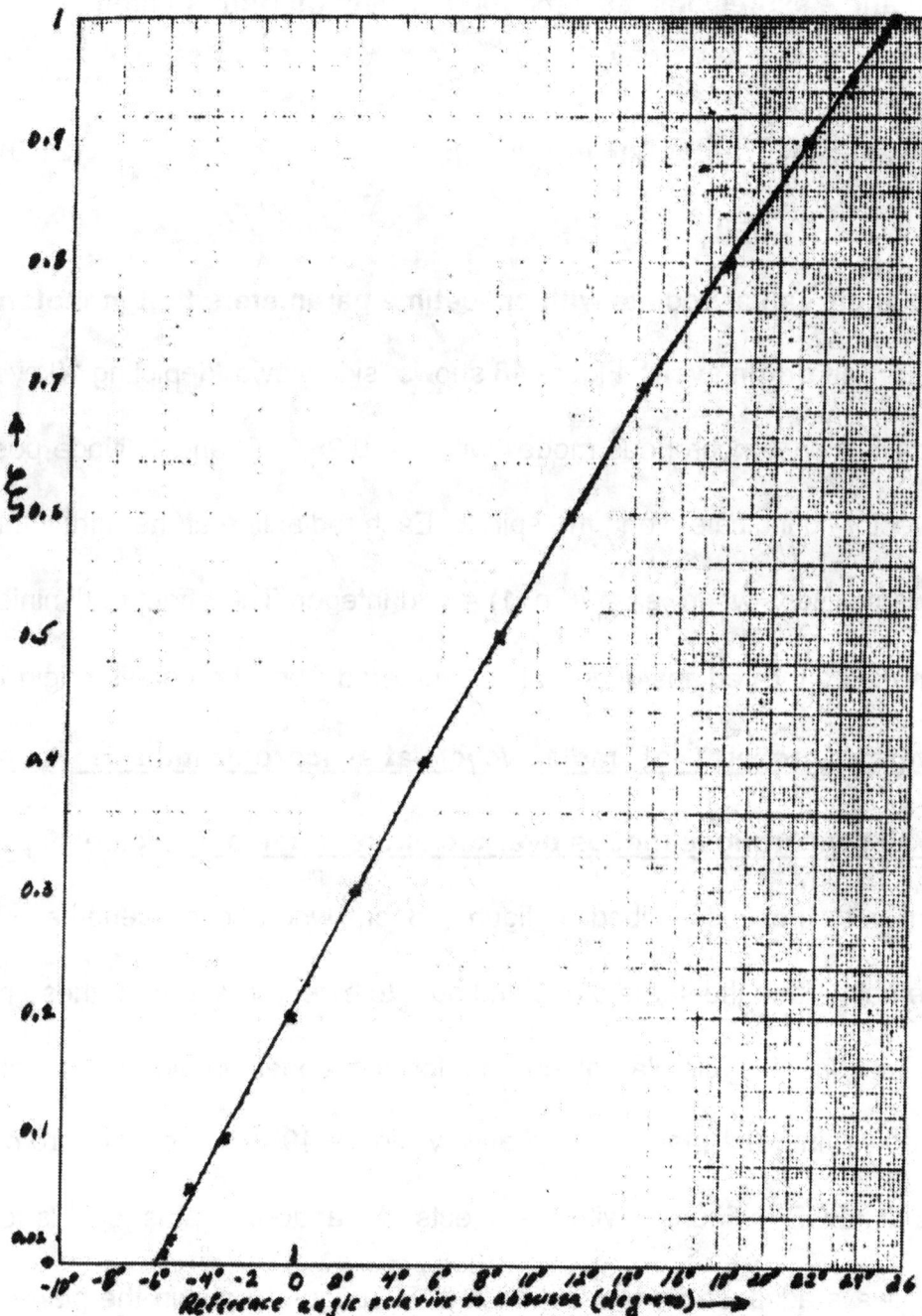

Figure 43 displays the empirically measured shift in polarization (symmetrical orientation) for the n = 2000 spirals of figures 39 and 40. Experimental accuracy was about ± 0.5° in establishing this almost-linear relationship over the range of about 30° for 1 ≥ ξ ≥ 0. With any specific ξ, absolute orientation is not particularly relevant to the cyclic nature of numerical advance in n.

281

There is always an n and ξ, and by straight-chord definition a cumulative angle expressible as a mapped projection onto a plane

$$\text{n}$$

as: $\psi(n,\xi) \equiv \{\pi/(2\sqrt{\xi})\} \sum \text{arcos}[(n+1-\xi)/(n+1)]$. (21*9)

$$\text{n=0}$$

That angle can associate with spacetime parameters that monotonically increase after an origin event. Figure 48 shows "side views" depicting 10 cycles of computer-calculated synchronous modes with ξ = 0.25, 0.5, and 1. Node positions of each n-vector-tip indicate along the spirals. Each radial line at the right intercepts synchronous positions where $s_n = \sqrt{(2n+1)}$ = odd integer. The effect of diminished ξ obviously reduces the radians/quadrant encountered from the cone's origin by $\sqrt{\xi}$. <u>**This universal property of radians/cycle-at-synchronization $\equiv \check{R}_s = 4\sqrt{\xi}$ describes all synchronous modes over the entire range of**</u> ξ. Figure 49 portrays modes similar to those described in figure 48 but where chord-lengths of each quadrant remain within the same plane and sum to extent $\sqrt{\xi}$. To form these modes one might imagine the circular quadrants for the cones in figures 37 and 48 "squished down" into flat planes. Alternatively, figure 49 can construct from right-angle folds at ABCD in figure 3 with the effects of ξ added. That is, spirals forming this mode always advance from direction A to B in figure 3 within the plane of the paper. Upon reaching B they invariably make a right-angle transition subsequently proceeding normal to the paper between B and C. After continuing in that normal plane from B to C they make another right

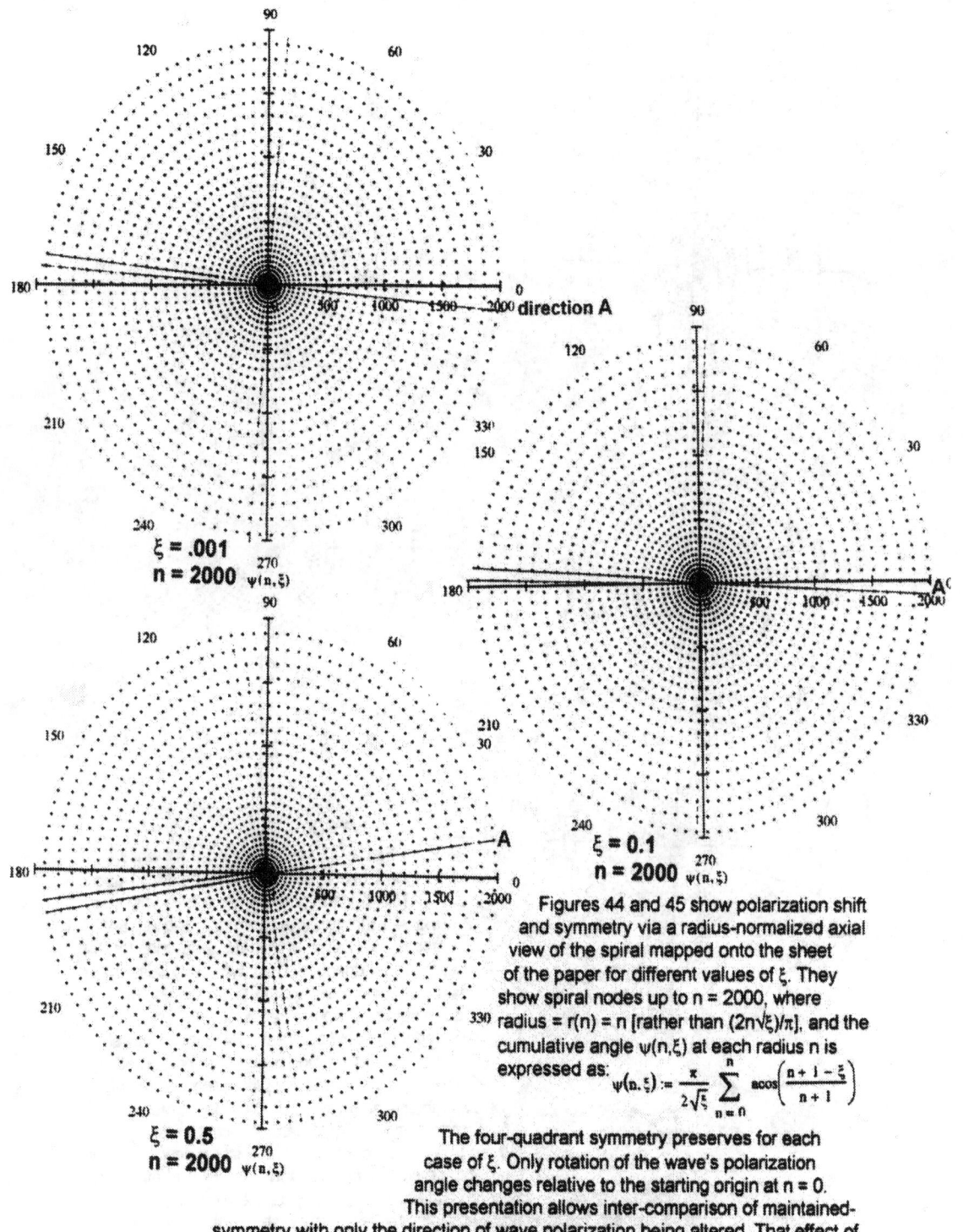

$\xi = .001$
$n = 2000$ $\psi(n,\xi)$

$\xi = 0.1$
$n = 2000$ $\psi(n,\xi)$

$\xi = 0.5$
$n = 2000$ $\psi(n,\xi)$

direction A

Figures 44 and 45 show polarization shift and symmetry via a radius-normalized axial view of the spiral mapped onto the sheet of the paper for different values of ξ. They show spiral nodes up to n = 2000, where radius = r(n) = n [rather than (2n√ξ)/π], and the cumulative angle ψ(n,ξ) at each radius n is expressed as:

$$\psi(n,\xi) := \frac{\pi}{2\sqrt{\xi}} \sum_{n=0}^{n} a\cos\left(\frac{n+1-\xi}{n+1}\right)$$

The four-quadrant symmetry preserves for each case of ξ. Only rotation of the wave's polarization angle changes relative to the starting origin at n = 0. This presentation allows inter-comparison of maintained-symmetry with only the direction of wave polarization being altered. That effect of diminished radians/cycle (an effectively smaller cone apex angle) is "compensated for" in these plots by putting √ξ in the denominator of ψ(n,ξ).

283

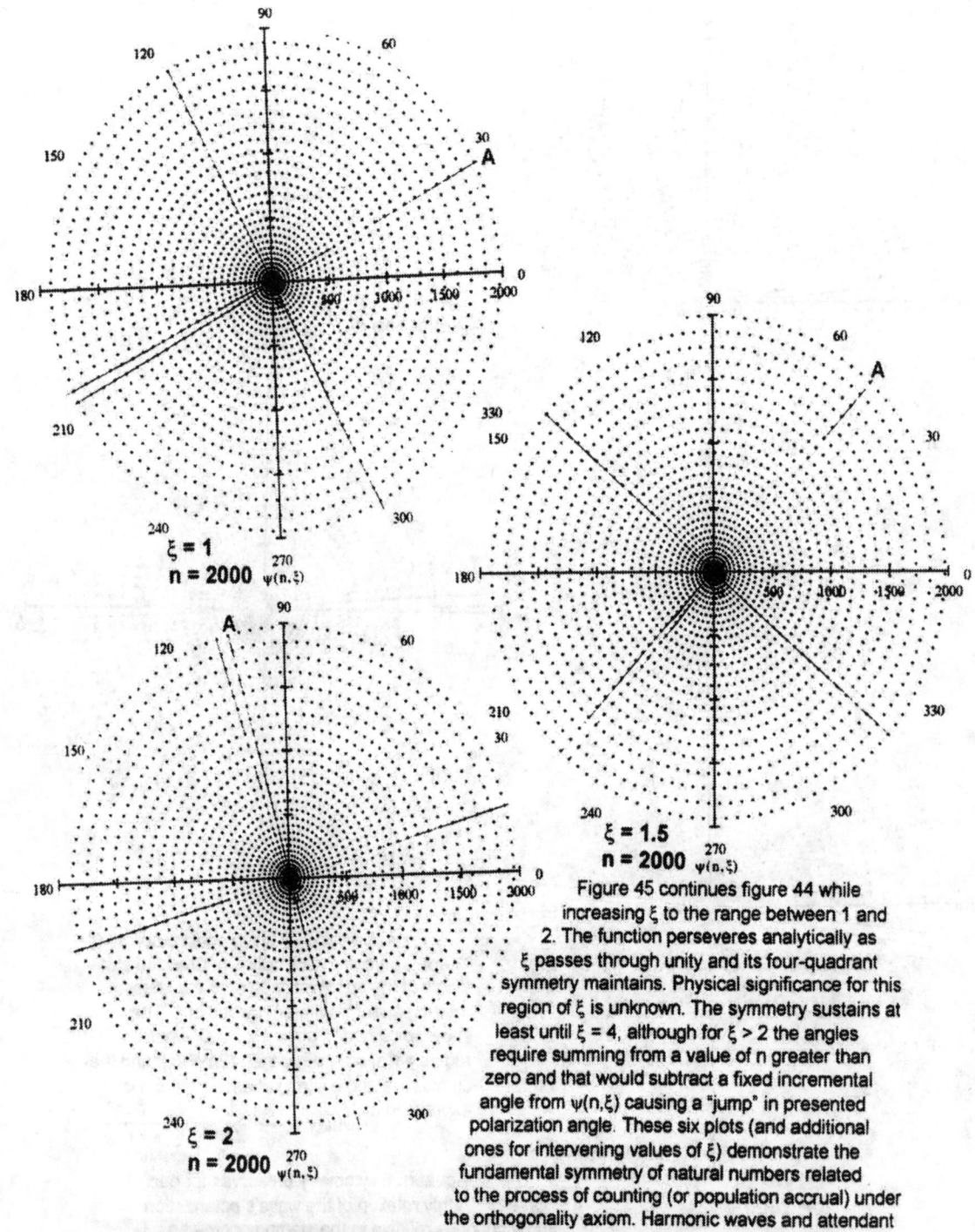

$\xi = 1$
$n = 2000$ $\psi(n, \xi)$

$\xi = 1.5$
$n = 2000$ $\psi(n, \xi)$

$\xi = 2$
$n = 2000$ $\psi(n, \xi)$

Figure 45 continues figure 44 while increasing ξ to the range between 1 and 2. The function perseveres analytically as ξ passes through unity and its four-quadrant symmetry maintains. Physical significance for this region of ξ is unknown. The symmetry sustains at least until $\xi = 4$, although for $\xi > 2$ the angles require summing from a value of n greater than zero and that would subtract a fixed incremental angle from $\psi(n, \xi)$ causing a "jump" in presented polarization angle. These six plots (and additional ones for intervening values of ξ) demonstrate the fundamental symmetry of natural numbers related to the process of counting (or population accrual) under the orthogonality axiom. Harmonic waves and attendant modes that result under the spiral construction method outlined in figure 12, (where $0 \le \xi \le 1$), constitute an intrinsic property for spacetime variables following an origin event. It is not a mathematical coincidence that these waves remain harmonic for all allowed values of ξ but a physical property of enumeration that has been overlooked by non-recognition of the orthogonality axiom. Harmonic waves and "frequencies" begin at the $t = 0$ origin, not at minus infinity in time as portrayed by the exponential function.

Figure 46. This graph illustrates the geometric formation of spirals for ξ values between zero and three. Only the transition from n = 13 to n = 14 is illustrated for different choices of ξ. For each respective ξ portrayed, the triangle formed by abscissa, angle-from-the-origin $\theta_{\xi=x}$, and its affiliated oblique sector $s_n \equiv s_{13,\xi=x}$, represents the triangle that would be added to the spiral for that-specific-ξ-case-transition from n = 13 to n = 14. Each $\theta_{\xi=x}$ itself actually represents a conic-surface-of-rotation-about-the-abscissa describing possible angles at which progression from the spherical surface at n = 13 (shown dotted) to the spherical surface at n = 14 can occur. Radians-at-synchronization $\equiv \check{R}_s$ characterizes the radians accrued within the spiral between perfect squares of (2n+1) = (odd #) for the respective ξ, as measured by the sum of the angles $\theta_{\xi=x}$ between those perfect squares. Precise interpretation for the region where ξ > 1 and $\check{R}_s$ > 4 is unknown but presented here for reasons of continuance in display of the concept. It is perhaps plausible that the maximum radians-at-synchronization would be $2\pi = \check{R}_s = 4\sqrt{\xi}$, thereby limiting ξ to $(\pi/2)^2 = 2.467$ – – –.

The transition from the concentric spherical surface at n = 13 to the spherical surface at n = 14 occurs forming an origin angle $\theta_{\xi=x}$. The transition is depicted here at different feasible values of ξ, with resultant values of $\check{R}_s$ = Radians-at-synchronization displayed at respective points, in the format: ξ, [$\check{R}_s$ = |

3, [$\check{R}_s$ = 6.928 –]

$s_{13,\xi=3}$

2, [$\check{R}_s$ = 5.656 –]

$s_{13,\xi=2}$

1.5, [$\check{R}_s$ = 4.898 –]

$s_{13,\xi=1.5}$

1, [$\check{R}_s$ = 4]

0.5, [$\check{R}_s$ = 2.828 –]

$s_{13,\xi=1}$

$s_{13,\xi=0.5}$

0.25, [$\check{R}_s$ = 2]

$\theta_{\xi=3}$

$\theta_{\xi=2}$

$\theta_{\xi=1.5}$

$\theta_{\xi=1}$

$\theta_{\xi=0.5}$

$\theta_{\xi=0.25}$

Spherical surface at n = 13

Spherical surface at n = 13

Spherical surface at n = 15

Spherical surface at n = 14

origin

n

← ξ

Irwin Wunderman

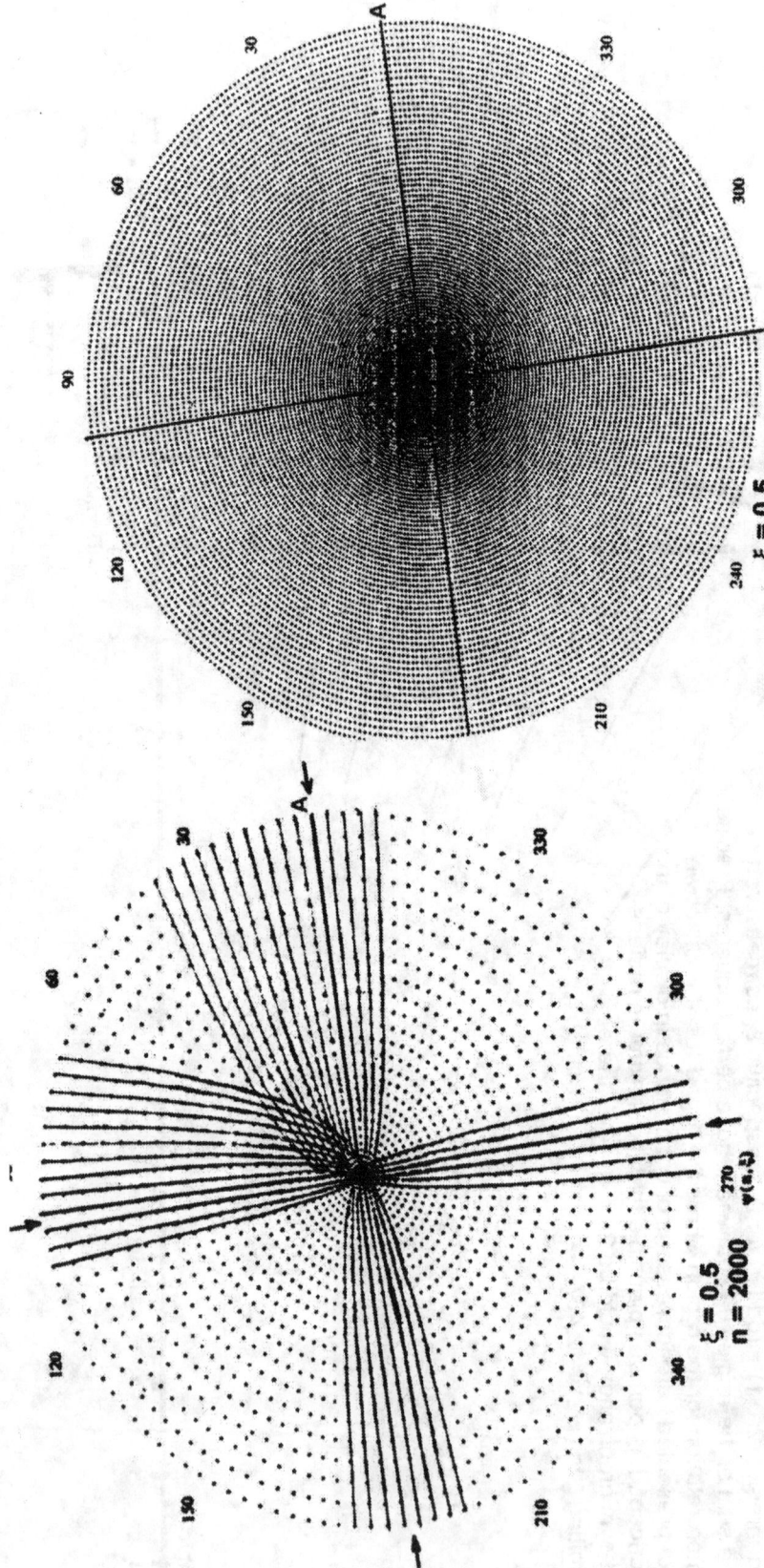

Figure 47. The symmetry in these plots derives from the spiral (or the natural numbers) having one additional sector per outward quadrant as noted in figure 3. This leads to perfectly symmetrical cyclic buildups portrayed by these examples, shown with ξ = 0.5 and n = 2,000 on the left and 20,000 right. A dot indicates each spiral node and in the left figure hand drawn "radial" lines connect outward-going dots to show the inherent symmetry. The outermost spacing between dots is about 3.2mm and a symmetry deviation of about 0.1 dot spacing would be discernable. The 41.5 cm perimeter thus has about 415/.032 ≈ 1300 resolvable parts per circumference. A mean angular deviation per spiral triangle of one part per 2000x1300 ≈ one part in 2.6 million could introduce a discernable deviation in symmetry. The overlapping hand-drawn lines exhibit "sunflower seed" topology common to many living objects of Nature. The right figure demonstrates similar tendencies when viewed obliquely.

286

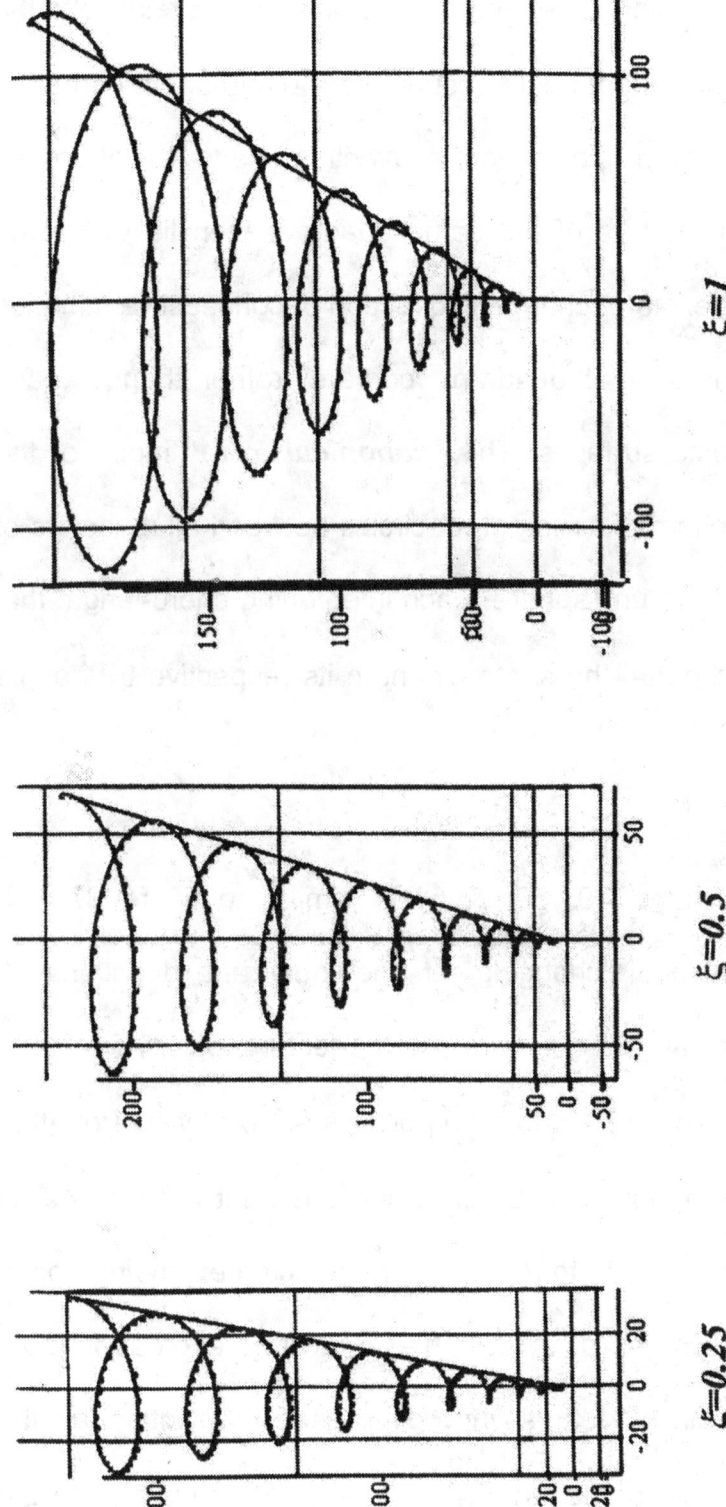

ξ=0.25   ξ=0.5   ξ=1

Figure 48. These views show ten cycles of three synchronous modes on respective conic surfaces using ξ = 0.25, ξ = 0.5, ξ = 1. These are the same three cases as illustrated in figure 35. Synchronization occurs on a conic surface of apex angle β = sin⁻¹(2√ξ)/π] and for ξ = 0.25 that angle is half the ξ = 1 case. Only the value of ξ differs for the three cases, with resulting differences in apex angle where synchronization occurs. For synchronous conditions, radians around each cone's cross-sectional perimeter will always be 4√ξ = Řₛ. Nodes where n-vectors terminate along the conic surface are shown along each spiral and they have matching respective distances from the origin in all three views. Only conic apex-angles and radians around the conic perimeter differ. In each case the view was aligned normal to the polarization axis so that n vectors along the right edge of each cone depict the perfect squares of (2n+1).

287

angle at direction C to progress within a plane that makes an angle $\sqrt{\xi}$ relative to the paper plane. At direction D another right angle occurs into a plane again normal to the page, which completes the cycle upon reaching direction E and A, both those directions being identical. All n-vectors of the same quadrant then lie within the same plane for each repeat cycle. Here, change in direction of consecutive triangles forming the spiral would occur only at quadrant "corners" rather than at each transition, as for modes on conic surfaces. This *"canonical form"* mode of the unified wave demonstrates as chords forming great-circles between all n-vectors of the spiral at their intersection of the unit-sphere. Each intervening chord-length that sums to $\sqrt{\xi}$ per quadrant would equal the same extent in its respective θ_n triangle between **J** and **K** in figure 15.

Mode arrangement in figures 49, 50, & 51 is similar to figure 37 with consecutive-triangles within planes, changing at the quadrant directions. It characterizes a wave of one radian per quadrant with perfect periodicity for $\xi = 1$. For three values of $\xi = 1$, 0.5, and 0.25, figure 50 portrays an axial view from-the-origin showing chord-lengths forming each straight-sided quadrant in figure 49. In each case about 2 ½ cycles (from P = 10 to P = 12.5) are depicted extending from n = 179 to n = 287. Respective values of s_n at corner directions express in ***bold italics***. All extents would actually appear slightly diminished in length than the presented values by foreshortening of the view. The inner mode "square" with $\xi = 0.25$ is "half the size" of the outer mode with $\xi = 1$ since $\sqrt{\xi} = \sqrt{0.25} = 0.5$. Quadrant lines for the center $\xi = 0.5$ mode were slightly broader than the other

A "Canonical Mode" Form for Representing Unified Waves

Directly from natural numbers without the use of π and with ξ =1, this figure defines both one absolute angular radian and one absolute steridian solid angle. It also precisely renders smaller angles and solid angles based solely on interpretation of ξ in figure 12.

Chord-lengths sum to √ξ along this arc, which is actually a quarter-section of each spiral's cycle.

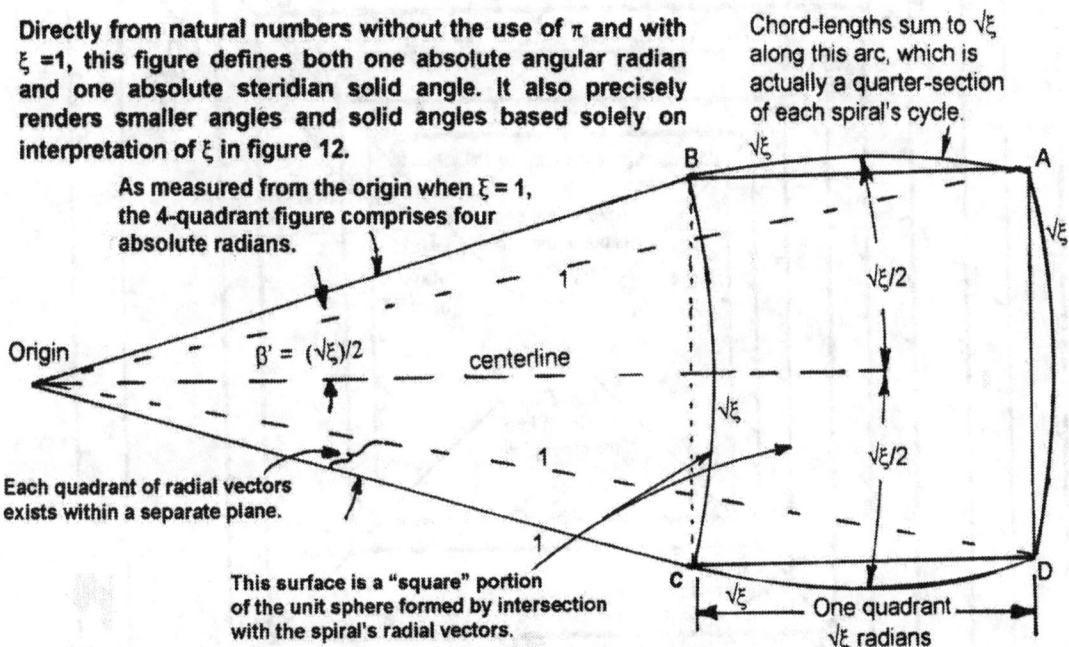

As measured from the origin when ξ = 1, the 4-quadrant figure comprises four absolute radians.

Origin

β' = (√ξ)/2   centerline

Each quadrant of radial vectors exists within a separate plane.

This surface is a "square" portion of the unit sphere formed by intersection with the spiral's radial vectors.

One quadrant √ξ radians

Figure 49. For the case of a specific ξ, this shows an alternate way of presenting a synchronous mode where adjacent triangles remain in a plane, except at directions ABCD. It is a variation of the lower part of figure 37 (divided by n) wherein the transverse projection of each quadrant's cumulative chord-lengths similarly sum to √ξ. All n-vectors in each quadrant constitute an arc within a plane as drawn in figure 3. The arc would more accurately be a spiral increasing radially in advance between quadrant corners so this picture signifies large n (and divided by n) where fractional advance in n per cycle would be negligible. Alternatively, it shows the intersection of the progressing n-vectors with a unit sphere (rather than the unit-circle). Optional to representation via progression around a cone's circular base with radius $(2\sqrt{\xi})/\pi$, it is depicted here as cumulative chord-lengths whose projections onto the page designate a "square" of side-length √ξ on a unit sphere. The four-sided cone-like figure from the origin has edges delineating directions ABCD each of length unity, with a "square" of sides √ξ as the surface segment of the unit sphere. If the figure were cut along direction A from the origin outward and unfolded onto a single plane, it would more appropriately simulate the fan of figure 3. That only applied there for ξ = 1, rather than for 0 ≤ ξ ≤ 1 here. Half of each side of the square has extent [√ξ]/2 so the half-angle formed at the figure's origin is [√ξ]/2 rather than $\arcsin[(2\sqrt{\xi})/\pi]$, for the cone. **This depiction avoids transcendental π in the analysis**. Each side of the square having transverse extent √ξ **represents one quadrant of wave and √ξ radian relative to the origin**. The radian representation shows-up within a plane as a "triangular-like-segment", rather than as chords bent around a curved cone and that may evade approximations of the cone analysis. In progressing around the square of projected side-length √ξ, orthogonal energy forms like electric and magnetic that constitute quadrants of the wave emerge at right angles. It is the freedom for orthogonal vectors-s_0 to progress in any direction that allows the abrupt transition-angle at each ABCD direction while experiencing no change in direction elsewhere. The cumulative sum of θ_n-angles form the synchronous harmonic and they are independent of the bearing of s_0. Diamond shape modes, (rather than a square) and arbitrary contours between ABCD intersections are also feasible.

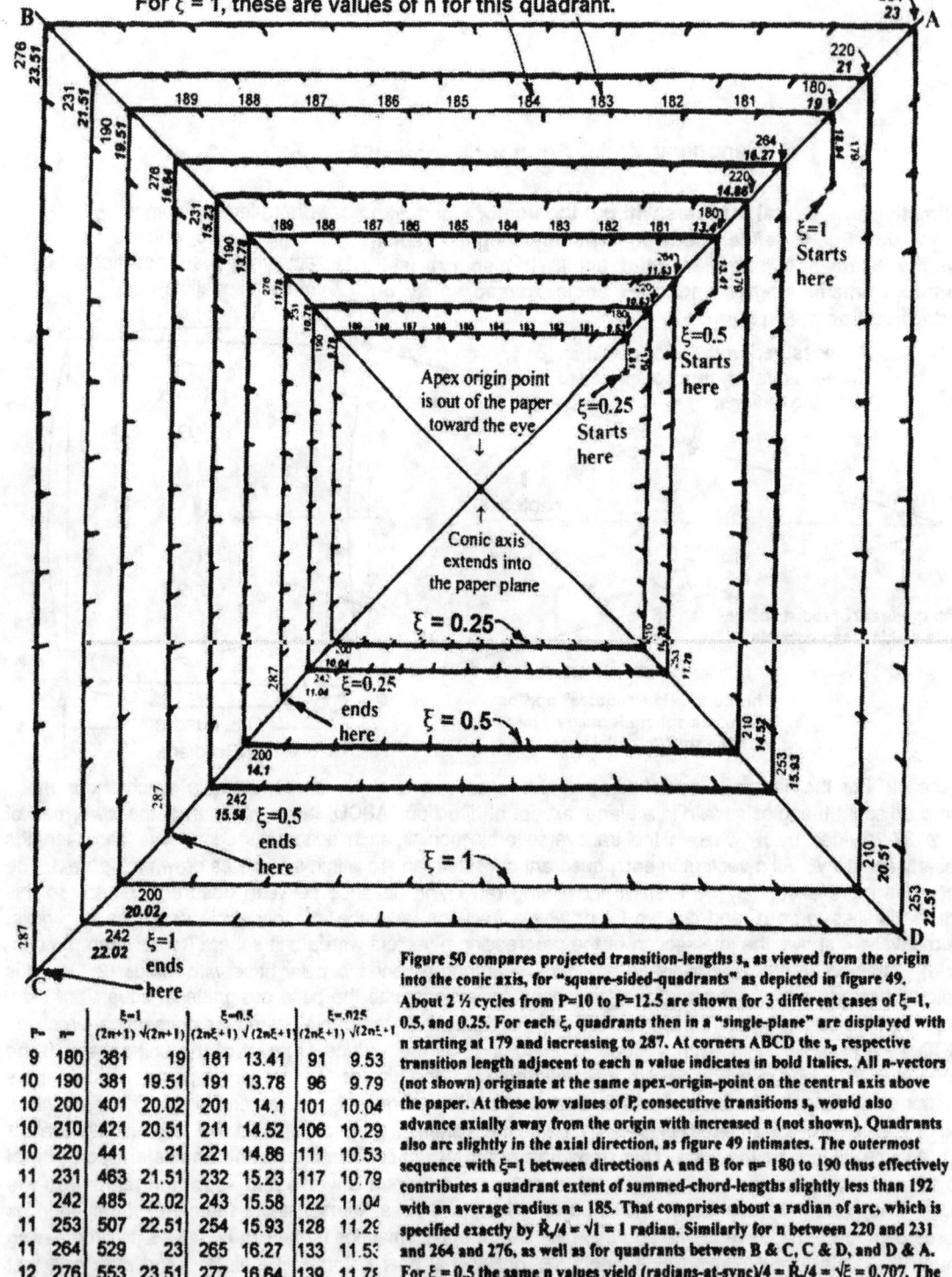

For ξ = 1, these are values of n for this quadrant.

P=	n=	ξ=1 (2n+1)	√(2n+1)	ξ=0.5 (2nξ+1)	√(2nξ+1)	ξ=.025 (2nξ+1)	√(2nξ+1)
9	180	361	19	181	13.41	91	9.53
10	190	381	19.51	191	13.78	96	9.79
10	200	401	20.02	201	14.1	101	10.04
10	210	421	20.51	211	14.52	106	10.29
10	220	441	21	221	14.86	111	10.53
11	231	463	21.51	232	15.23	117	10.79
11	242	485	22.02	243	15.58	122	11.04
11	253	507	22.51	254	15.93	128	11.29
11	264	529	23	265	16.27	133	11.53
12	276	553	23.51	277	16.64	139	11.78
12	288	577	24.02	289	17	145	12.04
12	300	601	24.51	301	17.34	151	12.28

Figure 50 compares projected transition-lengths s_n as viewed from the origin into the conic axis, for "square-sided-quadrants" as depicted in figure 49. About 2 ½ cycles from P=10 to P=12.5 are shown for 3 different cases of ξ=1, 0.5, and 0.25. For each ξ, quadrants then in a "single-plane" are displayed with n starting at 179 and increasing to 287. At corners ABCD the s_n respective transition length adjacent to each n value indicates in bold Italics. All n-vectors (not shown) originate at the same apex-origin-point on the central axis above the paper. At these low values of P, consecutive transitions s_n would also advance axially away from the origin with increased n (not shown). Quadrants also arc slightly in the axial direction, as figure 49 intimates. The outermost sequence with ξ=1 between directions A and B for n= 180 to 190 thus effectively contributes a quadrant extent of summed-chord-lengths slightly less than 192 with an average radius n ≈ 185. That comprises about a radian of arc, which is specified exactly by $\check{R}_c/4 = \sqrt{1}$ = 1 radian. Similarly for n between 220 and 231 and 264 and 276, as well as for quadrants between B & C, C & D, and D & A. For ξ = 0.5 the same n values yield (radians-at-sync)/4 = $\check{R}_c/4$ = $\sqrt{ξ}$ = 0.707. The RMS sum of chord-lengths there using n ≈ 220 would be √(ξ+s_n/n) ≈ √(0.5+.5/220) ≈ √0.5022 ≈ 0.7087. The axial unit-transition-length between integers contributes slightly to the RMS sum but axial advance does not effect radians/quadrant. The innermost sequence in the figure with ξ = 0.25 has $\check{R}_c/4$ ≈ √0.25 = √0.5. The RMS sum of chord-lengths there ≈ √(0.25+0.75/n) ≈ 0.5034. Convergence of the RMS sum to $\check{R}_c = 4\sqrt{ξ}$ becomes ever more exact at large n, (and large ξ). The drawing is approximately to scale within limitations of projection foreshortening, etc. Values derive from the tabulation to the left.

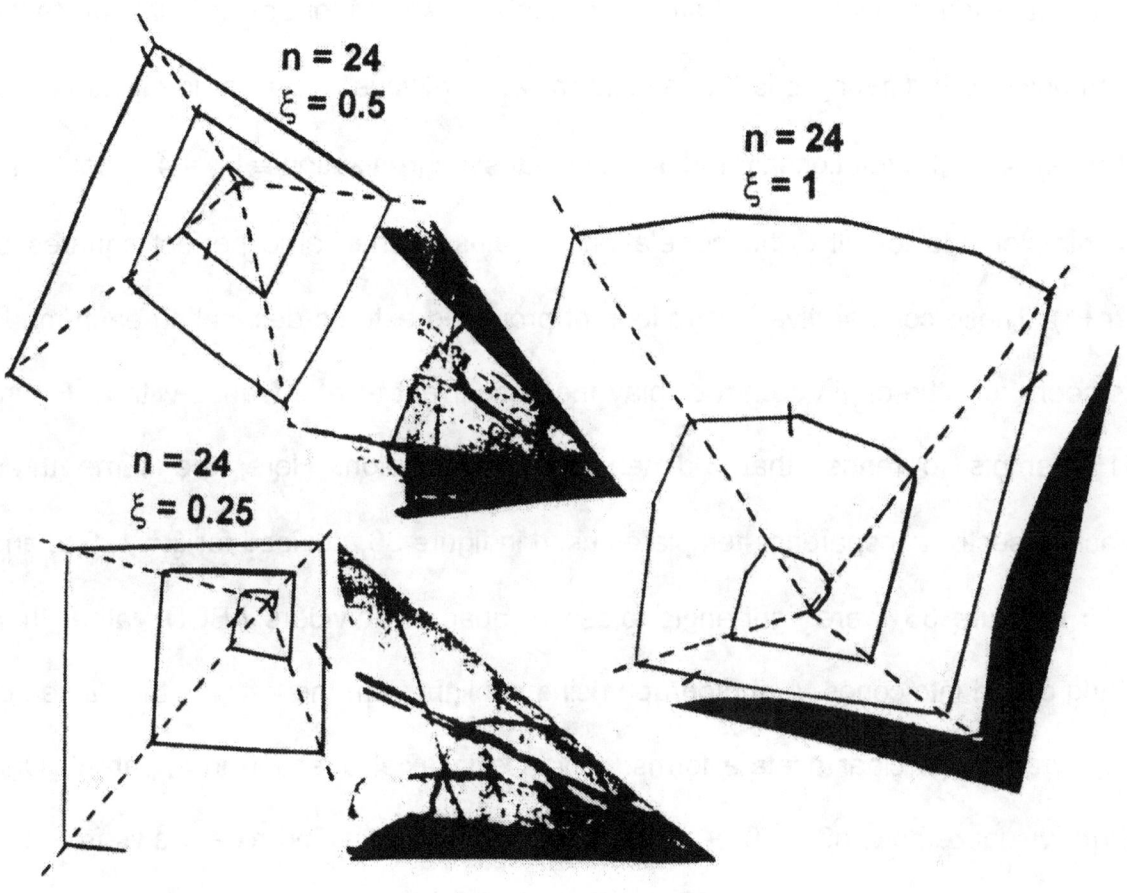

Figure 51 pictorially illustrates the formation of "square type modes" identified in figures 49 and 50. As described in figure 35, clear copies of all three stacked-sector templates for ξ = 1, 0.5, and 0.25 were folded at their dotted-line quadrant-directions B, C, D. Dotted-line exterior sides A and E of the templates were then joined to form three tetrahedron-like cones, each having a "square" base. This top-view photograph then shows the transition-chord-lengths in the plane of each quadrant plus a shadow from sun illumination on the left. Sector chord-lengths were redrawn to clarify the figure but shadow areas were not cleaned up. The three "square cones" reiterate how apex angle and radians/quadrant diminish with reduced ξ though synchronization remains ostensibly unchanged. As quantitatively portrayed in figure 50, the cones become more pointed at smaller ξ. Axial-direction advance with increased n is also more obvious here. Relative rate of axial advance monotonically diminishes with cycles P traversed.

two but such rendering is barely noticeable. An important feature of these harmonics with differing ξ is that synchronization always occurs at identical values of n for all ξ. The relationship radians/cycle-at-synchronization $\equiv \check{R}_s = 4\sqrt{\xi}$ could not be simpler and for all ξ the correlation sustains at the same perfect squares of (2n+1). These comparative illustrations of projected extents delineating each mode as seen from the origin do not display the concurrent axial advance with n. Figure 51 attempts to render that 3-dimensional progression. Here, the same three stacked-sector transparency-templates used in figure 36, devised for ξ = 1, 0.5, and 0.25 in figure 35, were right-angle-folded at quadrant dividers ABCD, rather than being curled into cones. A photograph using sunlight from the left was then taken of the pyramid-like clear acetate forms. Advance in axial direction is apparent. One might visualize how, after 10 or 12 complete cycles, projected axial views from the origin of these "pyramids" would appear as indicated in figure 50. Description of unified wave harmonic modes using four straight-sided planes simplifies analysis, revelation, and absolute standardization of angular plus solid-angle dimensional units.

Figures 44 and 45 demonstrate how 4-quadrant symmetry hardly changes over the range of ξ. Polarization angle shifts slightly but quadrant symmetry and synchronization sustain at respective n values. As n increases, superimposed graphs of figure 42 appear more-and-more like symmetrical four-leaf-clovers. Four quadrants repeatedly traversed in these figures are synonymous with the four consecutive quadrants every wave must progress through in its advance.

"Injection" of one additional node/quadrant with each successive cycle, exemplified in figure 3, engenders sinusoidal-type harmonics having four symmetrical regions. <u>In a simple harmonic oscillator, quadrature interchange in energy-form occurs four times during every cycle</u>. That is what the four quadrants of every pure wave signify. Such energy interchange associates with every quarter cycle of electromagnetic waves. They manifest as every rung of the DNA ladder. A weight on a spring exchanges kinetic and potential energy at the quadrant points. More than merely being a coherent part of every wave, their interpretation here in terms of natural numbers imparts new analytic understanding for all waves. Each synchronous wave defines as a three-dimensional entity expressible on a conic surface (or as in figure 49) rather than being comprised from a pair of two-dimensional sinewave **components** in quadrature phase separated using Cartesian Coordinates. <u>**Every discrete "frequency" has its origin at initiation of the causal impetus, not at minus infinity in time and space.**</u> Every consecutive cycle of every wave has numerical value n encrypted arithmetically into the sequence of that wave's progression. The numerous unified wave features mentioned herein should be attendant to all pure periodic waves that can be, or ever have been generated. For elementary spin-unity particles, such waves might represent a disturbance in the sea of sub-atomic quanta, the virtual particles of a vacuum. Elementary waves might be those resonances based on monads, though we associate such fluctuations with more familiar macroscopic processes.

Parameter $s_a = (1-\xi)$ indicates the axial inter-integer direction-component of transition s_n while ξ indicates the extent bypassed orthogonally. From the totality of all possible directions each transition from n to (n+1) can take place in, they both equivalently restrict that range to a particular compass of potentiality. Of all feasible directional possibilities, when or if vectors ABCD can align on a conic surface of apex angle $\beta_s = \sin^{-1}[(2\sqrt{\xi})/\pi]$, $\qquad\qquad$ (21*10)

an harmonic mode solution rather than a random mode can portray the spacetime aftermath of that origin event. Thus ξ articulates a mathematical parameter that influences when, how, and in what form standing wave solutions might emerge. Those possible synchronous-mode solutions exemplify on a conic surface whose apex angle and quadrature extents characterize via $\sqrt{\xi}$. Each conic-unit-circle quadrant-length-$\sqrt{\xi}$ formed from summed chord-lengths has a graphical equivalent as in figure 49 and as the $\sqrt{\xi}$ leg of a circle-inscribed triangle in figure 12. Parameter ξ of each synchronous mode as portrayed in figures 37 and 49 also delineate in figure 12.

Figure 46 shows the construction for $\xi > 1$, which region is "analytically continuous" through the $\xi = 1$ condition. Significance of that extended domain remains to be determined. Figures 33 to 46 reveal that **angle comprises an extremely fundamental property of the sequence of numbers.** In the 6000 consecutively different angles accrued for n = 6000 in figure 41, the wave traverses 217 quadrants while always reaching the appropriate point of phase for each one.

That process continues till n→∞. Additive and subtractive algebraic traits of the number system pale compared to the feat of maintaining that angular cadence over the full range of integers. Illustration emerges even more dramatically in figure 47 where n is increased to 20,000. That incredible attainment in the natural synchronization of summed θ_n angles for all $1 \geq \xi \geq 0$, constitutes an almost impossible truth. Such potential for synchronicity, combined with the facility for each transition to progress in a multiplicity of "directions", (rather than as illustrated on planes herein) allows harmonic modes of unbelievable diversity. However improbable that unique rhythm seems, it remains a truism and representative of the important role orthogonality plays in inter-integer transitions. In retrospect though, it is precisely what the natural numbers tell us, that only integers must be allowed for non-divisible things. In counting, the space between them need encompass a different dominion than the whole numbers themselves, a different zone than the allowed concentric-spherical-surfaces of integer-radius that surround zero. Angles must play a vital role in the enumeration sequence we have mistakenly imagined to take place along a one-dimensional Euclidean line. The compass of numbers is considerably richer than that one-dimensional likeness. Kronecker said, "God gave us the natural numbers and all else is man made". Were that so, the basic gift was far more substantial than comprehended, embracing a vast territory yet to be explored.

In figures 41 and 42 the pico-degree exactness of phase accumulation in summed θ_n angles with increased n (wherein each consecutive cycle always attains

295

precisely repeated cadence for the wave), extends to fempto-degrees, atto-degrees, etc. for ever large n. The exactness progresses to absolute accuracy and an absolute radian unit toward $n \rightarrow \infty$. That incredible angular precision is so easily disturbed that synchronization might be evaded through any angular perturbation attempting to determine the state of the system. But somewhere that repeatedly overlapping angular precision should be expected to run out. It should happen when accumulated Planck angles (one per cycle of spiral) have accrued to one full cycle at $n_{max} = 2/\Delta\theta^2 \approx 2P^2$, or at $P = 1/\Delta\theta$ cycles. Mathematically, (but not physically as n $\rightarrow\infty$) that absolute accuracy for repeated four-radian-cycles embedded within the natural numbers grants radians and cycles the status of **absolute rotational dimensional-units**, rather than their being dimensionless. It is that cyclic absoluteness in $h = E/\nu$ = energy-time/cycle = a Planck-angle fraction of a cycle/cycle = dimensionless = $\Delta\theta$ that bestows an unqualified property onto Planck's constant. That unconditional property thereby demands absolute dimensional units for energy-time, with such supreme-unit numerically equaling the actual excess-radian-angle/cycle of spacetime. What appears truly amazing is that such absolute accuracy endures throughout the gamut of all $0 \leq \xi \leq 1$. **The 4 symmetrical harmonic quadrants remain inseparable from enumeration virtually irrespective of the explicit process by which inter-integer transitions occur.** Traditionally-overlooked orthogonality in transitions is so fundamental that regardless of where, or when, or how it manifests in the counting accumulation progression, the four symmetrical quadrants sustain mathematically from n = 0 to n = ∞.

Figures 33 to 45 infer how synchronicity of the wave derives from consecutive sums of origin-angles repeatedly accruing to the same value though **no two of those angles are identical**. **Since complete cycles occur for every** $s_n{}^2 =$ **(odd integer), periodicity of the produced waves stem from the perfect algebraic repetition intervening all odd integers within the number sequence**! One could not have a more perfect basis for pure periodicity, which incidentally becomes **violated only at the final limit point of n = ∞, a point that is neither odd nor even**. Moreover, while for any fixed ξ each origin–angle differs, synchronicity from their sum sustains through all n and all ξ. Each periodic accumulation constitutes a quadrant of the wave that can yield exact repetition in radial direction every four quadrants. For an idealized mathematical world of Euclidean space without curvature, this process of recurring periodicity would continue for an infinite number of differing triangle angles. That may be difficult to believe but the source of skepticism would be unfamiliarity with a basic property of natural numbers. That those cyclic summations sustain for say a trillion trillion different triangles is truly amazing, but it almost defies belief that mathematically the perfect cycles go on indefinitely. Such circumstance insinuates a precarious condition however. As the number of triangles being summed gets ever larger, synchronization would be lost if origin-angles differed from what is exactly required of them to virtually-infinite-degree. The number of digits necessary to specify each different origin angle would tend to approach infinity. It is not possible for physical spacetime to track what mathematical ideals dictate to endless degree. There exists no compulsion for the physics to follow the mathematics to infinity. If space entailed

any warpage whatsoever, a deviation from mathematical perfection in accrued angles would result at some finite value of ever increasing n. Any non-compliance in origin-angles required for cadence, however tiny, would eventually jeopardize synchronization at sufficiently large n. For physical reality, perfection simply runs out after summing n_{max} triangles, which value confirms the prevailing excess Planck angle as $\Delta\theta = \sqrt{(2/n_{max})}$. Empirically, cadence ceases after n_{max} summed angles thanks to Planck's constant.

Use of the exponential function to model harmonic waves unknowingly infers that summed origin-angles as outlined herein track mathematical requirements to infinite degree. In so doing, such functions misrepresent real world harmonics. One indication of that inappropriate infinitude manifests in **the exponential's violation of causality since each discrete "solution frequency" evokes a sinusoidal response that begins at -∞ in time for a causal event that occurred after the response**. This violation of causality inherent within the expression provides a good example of a legitimate mathematical procedure that is outside the realm of objective physical reality. However commonplace, <u>**every application of the exponential harmonic function injects causality-violation into the analysis from the onset**</u>. The system's model is made non-physical by inadvertently invoking the requisite that response to an event has ability to precede the event. By contrast, even as n→∞, unified wave resonances begin at the n = 0 disturbance-origin and thus preserve causality.

For modes where ξ differs slightly from zero, "amplitude" of the wave characterized in terms of the four-quadrant perimeter (or <u>radius</u>) of the unit-cone's circular-base increases proportional to $\sqrt{\xi}$. In this region close to Euclidean axis conditions, amplitude (measured as radius of the cone's circular base) would increase with increased ξ in correspondence with $\sqrt{\xi}$. However, the unit-length radial extent of the unit-sphere or unit-cone's exterior side does not change with ξ, so based on that unit-distance from the origin amplitude would always be unity regardless of any other variables. Definable amplitude is thus moot. The picture entails one "extra dimension" than the conventional unit circle, which here forms with a unit-cone or unit-sphere as its base. Neglecting the additional dimension can add ambiguity when components are interpreted in context of a two (rather than three) dimensional figure; i.e., interpreted as the circle, at the base of the unit cone. Distinction between harmonics formed as paired X and Y sinusoidal-**components** from traversing a two dimensional circular figure (as with the exponential), and use of a three dimensional figure for the unified wave, is not trivial. The extra dimensional degree of freedom arises by retaining n, [which vanishes from the result in circling the conventional unit-circle, or within the definition $i\omega t \equiv \lim n\to\infty$ of $(1+i\omega t/n)^n$]. On the circle within the unit-sphere's surface, n embeds in all chord-lengths as $s_n = \sqrt{(2n\xi+1)}$. Analysis of a wave (that knowingly must physically be three-dimensional) but decomposed into a pair of two-dimensional components (e.g. as intertwined electric and magnetic sinusoids) a priori presumes Cartesian coordinate rectilinearly for spacetime. That lower-order inquiry does not encompass spacetime curvature, which however small can have a profound effect for high frequency resonances. The unified wave includes three dimensional portrayal without need for $i = \sqrt{-1}$ to distinguish man-made Cartesian components in deciphering it. Neither n nor ξ emerge in examination under Cartesian Coordinates and without n monadic properties cannot be accommodated in the analysis.

It has been suggested that integer natural numbers designate concentric spherical surfaces of radius n about an origin. The numbers themselves provide no information, constraint, or clue about how transitions can occur between those concentric surfaces. For indivisible monads possessing an information shroud, transit through the "unoccupyable" annular gap between spherical surfaces should be orthogonal to the existing origin-to-n direction. Fractional values along the axis cannot be encountered since further division of each tallied population member is impossible. Each unit-advance then entails maximum independence from, (or is linearly independent of) prevailing origin-to-n direction. Traversal through the annular transition region between spherical surfaces at radius n and (n+1) from the origin then possesses zero directional-component concordant with n. That expresses as orthogonality to the direction of n.

But what if the entity being enumerated were "partially divisible"? Postulate a population being counted with members having some discernable attribute or having an incomplete shroud. At an extreme of that possibility, suppose the relevant enumerated entities depicted large divisible macroscopic objects. In such cases one would expect the transition to occur entirely along the conventional Euclidean axis with fractional values feasible. In effect, a **range** of treatable "**divisibility's**" exists extending from **totally non-divisible monads** to **completely divisible large rest mass objects**. Harmonic modes for the extreme monadic case would entail maximally orthogonal transition directions and portray discrete indivisible particles like photons. Large rest mass objects would be highly divisible and such divisions could be signified by fractional values between integers. It is thus not unreasonable that a component of transition along the origin-to-n direction have some association with divisibility while the transition component orthogonal thereto affiliated with indivisibility. Of course the designation "divisibility" is applied somewhat metaphorically here to describe an unfamiliar attribute for a new parameter. The "degree of divisibility" quality could also be presented in terms of degree-of-directional-correlation between transition and n, or the penetrability of the shroud, etc. What is implied is, a **degree-of-orthogonality** can apply for transitions; a gradation-type property providing commensurate reduction in the effect of the shroud; i.e., "monadicity" as parameter ξ. Orthogonal transition components maximize when the shroud is impenetrable and minimize for cases of Newtonian continuity. The former yields resonances typical of unity spin, discrete, non-divisible, photon-like particles, whereas harmonics become more obscure for the latter case.

That the $\xi = 1$ case depicts photons might be questioned, but that waves with similar attributes emerge from the mathematics is undeniable. Symmetry and richness of the result is not an oddity of the situation, but a lack of prior awareness exposed by discovery. Derivation strictly from the sequence of integers n adds credence to the thesis. The upheld prevailing-picture about enumeration has simply been incomplete. Causality-violating sinusoids derived from exponential harmonics are plainly not the logical or mathematical origin for waves. Harmonics procure in the form of unified waves, from tallying discrete objects however small such that their summation appears continuous to us. **That peaks and valleys of all harmonics never occur at a prime number in the ordinal sequence is an innate connecting link between natural numbers and resonant mechanisms. In these waves zero-crossings always occur at <u>integer</u> values of s_n.** That enacts a basic inter-relationship between waves formed from non-zero entities. When new secrets of Nature become exposed in the course of science they might plausibly provide foundations for previously unexplained phenomena. Such oscillatory modes embedded within consecutive integers are conjectured to constitute viable descriptions for all quantized events. Allegedly the represented possibilities issuing as resonances distinguish from the array of particles and force carriers arrived at through other means. The orthogonal transition component that straddles integers, though unrecognized, plays an important role in enumeration and wave generation. In retrospect, necessity for that orthogonal component is totally logical. Its previous omission has been an oversight. These pages attempt to provide foundation reason and justification for its existence, but many concepts are new. The background glossary of prevailing words and ideas are often insufficient, or concepts are marginally understood. Proof of the pudding rests in the mathematical results however, and they display the waves very explicitly.

The question arises as to whether following an event, ξ remains an unchanging parameter (i.e., a single fixed value characteristic of the system), or unspecified with ability to take on any value at any n. The more general answer would allow it to can take on any value just as orthogonal transitions from n could be in any direction. Then, ξ merely grants another degree of freedom to the transition direction. As s_a (= 1-ξ) it would specify the extent of each axial-component transition-direction. Any allowed $s_a > 0$ would imply transit from n could occur anywhere within the **distal** hemisphere of the plane tangent to the spherical surface at n. It could progress outward from n at any angle up to 90° relative to vector n. Each such angle, called γ, would be the consequence of a different s_a at any n since $\gamma = \cos^{-1}(s_a/s_n)$. If s_a were permitted to go "negative" (or ξ >1), the transition between n and (n+1) could conceivably occur in any possible 360° spherical direction from vector n. All radial-direction vectors from n to (n+1) would be feasible and that would encompass the most generic description for how advance between integers might transpire. **Natural numbers stipulate no constraint concerning inter-integer transitions so permitting any value of $\xi \geq 0$ would accommodate that most general case, seemingly without violating restrictions.** That logic abides generically for monadic enumeration, which infers the transition-region between "spherical surface shells" of radius n and (n+1) remains indeterminate by virtue of the monad's shroud. How the transition passes through each one-unit of "indeterminate fractional-integer-territory" would ostensibly be unknown. Under that rational all angular directions emanating from wherever n locates on a spherical surface would be viable.

On the other hand, if ξ remained a fixed value throughout all n **it depicts the net fraction of Euclidean axis bypassed by orthogonal transition**. Transitions with ξ > 1 would then enter inter-integer regions smaller than n in getting to (n+1) and thereby might be disallowed. That provides justification to only consider the range where 0 ≤ ξ ≤ 1. Just as of all the transition-directions possible, harmonic solutions only result when directions ABCD align on the conic surface of apex angle β = $\sin^{-1}(2/\pi)$, with arbitrary ξ included, harmonic solutions only occur as chords on the conic surface of a lesser apex angle established by √ξ. From the totality of possible synchronous mode cones or harmonic-depicting surfaces, the cones delineating resonance become more "pointed", when encompassing the influence of √ξ. A simple elegance thus results when ξ remains a fixed retained value throughout all n. Any transition angle γ (from zero upward representing the n direction), to 90° (representing a tangent plane to the sphere at n), is permitted for 0 ≤ ξ ≤ 1. That freedom-of-possibility would suggest any or all of the harmonic solutions can be simultaneous solutions. Which or what combination might emerge will be differentiated by other factors imposed on the system.

The issue arises as to whether ξ is rational or can be irrational. **Were it rational, than all irrational values that could possibly have relevance along the Euclidean axis (radial direction) would be bypassed by tangential traverse**. The graphics illustrate a condition where the axis-segment ξ between every pair of integers effectively becomes "avoided" by orthogonal progression of s_o. Each axial region between $(n+s_a)$ and $(n+1)$ is evaded so the only "encounterable" irrational value in the radial direction would be ξ, (or s_a). Now ξ's mathematical role is exclusively **a multiplier of all even numbers** in $s_n = \sqrt{(2n\xi+1)}$. **Its algebraic sum with s_a is always unity** suggesting the point of partition it defines is a ratio or rational number. Moreover, **since $\xi - s_a = \xi^2 - s_a^2$**   (21*11)

the difference between their squares would have to be irrational for them to be irrational. Their irrationality could not stem from being an irrational square root of a rational number. [Possibly, ξ quantizes into smallest increments as result of γ_n in figure 12 being restricted to discernable angular increments of $\Delta\theta$.] While these arguments do not provide definitive proof, ξ will be treated as rational based on further justification.

In figure 49, each quadrant's n-vectors would intersect a unit-sphere-surface centered at the origin along an "equatorial-type-curve". The four closed quadrants would characterize a "square chunk" of that unit sphere. Since radius of the unit-sphere is always unity and cumulative chord-lengths of each quadrant sum to $n\sqrt{\xi}$, when divided by n they sum to $\sqrt{\xi}$. Total radian angle $2\beta'$ between opposite-side planes of the figure would be $\sqrt{\xi}$. This method of mode characterization in terms of $\sqrt{\xi}$ is very fundamental, simple, and useful. Chord-lengths always follow the curved spherical surface along an equatorial-type path. Moreover, for $n \to \infty$ and $\xi = 1$, **<u>each quadrant angle from the origin is exactly one absolute radian defined</u>**

exclusively through natural numbers. This endows a wave format based on quadrants that equal a standard absolute radian of rotation not dependent upon π. The solid angle formed by the figure is then **exactly one steridian**, which can equally serve as an **absolute solid-angle standard based on natural numbers**. Since the intersection shape on the unit-sphere is a "convex square" of sides $\sqrt{\xi}$, its surface area proportions to ξ, thereby furnishing a rational-number fractional-steridian solid angle. The resultant mathematical description could not emerge in plainer terms. Another feature of this configuration is that energy-forms participating in consecutive quadrants of the wave (like electric and magnetic), appear orthogonal commensurate with actual harmonic interchanges in energy form per quadrant.

 A most significant emergent parameter specifying mode characteristics is ξ, requiring neither π, e, $\sqrt{-1}$, ∞, a Planck length or a Planck time. The latter two can be defined on the unit sphere exclusively as chord-lengths at n_{max}, at the limit Planck angle $\Delta\theta$. Moreover with $\xi = 1$, each side of the "square" with summed chord-lengths of $\sqrt{\xi}$ represents exactly one radian. The solid angle of the "square" of spherical surface area $(\sqrt{\xi})^2 = \xi$, at $\xi = 1$ comprises exactly one steridian toward large n and ξ steridians for arbitrary ξ. Parameter ξ defined in figure 12 as the point of orthogonal transition **turns out to also equal the "steridian spreading" of the wave.**

A	B	C	D	E	F	G	H	I	J	K	L
n	quadrant	cycle	Sn (2n+1)^.5	angle theta	sum thetas	quadrant data-points (n+1/4)^.5	added angle @ (n+1/2)	full sum of thetas	sum theta * Pi/2	n*COS J	n*SIN J
0			1	0	0	0.5		0	0	0	0
1		1	1.73205	1.570796	1.570796	1.118034		1.570796	2.467401	-0.78121	0.624266
	1.5						0.8411	2.411865	3.788549	-1.19688	-0.90414
2			2.23607	1.047198	2.617994	1.5		2.617994	4.112335	-1.12937	-1.65061
	2.5						0.6435	3.261495	5.123144	0.998255	-2.29205
3			2.64575	0.841069	3.459063	1.802776		3.459063	5.433483	1.98062	-2.25325
4		4	3	0.722734	4.181797	2.061553		4.181797	6.568751	3.83801	1.126801
5			3.31662	0.643501	4.825298	2.291288		4.825298	7.57956	1.35495	4.812911
	5.5						0.4297	5.254998	8.254531	-2.14458	5.064658
6			3.60555	0.585686	5.410983	2.5		5.440088	0.469558	-3.00393	-7.792557
7			3.87298	0.5411	5.952083	2.692582		5.952083	9.34951	-6.98018	0.526378
	7.5						0.3672	6.319291	9.926319	-6.57632	-3.60583
8			4.12311	0.505361	6.457443	2.872281		6.457443	10.14333	-6.02209	-5.26635
9			4.3589	0.475882	6.933326	3.041381		6.933326	10.89084	-0.94086	-8.95069
	9.5						0.3259	7.259209	11.40274	3.762065	-8.72335
10			4.58258	0.451027	7.384353	3.201562		7.384353	11.59931	5.677249	-8.23218
11			4.79583	0.4297	7.814052	3.354102		7.814052	12.27428	10.5341	-3.16746
12		12	5	0.411138	8.22519	3.5		8.22519	12.9201	11.25706	4.156766
13			5.19615	0.394791	8.619981	3.640055		8.619981	13.54023	7.307402	10.75183
14			5.38516	0.380251	9.000232	3.774917		9.000232	14.13753	-0.00511	14
	14.5						0.2634	9.263606	14.55124	-5.83393	13.27461
15			5.56776	0.367208	9.36744	3.905125		9.36744	14.71434	-8.18487	12.57012
16			5.74456	0.355421	9.722862	4.031129		9.722862	15.27264	-14.5077	6.747324
17			5.91608	0.344701	10.06756	4.153312		10.06756	15.81409	-16.9044	-1.80078
	17.5						0.2396	10.30718	16.19048	-15.502	-8.12022
18			6.08276	0.334896	10.40246	4.272002		10.40246	16.34014	-14.5213	-10.6363
19			6.245	0.325883	10.72834	4.387482		10.72834	16.85204	-7.86385	-17.2962
20			6.40312	0.31756	11.0459	4.5		11.0459	17.35086	1.440816	-19.948
	20.5						0.2213	11.26722	17.6985	8.354281	-18.7205
21			6.55744	0.309845	11.35575	4.609772		11.35575	17.83757	11.13366	-17.8057
22			6.7082	0.302665	11.65841	4.716991		11.65841	18.31299	18.90834	-11.2461
23			6.85565	0.295963	11.95438	4.821825		11.95438	18.77789	22.94096	-1.64694
24		24	7	0.289687	12.24406	4.924429		12.24406	19.23293	22.25781	8.977195
25			7.14143	0.283794	12.52786	5.024938		12.52786	19.67871	16.88748	18.43402
26			7.28011	0.278247	12.8061	5.123475		12.8061	20.11578	7.79702	24.80336
27			7.4162	0.273013	13.07912	5.220153		13.07912	20.54463	-3.34679	26.79177
	27.5						0.191	13.2701	20.84462	-11.3205	25.06184
28			7.54983	0.268063	13.34718	5.315073		13.34718	20.9657	-14.524	23.93852

Table 5. In this tabulation, column B inserts an additional half-integer data-point at the quadrant dividers. It results in a data-point removed from vector n an extent √[n+1/2]) along s_n, as registered in column G. The added angle of that data-point at quadrant-dividers is shown in column H making the total accrued angle at each quadrant divider as column I indicates. On a conic surface of apex angle β = arcsin(2/π), all the summed angles would appear increased by π/2 when viewed from the origin. Column J thus denotes the angle of all data-points at a distance n [or (n+1/2) respectively] from the origin. Columns K and L calculate the x and y components of those n and (n+1/2) data-points and these data are plotted as spirals in figures 53 and 54.

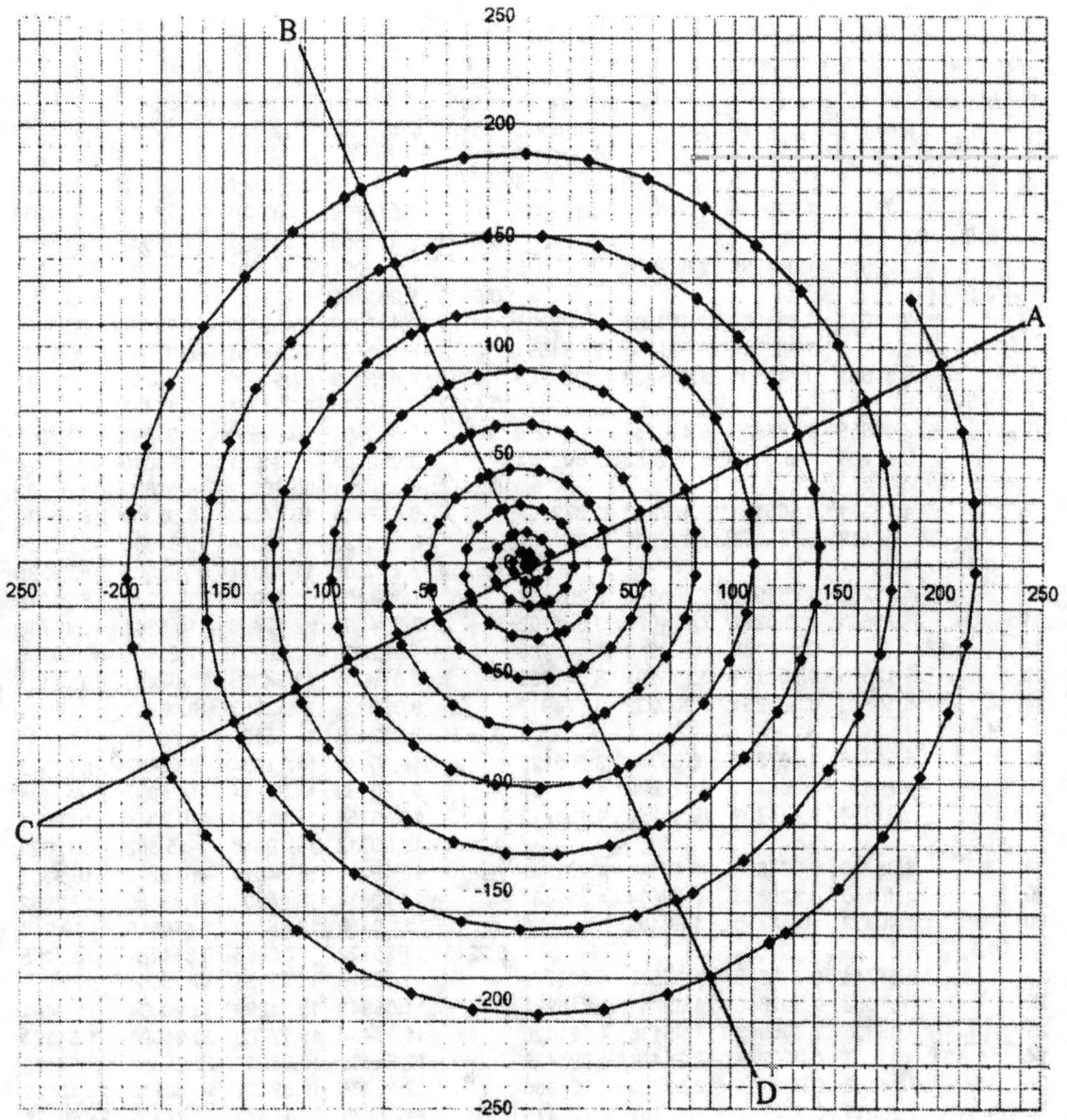

Figure 53 plots the data of Table 5 extended to ten cycles = P, where n = 221. The "telephoto view" along the conic axis clearly exhibits the inherent symmetry of the spiral. The included half integer data-points at the quadrant dividers are seen to be very close to exact rectilinear dividers. Each such data-point occurs a tangential extent $\sqrt{(n+1/4)}$ from its respective radial vector n along the associated tangential-going $s_n = \sqrt{(2n+1)}$. The rectilinear symmetry becomes perfect at large n but is slightly imperfect near n = 0. An oblique view along directions A to C indicates that minute deviation, which shows up magnified in figure 54. Always referencing the half-integer n-values in connection with quadrant dividers becomes cumbersome in discussions, so mention of their germaneness is generally avoided. However, the inherent harmonic symmetry within natural numbers entails midpoints between integers in addition to the integers.

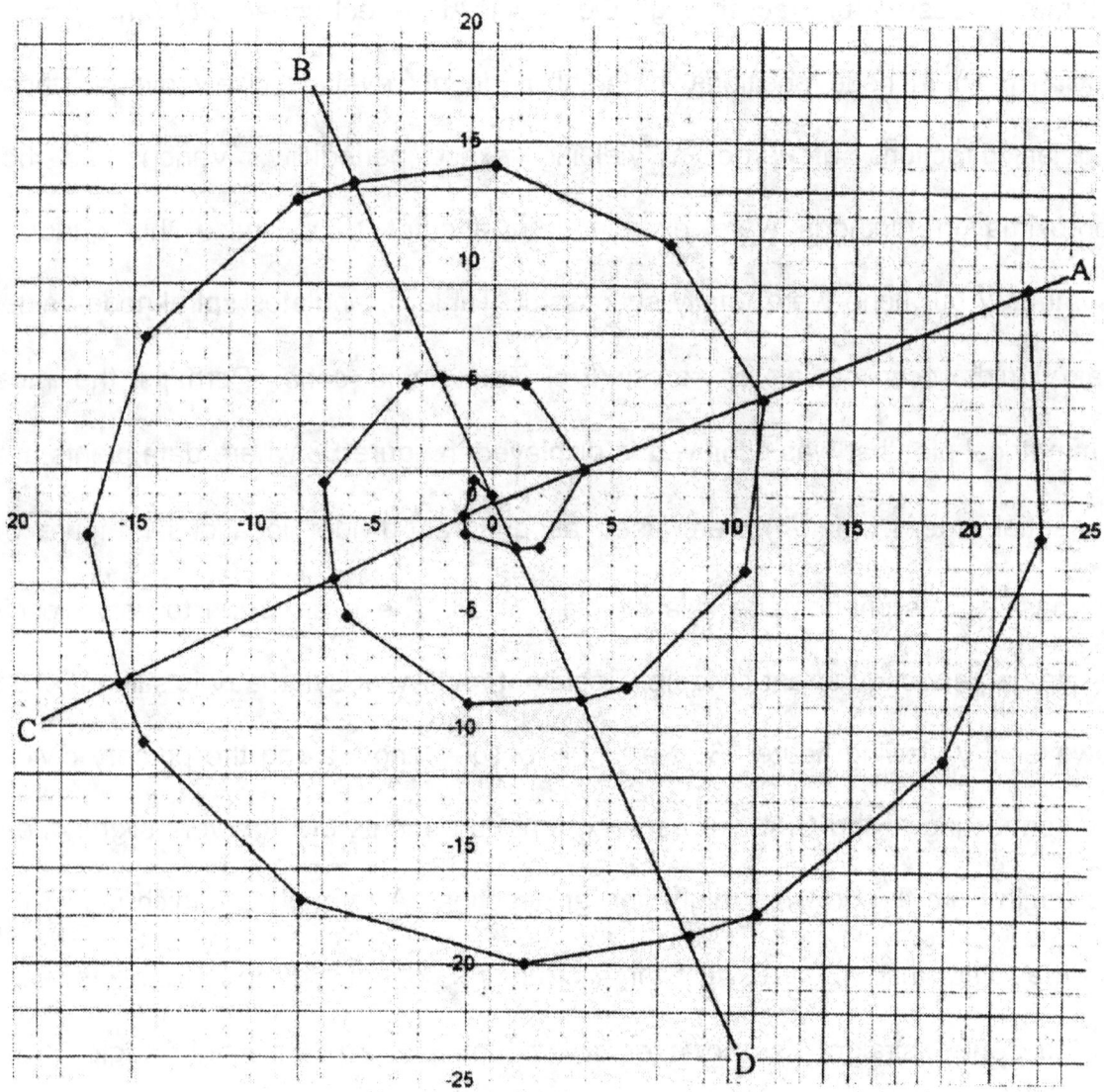

Figure 54 plots the initial three cycles of the Table 5 data depicting a conic spiral. It can be seen that data-points do not converge in a straight line to the origin, but slightly above. Directions A and C are not precisely 180° apart at the origin, although they asymptote toward the same straight-line at large n. Significance of this imperfect rectilinearly is unknown.

As presented in figures 38 through 45, four-quadrant rectilinear symmetry appears slightly "inexact" in the spirals, particularly at small n. Perfect 4-quadrant symmetry does not protract through the origin. At perfect squares of $(2n+1) = s_n^2$, ABCD direction node positions for small n do however precisely overlap those angular directions at large n yielding exact periodicity. Various ad hoc approximation methods were examined in attempt to minimize this quadrant asymmetry but none were totally successful. Table 5 tabulates spiral-node-values that would occur on a $\beta = \arcsin(2/\pi)$ apex-angle cone. Perhaps the most symmetrical plot that was achieved is displayed in figure 53 where data points at a tangential extent $\sqrt{(n+1/4)}$ were used as quadrant-divider locations. Figure 54 emphasizes this phenomena in a plot to n = 26. These are seen to very closely designate perfect quadrant dividers. Resultant rectilinear symmetry is still not exact however. By viewing figure 53 at a highly oblique angle along the primary divider line connecting A and C, it can be seen that data points diverge very slightly near the origin from the drawn straight line at directions A through C, which insect the origin at other than 180°. Toward large n the line connecting the data-points tends toward perfect straightness however, and become at a precise right angle to the straight line connecting directions B and D.

Consequences of this seeming deviation from perfect symmetry at small n are derived from straight chords. It is feasible that some simple unknown adaptation would enable increased or exact symmetry. Perfect rectilinear symmetry near small n may be incompatible with straight chord models, however familiar we have become with that rectilinear concept within the paradigm of Euclidean space. This shortcoming might be physically tested by examining symmetry of initial quadrants for elementary oscillatory processes. Also, as outlined in Appendix B, the condition establishing that between $0 \leq n \leq \infty$ no two straight chord lengths forming the spiral (or chord-lengths defined by its intersection with the unit sphere) are identical, hints of discordance between rectilinear Euclidean space and the space curvature educed through natural numbers.

22) ROOT MEAN SQUARE SUMMATION OF INTER-INTEGER TRANSITIONS

Science is the process of finding order in the universe. Whatever order prevails in natural numbers would likely infiltrate a multitude of physical and biological specialties. It is therefore worthwhile to search for patterns in the numbers, particularly where such structure plausibly takes part in fundamental processes like waves and quantized procedures. Besides the credible influence those patterns might have on elementary particles, they should be useful in a host of other disciplines. The following extends the systemization of integers in connection with their cyclic quadrants. Every integer n is defined as having a **Veil** [designated $V(n,\xi)$] that is the RMS sum of all transition magnitudes from zero upwards through every lesser positive integer to reach that integer n under a fixed ξ. For any specific ξ that would be the RMS sum of the relevant spiral's chord-lengths to reach arbitrary n. No account is taken of the directions or projections of those chord-length magnitudes in taking the RMS sum. Because the last transition leading to n is s_{n-1}, we can write

$$V(n+1,\xi) \equiv \sqrt{\left[\sum_{\check{n}=0}^{n} s_{\check{n}}^2\right]} = \sqrt{\left[\sum_{\check{n}=0}^{n} \{\sqrt{(2\check{n}\xi+1)}\}^2\right]} = \sqrt{\left[\sum_{\check{n}=0}^{n}(2\check{n}\xi+1)\right]} \qquad (22*1)$$

Here, $\check{n}$ takes on the values 0, 1, 2, 3, 4, 5, - - - to (n), respectively yielding summation terms for $\xi = 1$ equaling 1, 1+3, 1+3+5, 1+3+5+7, etc. These become $\sqrt{1}$

313

for summation limit n = 0, √4 for n = 1, √9 for n = 2, √16 for n = 3, etc. resulting in Veil V(n+1,1) = 1, 2, 3, 4, etc., this sequence equating to the natural numbers (n+1). With the understanding that the largest transition to reach n is always s_{n-1}, this property can of course also be stated as: V(n,1) will always equal the natural number n. Veil V(n, ξ) will have a value equal to n (when ξ = 1), and a minimum equaling the RMS of n unit-lengths, or √n for ξ = 0. The definition can extend to greater values of ξ so for example when 1 ≤ ξ ≤ 2, Veil would range between n and the result of equation 22*1 with substituted ξ between 1 and 2.

Table 3 and 3a tabulate values compiled from sub-terms in equation 22*1 to express V(n,ξ) for ξ = 0, .05, 0.25, 0.5, 1 for specific n values up to 92. Table 3a is a continuation of columns to the right in Table 3. Alphabetical column headings are horizontally referenced just below, indicating how other column values mathematically insert. The vertical expressions delineate functional relations for values tabulated in that column. The right-hand eight columns of Table 3a consolidate results in Table 3 to its left with its five rightmost columns indicating Veil for values of ξ. Figure 52 plots those Veil V(n,ξ) values as ordinate, for abscissa values of n-1 up to 92, with ξ as a parameter. For clarity, curves are only presented for ξ = 0, .05, 0.25, 0.5, 1, but tendencies for intervening values are obvious. Now the prefect squares of (2n+1) being 1, 9, 25, 49, 81, 121, 169, - - -, where n = 0, 4, 12, 24, 40, 60, 84, —— respectively, articulate cycles within the natural numbers. As shown in Table 1, quadrants of those cycles evenly partition the intervals in n

314

between those perfect squares. The data points of figure 52 similarly partition cycles into quadrants. These graphs, mathematically expressed as a function of n, highlight the uniformity and the symbiotic-relationship between cycles and integers. That quality continues out to $n \rightarrow \infty$.

Veil provides a measure of how indivisible each ensemble member that accrues to a population of n integer things would be. The Veil of integer 10 for example communicates non-divisibility for each of the ten amassed items referenced. Normalized Veil $\equiv V(n,\xi)/n$ is a fraction describing the "monadicness" of each member of the population of objects signified by integer n. Various analogous meanings can also apply. The Veil of an integer is the RMS sum of all transition magnitudes from zero upwards to that integer. It also elucidates the gradation to which all-prior inter-integer transition-directions remained un-correlated in reaching that integer. It furthermore makes known the extent new population-members remain statistically-un-correlated to the existing population upon accrual. Figure 52 shows that for large integers where $2n\xi \gg 1$,   (22*2)

Veil graphically equals the product of any respective n and $\sqrt{\xi}$. "Coincidentally", the right-most vector inscribed in each unit hemi-circle of figure 12 is also of length $\sqrt{\xi}$. From the intersecting straight lines of figure 52, normalized Veil can then approximate as $V(n,\xi)/n \approx \sqrt{\xi}$ for large n.   (22*3)

$$\text{Veil of }(n+1,\xi) = V(n+1,\xi) = V(n,\xi) = \{\textstyle\sum_{\tilde n=0}^{n} s_{\tilde n}{}^2\}^{1/2} = \sqrt{[\textstyle\sum_{\tilde n=0}^{n}(2\tilde n\xi+1)]} = \sqrt{[\textstyle\sum_{\tilde n=0}^{n}(2\tilde n\xi+1)]} \quad\text{and}\quad \text{Veil}(n,\xi) = V(n,\xi) = \sqrt{(n+2\xi\Sigma)}$$

A	B	C	D	E	F	G	H	I	J	K	L	M	N	O	P
1	n-1 ($\tilde n=n-1$)	n	sum $\sum\tilde n$ from $\tilde n=1$ to $(n-1)$	2*D	E+C	Sq.Rt($\tilde F$) V(n,1)	D+C	root of H V(n,0.5) $\sqrt{(n^2+\xi)}$	$\delta\text{Veil}_{.5}=V_T-V_t$ Delta I	$\delta^2\text{Veil}_{.5}=\delta(\delta\text{Veil})$ Delta J	D/2 $\xi/2$	L+C $n^2+\xi/2$	root of M V(n,0.25)=$\sqrt{(n^2+\xi/2)}$	$\delta\text{Veil}_{.25}=V_T-V_t$ Delta N	$\delta^2\text{Veil}_{.15}=\delta(\delta\text{Veil})$ Delta O
1	1	2	1	2	4	2	3	1.732051			0.5	2.5	1.581139		
1	2	3	3	6	9	3	6	2.44949	0.717439	0.717438935	1.5	4.5	2.12132	0.540182	0.540181513
1	3	4	6	12	16	4	10	3.162278	0.712788	-0.004651018	3	7	2.645751	0.524431	-0.015750546
1	4	5	10	20	25	5	15	3.872983	0.710706	-0.002082231	5	10	3.162278	0.516528	-0.007904618
1	6	7	21	42	49	7	28	5.291503	1.418519	0.70781359	10.5	17.5	4.1833	1.021022	0.504496123
1	8	9	36	72	81	9	45	6.708204	1.416701	-0.001817966	18	27	5.196152	1.012852	-0.008170182
1	10	11	55	110	121	11	66	8.124038	1.415834	-0.000866338	27.5	38.5	6.204837	1.008684	-0.00416789
1	12	13	78	156	169	13	91	9.539392	1.415354	-0.000480863	39	52	7.211103	1.006266	-0.002418572
1	15	16	120	240	256	16	136	11.6619	2.122512	0.707158166	60	76	8.717798	1.506695	0.500429608
1	18	19	171	342	361	19	190	13.78405	2.122145	-0.000366813	85.5	104.5	10.22252	1.504726	-0.001969073
1	21	22	231	462	484	22	253	15.90597	2.121925	-0.000219994	115.5	137.5	11.72604	1.503515	-0.001211014
1	24	25	300	600	625	25	325	18.02776	2.121783	-0.000142312	150	175	13.22876	1.502717	-0.000798094
1	28	29	406	812	841	29	435	20.85665	2.828897	0.707114581	203	232	15.23155	2.00279	0.500072501
1	32	33	528	1056	1089	33	561	23.68544	2.828785	-0.000112287	264	297	17.23369	2.002142	-0.000647929
1	36	37	666	1332	1369	37	703	26.51415	2.828709	-7.63476E-05	333	370	19.23538	2.001696	-0.000445606
1	40	41	820	1640	1681	41	861	29.3428	2.828654	-5.42874E-05	410	451	21.23676	2.001377	-0.000319802
1	45	46	1035	2070	2116	46	1081	32.87856	3.535763	0.707108608	517.5	563.5	23.73815	2.501394	0.500017839
1	50	51	1275	2550	2601	51	1326	36.41428	3.535718	-4.44768E-05	637.5	688.5	26.23928	2.501129	-0.000265773
1	55	56	1540	3080	3136	56	1596	39.94997	3.535686	-3.26677E-05	770	826	28.74022	2.500932	-0.000196386
1	60	61	1830	3660	3721	61	1891	43.48563	3.535661	-2.46994E-05	915	976	31.241	2.500783	-0.000149222
1	66	67	2211	4422	4489	67	2278	47.7284	4.242768	0.707107345	1105.5	1172.5	34.24179	3.000789	0.500005656
1	72	73	2628	5256	5329	73	2701	51.97115	4.242748	-2.08592E-05	1314	1387	37.24245	3.000682	-0.00012707
1	78	79	3081	6162	6241	79	3160	56.21388	4.242731	-1.61361E-05	1540.5	1619.5	40.24301	3.000563	-9.86459E-05
1	84	85	3570	7140	7225	85	3655	60.4566	4.242719	-1.27389E-05	1785	1870	43.2435	3.000485	-7.81115E-05
1	91	92	4186	8372	8464	92	4278	65.40642	4.949826	0.707106992	2093	2185	46.74398	3.500487	0.500002144

Table 3 shows the calculated values used to form the chart of figure 51 and display the Veil of all numbers up to 92. The Table is two pages wide and continues horizontally as Table 3a. After the vertical break in Table 3a, important columns are repeated. The Veil of a number n is defined as the RMS sum of all transition magnitudes $s_n = \sqrt{(2n\xi+1)}$ for each consecutive integer from zero upwards to the number n. The symbol $\tilde n$ is used for the summation in the Table, it being understood that $\sum\tilde n$ accumulates from $\tilde n = 1$ to $\tilde n = (n-1)$, which limit is expressed as $(n-1)$ so that data points and vertical markers fall at the values 4, 12, 24, 40, 60, 84, ---etc., which markers represent perfect-square cycles. Data points denoting four quadrants per each cycle were selected as evenly spaced horizontally. At large in they emerge very close to evenly spaced vertically as the $\delta^2\text{Veil}$ columns signify. That is particularly true toward larger ξ. Clustering the data into groups of four rows also allow recognition of each quadrant and each cycle. One immediate observation of table 3 and figure 51 is that linearity of summed RMS transition magnitudes improves markedly with both ξ and n, being perfectly linear for $\xi = 1$. Natural numbers included by extending the calculations downward on the table to $n \to \infty$ and to the right in the chart, would tend toward perfect straight lines for $\xi > 0$. It demonstrates how orthogonal components of inter-integer transitions portray a fundamental property of numbers that can be used in all disciplines of science and engineering. Is of particular interest that with increased n each $\delta^2\text{Veil}$-column approaches $\sqrt{\xi}$ at each four-quadrant gap. It confirms how the sum chord-lengths for each quadrant of every wave sum to $\sqrt{\xi}$.

$$\text{Veil of }(n+1,\xi) = V(n+1,\xi) = \sqrt{\sum_{\check n=0}^{n} s_n^2} = \sqrt{\sum_{\check n=0}^{n}\{\sqrt{(2\pi\xi+1)}\}^2} = \sqrt{\sum_{\check n=0}^{n}(2\pi\xi+1)} \quad\text{and}\quad \text{Veil}(n,\xi) \equiv V(n,\xi) = \sqrt{(n+2\xi\Sigma)}$$

$$V(n,0) = \sqrt{n}$$

D/10 (Σ/10)	Q+C	V(n,0.05)=√(n+Σ/10) [root of R]	δVeil₀₅ [Delta S]	δ²Veil₀₅ [Delta T]	V(n,0)=√G [root of G]	n	Σ	n−1	V(n,1)=√(n+2·Σ)	V(n,0.5)=√(n+Σ)	V(n,0.25)=√(n+Σ/2)	V(n,0.05)=√(n+Σ/10)	V
0.1	2.1	1.449138			1.414214	2	1	1	2	1.73205	1.581139	1.449138	1.414214
0.3	3.3	1.81659	0.367453	0.367453	1.732051	3	3	2	3	2.44949	2.12132	1.81659	1.732051
0.6	4.6	2.144761	0.328171	−0.03928	2	4	6	3	4	3.16228	2.645751	2.144761	2
1	6	2.44949	0.304729	−0.02344	2.236068	5	10	4	5	3.87298	3.162278	2.44949	2.236068
2.1	9.1	3.016621	0.567131	0.262402	2.645751	7	21	6	7	5.2915	4.1833	3.016621	2.645751
3.6	12.6	3.549648	0.533027	−0.0341	3	9	36	8	9	6.7082	5.196152	3.549648	3
5.5	16.5	4.062019	0.512371	−0.02066	3.316625	11	55	10	11	8.12404	6.204837	4.062019	3.316625
7.8	20.8	4.560702	0.498682	−0.01369	3.605551	13	78	12	13	9.53939	7.211103	4.560702	3.605551
12	28	5.291503	0.730801	0.232118	4	16	120	15	16	11.6619	8.717798	5.291503	4
17.1	36.1	6.008328	0.716825	−0.01398	4.358899	19	171	18	19	13.784	10.22252	6.008328	4.358899
23.1	45.1	6.715653	0.707326	−0.0095	4.690416	22	231	21	22	15.906	11.72604	6.715653	4.690416
30	55	7.416198	0.700545	−0.00678	5	25	300	24	25	18.0278	13.22876	7.416198	5
40.6	69.6	8.342661	0.926463	0.225918	5.385165	29	406	28	29	20.8567	15.23155	8.342661	5.385165
52.8	85.8	9.262829	0.920167	−0.0063	5.744563	33	528	32	33	23.6854	17.23369	9.262829	5.744563
66.6	103.6	10.17841	0.91558	−0.00459	6.082763	37	666	36	37	26.5141	19.23538	10.17841	6.082763
82	123	11.09054	0.912128	−0.00345	6.403124	41	820	40	41	29.3428	21.23676	11.09054	6.403124
103.5	149.5	12.22702	1.136483	0.224355	6.78233	46	1035	45	46	32.8786	23.73815	12.22702	6.78233
127.5	178.5	13.36039	1.13337	−0.00311	7.141428	51	1275	50	51	36.4143	26.23928	13.36039	7.141428
154	210	14.49138	1.130988	−0.00238	7.483315	56	1540	55	56	39.95	28.74022	14.49138	7.483315
183	244	15.6205	1.129123	−0.00186	7.81025	61	1830	60	61	43.4856	31.241	15.6205	7.81025
221.1	288.1	16.97351	1.353009	0.223887	8.185353	67	2211	66	67	47.7284	34.24179	16.97351	8.185353
262.8	335.8	18.32485	1.351338	−0.00167	8.544004	73	2628	72	73	51.9711	37.24245	18.32485	8.544004
308.1	387.1	19.67486	1.350011	−0.00133	8.888194	79	3081	78	79	56.2139	40.24301	19.67486	8.888194
357	442	21.0238	1.348939	−0.00107	9.219544	85	3570	84	85	60.4566	43.2435	21.0238	9.219544
418.6	510.6	22.59646	1.572664	0.223725	9.591663	92	4186	91	92	65.4064	46.74398	22.59646	9.591663

Table 3a continues table 3 with repeat columns for the Veil of (n, ξ) ξ-grouped together on the right. That rightmost tabulated data is the basis for figure 51 for the values of ξ shown. Column AA being (n−1) is abscissa with the five plots for respective values of ξ to the right. For ξ = 0 transition magnitudes between the integers are all unity and V(n, 0) = √n. For ξ = 1, V(n, 1) = n so that curve has a slope of unity. The veil of each integer is then precisely one unit greater than its predecessor and δVeil, would be zero everywhere so it was not listed in the Table. As can be seen from the data and figure 51, all lines are quasi-linear with slope ≈ √ξ and pass through (1, 1) so that normalized Veil can be approximated as V(n, ξ)/n, or √ξ ≈ V(n, ξ)/n.

That means $\sqrt{\xi}$, which was shown to have other interpretations, is also $\approx$ (RMS sum of transition magnitudes to reach n)/n. Squaring both sides of 22*3 after multiplying by n; $\xi n^2 \approx 1 + s_1^2 + s_2^2 + s_3^2 + s_4^2 + - - - s_{n-1}^2 =$

$$1 + (2\xi + 1) + (4\xi + 1) + (6\xi + 1) + (8\xi + 1) + - - - (2[n-1]\xi + 1) \qquad (22*4)$$

The tabulation below allows evaluation of the low-n errors incurred in this equation under the approximation condition $2n\xi > 1$ of equation 22*2.

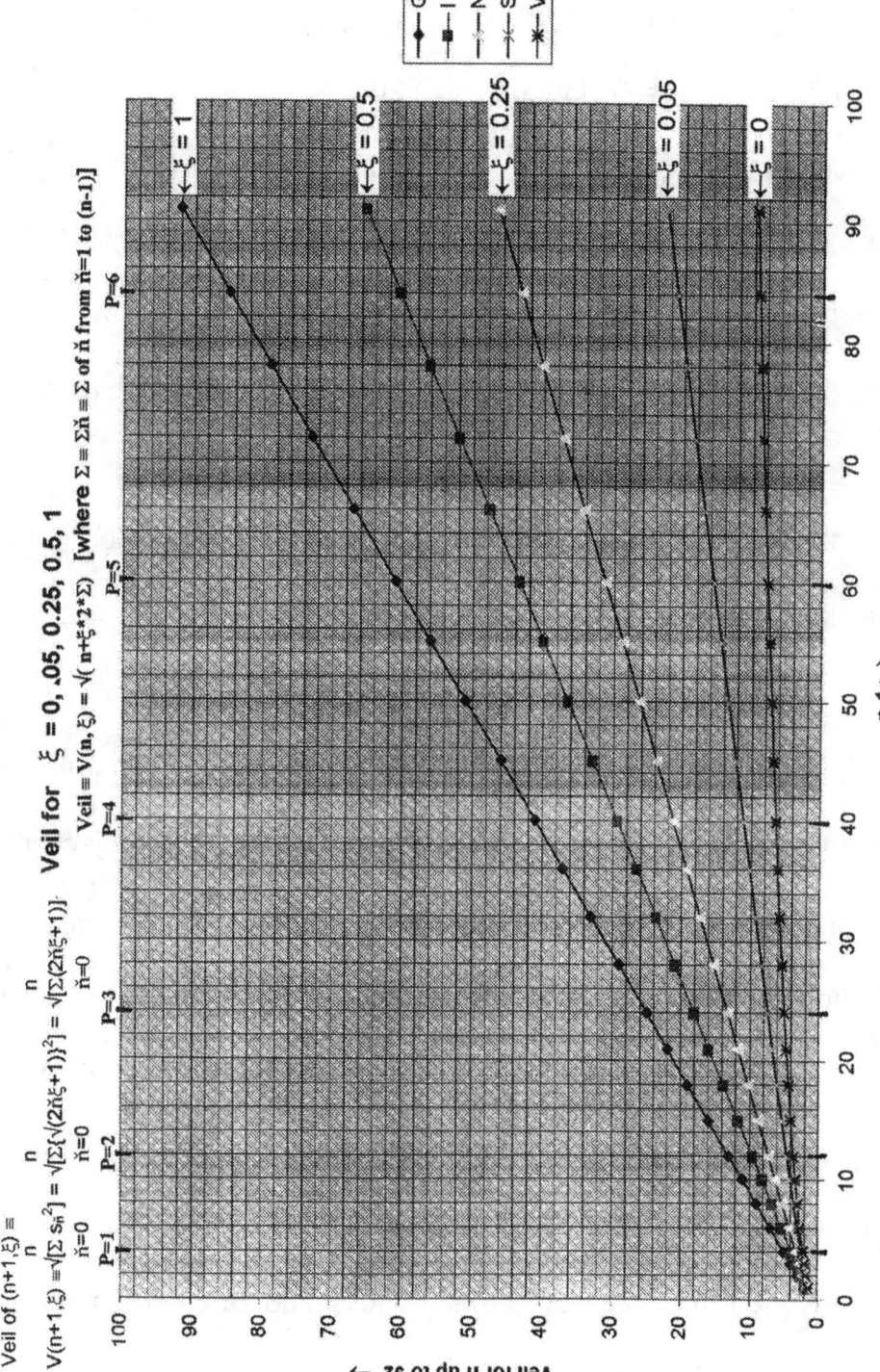

Veil of $(n+1, \xi) \equiv$

$$V(n+1,\xi) \equiv \sqrt{[\Sigma \sqrt{s_{\check{n}}^2}]} = \sqrt{[\Sigma \{\sqrt{(2\pi\check{n}\xi+1)}\}^2]} = \sqrt{[\Sigma(2\pi\check{n}\xi+1)]};$$

Veil for $\xi = 0, .05, 0.25, 0.5, 1$

$$\text{Veil} = V(n, \xi) = \sqrt{(n+\xi*2*\Sigma)} \quad [\text{where } \Sigma \equiv \Sigma\check{n} \equiv \Sigma \text{ of } \check{n} \text{ from } \check{n}=1 \text{ to } (n-1)]$$

Figure 52 plots the Veil $V(n, \xi)$ of any integer n up to n = 92 using data derived from tables 3 and 3a. Spacing uniformity is apparent. Because the charts are almost linear and converge to unity, the veil of all large integers will ostensibly be the same fixed fraction of the integer for a specific $\xi > 0$. The slope of each line is essentially $\sqrt{\xi}$ for large n, so that the Veil of any integer is well approximated as $n\sqrt{\xi} \approx V(n, \xi)$. Thus, besides being the chord-lengths per quadrant, and the radians at synchronization per quadrant, $\sqrt{\xi}$ approximately equals [(RMS sum of transitions to reach n)/n] $\approx \sqrt{\xi}$. For a parameter so closely affiliated with every integer, every quadrant and fractional radian of every wave, every orthogonal component of inter-integer transition, and every discrete chord or our granular world, $\sqrt{\xi}$ is a very neglected variable of science and engineering.

For n =	Evaluation of Equation 22*4	Comparative error term for $\xi=0$	Maximum relative % error	$\sqrt{\%}$ for RMS
5	$25\xi = 5 + 20\xi$	$5\xi = 5$ in 25	20%	44.7%
6	$36\xi = 6 + 30\xi$	$6\xi = 6$ in 36	16.7%	40.8%
7	$49\xi = 7 + 42\xi$	$7\xi = 7$ in 49	14.3%	37.8%
8	$64\xi = 8 + 56\xi$	$8\xi = 8$ in 64	12.5%	35.3%
100	$10{,}000\xi = 100 + 9900\xi$	$100\xi = 100$ in 10000	1%	10% (22*5)

Maximum error $\approx 1/\sqrt{n}$. Typical error $\approx \sqrt{(s_a/n)}$. For $\xi > 0$ the error will be $< 1/\sqrt{n}$. This error can be almost accommodated by $V(n,\xi)/n \approx \sqrt{(\xi + s_a/n)}$ $\qquad$ (22*6) That expression certainly provides exact results at extremes of ξ, namely $1/\sqrt{n}$ at $\xi = 0$ and $\sqrt{\xi}$ at $\xi = 1$. For large n, this relationship simulates radians/quadrant = $\sqrt{\xi}$ as seen from the origin, or $\check{R}_s/4 = \sqrt{\xi}$. In figures 37 and 49, the projection of tangential chord-lengths in each quadrant sum to $n\sqrt{\xi}$ and at large n so does the RMS sum of all chord-lengths up to n. In figure 49 with square cross section, the "apex angle" $\sqrt{\xi}$ in each quadrant of the square cone would then also be determinable by the slope in figure 52 and that slope approximates as $\sqrt{\xi}$, which equals $\check{R}_s/4$. The more accurate value for $V(n,\xi)/n$ at small n is $= \sqrt{(\xi + s_a/n)}$. **Along with the wave's absence of primes at peaks, valleys and zero-crossings and radians–at-sync $\equiv \check{R}_s = 4\sqrt{\xi}$, the RMS sum of all transition magnitudes constitute a comparable relevant property of all waves.** Veil characterizes the non-divisibility of elemental constituents that comprise the wave and at large n (or

large ξ), approximates $\check{R}_s/4$.

Veil and ξ reveal "hidden variables" within the process of enumeration and the interrelated cycles. All the variables are however eminently connected and aesthetically simple, however unfamiliar their existence. The Veil of every integer intrinsically conveys an additional meaning to that integer when representing a population, (which is what the integer signifies). It provides a contextual mathematical attribute for something that integer symbolizes concerning the association of objects represented. The name Veil of integer n was chosen because these characterizations depict manifestations of monadic-shroud prominence for whatever entities are counted to n. That ancillary property of every number n elaborates how "Veiled" the members comprising collection n must be. In linking to the radians per quadrant of every wave Veil also exhibits a commensurate monadic-attribute of resonant quanta.

To reiterate, any and every arbitrary natural number (n+1) has a Veil designated $V(n+1,\xi)$ that depends upon circumstances population (n+1) represents. Such parameter identifies an "indivisibility" property associated with each of the objects being signified. The same statement applies with (n+1) everywhere replaced by n. For any given ξ in figure 12, Veil $V(n+1,\xi)$ of number (n+1) along the abscissa is defined as the RMS sum of all transition magnitudes from n = 0 upwards to the number n. As previously shown in 21*1, that is the simple summation:

$$V(n+1,\xi) \equiv \sqrt{[\sum_{\check{n}=0}^{n} s_{\check{n}}^2]} = \sqrt{[\sum_{\check{n}=0}^{n}\{\sqrt{(2\check{n}\xi+1)}\}^2]} = \sqrt{[\sum_{\check{n}=0}^{n}(2\check{n}\xi+1)]}$$

Such summation as Veil is an efficacy measure of the information shroud associated with sub-attributes of each population member. When Veil = n, (or ξ = 1), no sub-attributes can associate with each population member while for $\xi \to 0$ many sub-attributes are possible. Since Veil represents the root mean square sum of s_n extents, it is also the RMS sum of s_0 and s_a transition components because $s_n^2 = s_0^2 + s_a^2$. Veil also ends up equaling each quadrant's summed-chord-lengths whose accrued value is $n\sqrt{\xi}$, as figures 37, 49 and equation 22*3 indicate. Figures 49 and 50 thus elaborate Veil as equivalent to each "quadrant side" for large n. **Presented in that form, there is no need for π in mathematically representing harmonics. The alternating interchanged energy-forms that comprise the wave distinctly emerge as orthogonal to each other as occurs in every pure periodicity**. Energy-forms like electric and magnetic alternatively traverse opposite sides of the figure 49 "square". **For ξ = 1 and n $\to \infty$, that configuration also provides the absolute definition of a radian and a steridian.** The surface area within the "square" is then exactly 1/4π'th the surface area of the unit sphere and it diminishes in proportion to ξ for ξ < 1. Each quadrant angle from the origin is exactly one radian, or $\sqrt{\xi}$ radians for ξ < 1. Though absent from our

repository of scientific concepts, Veil and ξ provide a fundamental characterization for each quadrant of every wave.

In figures 39 and 40, viewed from the cone's apex under conditions of synchronization, each consecutive radial vector n appears at a different "clock-hand angle". **Those angles from the origin derive from projections of each s_n onto the plane of the paper normal to the central axis.** The n vectors form a sequence of 4P such adjacent angles per each 360° cycle. We loosely call such projected incremental angles "clock angles" α_{cn} for purposes of discussion. They are also the intercepted angles around the $(2\sqrt{\xi}/\pi)$-radius circle at the base of the unit cone. That provides terminology for consecutive angular increments that produce net **rotation** about the unit-cone on the $(2\sqrt{\xi}/\pi)$-radius circle, counterclockwise being positive. Just as any right angle direction from n is feasible for each transition, it might be speculated that there are circumstances where any value of ξ is similarly feasible at any point. That would assume the gamut of all possible transition-directions "remain open" regarding how the spherical annular-regions between integer spherical surfaces at n and (n+1) get traversed. Nothing seems to dictate how that inter-integer annular gap is crossed so variations in transition angles between integers during a single mode (including variations in α_{cn} and ξ) could conceivably be feasible. That annular region between the spherical surface shells at integers, associates with indeterminism due to the monadic shroud and so internal details there are not totally ascertainable. In a "strobed" manner, parameter α_{cn} quantifies circulatory advance within the 360° circle viewed normal to the axis. Those joint variables, α_{cn} and ξ effectively specify transition-direction from each vector n into the hemisphere distal from the origin and terminating at (n+1). Various spherical-coordinate descriptions could apply to characterize advance into the distal hemisphere from vector-tip n.

Figure 49 shows that **every cycle references from the cone's apex, the origin of the wave**. The figure generated through intersection of the radial n-vectors at conditions of synchronization and the surface of an origin-centered **unit sphere** provide a visual analysis tool in analogue of how exponential harmonics visualize on the unit-circle. Transition extents map uniformly onto that unit-sphere surface granting it particular utility for comprehension. Figure 49 displays that condition. The range of the unit sphere's surface-area encompassed per cycle would extend from a very small area near $\xi = 0$ (small steridians/cycle related to squaring side-length $\sqrt{\xi}$, which area equals ξ), to one radian per quadrant and one steridian of area per cycle at $\xi = 1$. The control parameter $0 \leq \xi \leq 1$ governs how many steridians the "square" intercepts on the unit sphere. Each quadrant-length of the figure 50 "square" devised from summed s_n chord-length-projections" varies as $\sqrt{\xi}$. The area of that "convex square" having sides $\sqrt{\xi}$ is ξ. That additionally portrays the solid-angle steridians represented by each such "square". Extent ξ initially articulated in figure 12 as the inter-integer location of orthogonal traverse, thus also signifies "steridians of the wave's directional dispersion". With smaller ξ, transitions experience more relative advance in the cone's axial direction. That grants a narrower amount of tangential circulation (less γ) with each respective elemental advance.

Several procedures mentioned herein allow a summary examination from n = 0 to any n of inter-integer transitions $s_n = \sqrt{(2n\xi+1)}$, where $0 \le \xi \le 1$. When $\xi = 1$, the totally orthogonal transition magnitudes comprise the sequence of square roots of all odd numbers, as $s_n = \sqrt{1}, \sqrt{3}, \sqrt{5}, \sqrt{7}, \sqrt{9}, \sqrt{11}$, - - - etc. Each such magnitude represents a different irrational value, except for the perfect squares of (2n+1), namely $\sqrt{1}, \sqrt{9}, \sqrt{25}, \sqrt{49}$,- - - etc., which magnitudes themselves form the sequence of all odd numbers 1, 3, 5, 7, - - - etc. Cyclic rectilinear synchronicity associates with that odd-number series for s_n by virtue of summed concatenated triangles of angle $\theta_n = \tan^{-1}(s_n/n)$. Intervals between each such rational-odd-number-magnitude for s_n constitute a cycle, with 4 symmetrical quadrants/cycle exhibited when viewed from the origin. Projection of those transition magnitudes onto the plane normal to the central axis displays that cyclic periodicity as in figures 37, 50. Values of n affiliated with quadrant conditions are never a prime number. The RMS sum of transition-magnitudes-s_n upwards from n = 0 will equal to n when $\xi = 1$ and each transition from integer n then takes place orthogonal to the zero-to-n direction. When orthogonal transitions occur a fractional distance s_a from n along the axis, $\xi \equiv 1 - s_a$ and transition-lengths s_n become $\sqrt{(2n\xi+1)}$, instead of $\sqrt{(2n+1)}$ for $\xi = 1$. Projections of those shorter transition lengths onto the plane normal to the axis **retain the same**

periodicity and quadrant values n as with $\xi = 1$, but cycles then progress within a smaller "conical" range of circulation about the axis. Synchronization then occurs on a cone of smaller apex angle $\beta' = \arcsin\,[(2\sqrt{\xi}/\pi]$. Viewed along the axis from the origin at synchronous conditions, the radians accrued per cycle will equal $\check{R}_s \equiv 4\sqrt{\xi}$ for all ξ. Radians/quadrant will be $\sqrt{\xi}$, so for any mid-quadrant n that sum of projected chord-lengths will approach $n\sqrt{\xi}$. When normalized through division by n, (or by examining the intersection of n-vectors at the origin-centered unit-sphere), $\sqrt{\xi}$ radians/quadrant result from the projected sum of s_n vectors. Such repeated synchronization to $n \to \infty$, exposed by invoking the orthogonality axiom, is a most remarkable property of natural numbers. For $\xi \leq 1$, the RMS sum of all transition magnitudes leading up to n is approximately $n\sqrt{(\xi+s_a/n)}$. For large n this becomes $n\sqrt{\xi}$, and when normalized by n [which essentially converts to angles], can also equate to radians/quadrant as $\sqrt{\xi}$. In effect, parameter ξ as defined in figure 12, permeates the process of enumeration, particularly when the entity being counted is not infinitely divisible.

In accommodating *degree-of-divisibility*, the generated spiral incorporates an ξ less than unity, and, referencing figure 12, ends up with "diagonal" transition lengths $s_n = \sqrt{[2n\xi+1]}$, instead of orthogonal extents $s_n = \sqrt{[2n+1]}$ when $\xi = 1$. An elegant simplicity engrains within natural numbers to encompass enumeration of "partially divisible" entities as distinguished by ξ. The case of $\xi = 0$ represents maximum or total divisibility while $\xi = 1$ (where transition orthogonality maximizes), depicts minimum divisibility. The range $0 \leq \xi \leq 1$ covers the gamut of intermediary possibilities. **This "modifier" ξ on the efficacy of the information shroud surrounding each population member constitutes an unfamiliar variable. It is however inextricable from natural numbers and enumeration procedures**. It is observed to characterize absolute rotation as radians and fractions thereof, plus solid-angle steridians and fractions thereof. **Four-quadrant symmetry intrinsic to the $\xi = 1$ case preserves for all values of** ξ. Values of n denoting quadrants of the wave **also remain invariant with** ξ. Axis directions ABCD of figure 3 that articulate the quadrants never associate with a prime value of n and **retain that same feature throughout all** ξ. **Arbitrary values of ξ do not alter the intrinsic 4 quadrant rectilinear synchronization generated by natural numbers**. Phenomena identified through ξ affiliate with important properties of numbers, waves, enumeration, and populations.

Figure 12 infers why for $\xi = 1$, transition-lengths between integers create wider circulation about the axis than for lesser ξ and how reduced ξ diminishes the summed transition-length/quadrant in proportion to $\sqrt{\xi}$. At diminished ξ each quadrant of the wave comprises from shorter chords in proportion to $s_n = \sqrt{(2n\xi+1)}$ and the projection of those lengths in the paper plane normal to the axis diminishes as $\sqrt{\xi}$. **At the extreme of very small ξ each transition-length itself (not its projection), diminishes toward unity while its projection approaches zero**. The RMS sum of those transition-lengths from zero to any n then proportions to $\sqrt{n}$. That is the root mean square of n "unit-lengths". At P cycles from the origin any quadrant

would have P such unity-transition-lengths with a total of 4P transition-lengths/cycle.

However as $\xi \to 0$ where transition-lengths approach unity, "radians as seen from

the origin" forming each "square quadrant side" on the unit sphere of figure 49 will

approach zero in accord with $\check{R}_s = 4\sqrt{\xi}$. $\qquad\qquad$ (22*7)

Origin-angle radians/quadrant delineated on the unit sphere approach zero as

$\sqrt{\xi}$ even though each chord-length approaches unity in the axial direction as

$s_n = \sqrt{(2n\xi+1)}$. When $\xi \to 0$, the side of each square comprised from the projections

in the paper plane of summed quadrant-lengths approaches $\sqrt{\xi} \to 0$ though each

chord of those quadrants becomes of unit-length. **That property of retaining a**

fixed relationship for radians-at-synchronization of $\check{R}_s = 4\sqrt{\xi}$, applicable for all

ξ constitutes an amazing feat of the natural numbers.

Parameter ξ plays a useful role delineating a host of mode properties and α_{cn}

complements it as the other degree-of-freedom to signify progression between

integers. Synchronous modes have quadrants of extent $\sqrt{\xi}$ defined from projections

of summed chord-lengths normal to the axis. Smaller ξ signifies "more direct"

transition in the axial direction with shorter projected chord-lengths comprising each

quadrant. Each degree-of-freedom might conceptually vary separately but to retain

synchronization would generally require their joint inter-related participation. The α_{cn}

angles for example, could relate to whether the intersection figure during

synchronization was circular or square as in figure 37 vs. 49. Axial contribution of ξ

establishes the fraction of the unit sphere's surface area encompasses by that circle or square. Synchronization may perhaps also be possible for shapes alternative to a circle or square, as determined by α_{cn} and ξ. Nothing precludes periodicity for any perimeter path-length of $4\sqrt{\xi}$. The shape and size of synchronous waves could conceivably change as the mode progressed, with attendant alterations occurring in angle α_{cn} and magnitude ξ.

Typically, synchronization mandates a changing α_{cn} during the course of the mode. On a conic surface for example, when the cross-sectional circular base of the cone is small (for small n), angle α_{cn} between consecutive chords differs significantly. For large n by contrast, the cross-sectional circle becomes large and change in each angle α_{cn} per transition diminishes. To achieve synchronous modes it is however unnecessary for ξ to change with n, although no known mechanism appears to stipulate it must be constant. For synchronous modes on a conic surface, angle α_{cn} will appear to rotate like the second-hand of a clock (but counterclockwise for these discussions). For fixed ξ throughout the mode it would make 4P "almost uniform increment" jumps per cycle. Figures 39 and 40 reveal such exemplary jumps per cycle. If in fact both α_{cn} and ξ could vary during the course of a synchronous mode, than variation seems possible in the shape, size or envelope of the resultant wave. That might permit "wavepacket-type" modes that grow and then decay analogous to conventional wavepacket descriptions. While properties of such envelope variations embed in α_{cn} and ξ alterations, such characterization might be hidden within the realm of indeterminism. No outwardly apparent reason appears to preclude such modes however. If ξ changed during the course of a mode than each inter-integer transition would contribute to the RMS sum in proportion to its respective transition-length squared. The square root of the sum of each separate $s_n^2 = [\sqrt{(2n\xi+1)}]^2 = 2n\xi+1$ would define the RMS sum and such sum would typically differ for fixed or variable ξ. It remains total speculation as to whether ξ can vary during the course of a single mode.

This thesis shows that waves are **an intrinsic property of natural numbers**. These unified waves do not derive from an artifact of mathematics or from **proffered progression around a unit circle**. They do not emerge from a synthesized model, but **from the essence of enumeration under the orthogonality axiom**, a most fundamental physical process. They require **a minimal number of arbitrary operators and constants** from mathematics or physics. They are the inextricable consequence of **potential synchronizations** that can occur when counting small things. They **mathematically reference the impetus origin event** that caused the wave, not an arbitrarily selected figure of geometry like the circle on a perfect plane. This unfailing reference to the origin is an important distinction between unified and exponential waves. When $\xi = 1$ each quadrant occupies <u>**one radian**</u> not $\pi/2$ radians. When ξ is smaller, the sum of projected chord-lengths per quadrant diminishes as $\sqrt{\xi}$ and each quadrant seen from the origin occupies $\sqrt{\xi}$ radians. That characterizes a wave **relative to its**

origin with **rectilinear symmetry** based on the **absolute radian** as a "degree of rotation" and independent of "**surface flatness**" over that radian of rotation. **Any shaped line of extent-4 on the unit-sphere surface constitutes 4 radians of orthogonal traverse from the origin**. The waves exhibit natural synchronization having four quadrature components that inherently **affiliate with two alternately interchanging energy-forms**. The composite wave comprises from **discrete incremental changes** rather than infinitesimal continuity. **Toward larger n they exhibit perfect periodicity and do not violate causality**. Even if $\sqrt{\xi}$ as radians/quadrant were unrelated to physical mechanisms, the analysis method **provides a useful tool to mathematically quantify the directionality of waves, or to quantify any other annular dispersive process in terms of integers**. It micro-identifies how directionality emerges in progression by interrelating comparative component amounts of axial bypass ξ and angular circulatory advance α_{cn}. **They relate to all the other parameters embedded within basic wave processes**.

Although particles and force carriers knowingly have equivalent wave depictions, precise mathematical details of how such waves arise has not been imminent. This theory postulates a mechanism. Traditional exponential-harmonic wave processes are here alleged to be ad hoc, reverse engineered for synthesizing sinusoids rather than a logical model for wave development. Besides at least ten other listed shortcommings[1] the exponential suffers a fundamental problem of violating causality. It generates a discrete-frequency solution that begins at minus infinity in time from a later-origin event that creates impetus for the wave.

23) SPECULATIONS, IMPLICATIONS & ATTRIBUTES OF UNIFIED

WAVES

This section discusses various highly speculative issues. Those conjectures may miss the mark slightly or be totally inaccurate, but are included here as possibilities anyhow. Such speculations are inferred from the mathematics, which may well have alternative interpretations. The questionable explicitness of these attempts to decipher the mathematics should not detract from the exactitude of that mathematics, which is here presented as being totally appropriate and the essence of this theory. Further possible phenomenological interpretations were reckoned worthy of offer and are thus included. They provide an attempt to clarify what all this may mean. Harmonic states described on conic surfaces herein constitute symmetrical solutions. As such they represent resonances, stable state, or standing wave conditions. The analytic description leading to the spiral is one that portrays possibilities. The resonances depict realizable harmonics within space and time based on smallest monadic constituents and with multiple simultaneous solutions feasible. When smallest quanta accrue characterizing a spacetime variable concerning the aftermath of an origin-event, potential repetitive symmetry remains inextricably engrained in the process. Resultant harmonics possess periodicity, which connotes energy and mass. Circumstances where this symmetry is "broken" or doesn't exist could signify asymmetric, non-interference, random, or undefined conditions. Field equations depicting stable state solutions would typically describe a symmetric world, whereas phenomena in the world of non-interference solutions would likely be asymmetric. Accordingly, conditions of symmetry breaking plausibly relate to criteria that determine when and how these conic-surface solutions occur. Reasonable conjecture could furthermore speculate these wave-representations in spacetime might be alternatively portrayed in terms of conventional fields, particles and criteria of force. For an interaction between particles, were one fully versed on all possible conic surface solutions over all ranges of spin, ξ, and figures-6-to-11-type folded-surfaces, it might be possible to affiliate analogous unified waves with respective conventional equivalents of that interaction. Were that true, systems could characterize on a relatively few parameters; Planck's constant $\Delta\theta$ and velocity c being primary fixed constants of Nature. Quantitative delineation of ξ and how these conic-depicted surfaces can "fold" would represent "hidden variables" of physics. S_a might connote rest mass.

Consider the following speculative viewpoint. Spacing between each n-vector is always unity. Thus slope, $[\sqrt{(2n+3)}-\sqrt{(2n+1)}]$ signifying change in s_n is the "derivative" with respect to n of accumulated-time, (i.e., of s_n or of transpired half-periods). Now s_n itself represents time since n = 0, being the integral of that derivative with respect to n. Rather than time being independent variable as in Newtonian systems, it "depends upon" independent variable n and the n = 0 origin reference. Parameter n portrays quanta (of action) participating in the aftermath of an event. As a wave, n monads transport or portray effects of the event through time and space. The derivative or slope of s_n also equals half the radians traversed for each increment in n, or radians are traversed at twice the derivative of accumulated-time with respect to n.

The question emerges concerning a reason these systems should necessitate description in terms of action units (energy-time) or (momentum-position). Why should the variable of relevance be "action" instead of "time"? Indeterminism from the shroud of each smallest entity allows multiple resonances as simultaneous possible solutions. Each orthogonal right angle can take place in many workable directions, all of which are independently feasible. Potentiality of one direction does not preclude the utilization of others. Fulfillment of one mode does not prevent the existence of others. Emergence of a harmonic mode, (say a specific conic-surface fold-configuration as depicted in figures 7 to 11) does not prevent another resonance that may simultaneously subsist. Like two possible simultaneous solutions to a quadratic algebraic equation expressed in terms of $i = \sqrt{-1}$, any of the available right angles each constitute one of many viable options for each transition from n to (n+1), as parallel choices so to speak. The specific method of getting from one spherical surface at n to spherical surface (n+1) remains unrestricted (not articulated or defined), so long as it remains orthogonal to the prevailing origin-to-n direction. Therefore, a band or range of possibilities that can yield A,B,C,D, direction overlaps may always allow jointly acceptable resonant-mode solutions. Possibility of one viable integer-spin resonant mode solution does not exclude other modes from potentially occurring. (Half-integer spin modes may conceivably impose restrictions on other simultaneous modes, which process might

335

manifest as the Pauli exclusion principle). By having the "variable" and its solutions expressed in terms of a conjugate product as (energy x time), different "frequency" solutions become simultaneously feasible. That is, time is not restricted to a single-valued function as it is when we express f(t) = - - - . Here, one resonant mode in energy-time can have a large energy and a short time (a high frequency), while another mode simultaneously exists with a small energy and a large time (a low frequency). That would explain why these variables with multiple possible simultaneous harmonic-mode solutions should express in terms of the conjugate product (energy-time). Solution modes occur in the mutually exclusive energy-time-product variable. Simultaneous solutions grant interactions we designate as "amongst many particles" of differing energies, with net conservation governing the result. It also tells us why in physical reality, time is not the "independent variable"; a status unfortunately promulgated by the insertion of functional solution elements like the harmonic exponential $e^{i\omega t}$.

The question of what "medium" these or generic waves might be disturbances in is worthy of conjecture, however speculative that supposition may be. Under this analysis nothing prohibits two counter-rotating spirals from generating a zero-spin pair. Such circumstance is suggestive of depicting Bose condensate and Higgs particle conditions. Paired harmonics of this type would also remain at least logically consistent with a model for vacuum entities derived from smallest repeated constituents upon which this theory bases. Accordingly, the conjecture that these waves may represent disturbances in the sea of vacuum quanta is not entirely illogical. That medium is empirically known to support wave propagation and such mechanism could seemingly apply for all traversing particles and force carriers passing through vacuum. Moreover, for a zero rest mass particle of $\xi = 1$, **population n was shown to equal the RMS sum of all prior transitions**. That infers waves sustained under those conditions constitute a disturbance wherein all "prior-participating" monadic entities must partake in (or "reshuffle") upon addition of each new population member. At least the time associated with transition vectors infer that, as well as continued reference to the origin by variable n that permeates all solutions. All prior transitions have influence upon each next population value. Higgs particles in a "lowest energy sea" might be candidate for the medium in which that type disturbance is feasible. A crude analogy supporting that rational would be waves on water or a bubble's rising motion, wherein all prior-effected singletons of the system participate in the propagation of a discrete identifiable response (disturbance). That type of propagation also entails "retardation delay" which would be consistent with a limiting velocity.

Analysis here avoids the $\sqrt{-1}$ imaginary-vector mathematically used to signify rectilinear space. That was a complex variable tool supplanted herein by triangles with specified side-lengths. For these solutions by contrast, symmetry derived from a superfluous angle accommodates curved space. Resultant rectilinearity of the wave emerges as a matter of course exclusively from enumerated natural numbers. Applicability of the unified wave in excess-angle-space cleaves a distinction between conventional mathematics and physics. It highlights dissimilarity[13] between traditional mathematical practice where n may proceed-as-a-limit toward infinity, and physics, where for second-order systems, integers greater than $2/h^2$ vanish. At the same time, these solutions of integer-based spirals (or perfect square repetitions) can depict waves of physics, while those same integer numbers constitute the sub-structure of mathematics. Both disciplines utilize the identical foundation basis, the set of sequential numbers. **That mutual groundwork of consecutive natural integers infers a "reason" why mathematics might so well describe physical laws. They both utilize the indistinguishable substructure of natural numbers.** Spiral harmonics may further provide the vehicle or a descriptive procedure by which "space curves". No guarantee prevails that the originating "Euclidean axis" now bypassed and shrouded by non-local surrounding vectors follow "straight Euclidean lines". They might bend slightly within the emergent unfamiliar compounding processes.

Uncertain angle $\Delta\theta$ can then be visualized to depict that minimum net ambiguity in bending, returning full circle toward justifying its own existence. It should be recognized that <u>**spacetime with an extraneous angle is invariably more embracing than the special case when that extra angle equates exclusively to zero**</u>. That logic-of-generality [and the associated accommodation of angular curvature] applies to spiral harmonics compared to its special-case limit-condition that defines exponential harmonics, when exponent n_{max} effectively equals infinity. An overall "higher order self consistency" dominates for $\Delta\theta = h$, above the deterministic rectilinear characterization of spacetime with $\Delta\theta = 0$.

Many physicists argue that quantum mechanics works so well for solving problems it is acceptable to forgo understanding why it works. They may even assert it is sufficiently accurate to defy improvement and so one should not bother trying. Such attitude to sacrifice comprehension to the altar of computation stands totally inappropriate to other aspects of science. Why is it tolerated here; particularly when QM retains a host of enigmas like reversible time, unchanging entropy, irrationality, normalization singularities, relativistic incompatibility, and quandaries of wave function collapse like Schrodinger's Cat measurement-problems. One reason for full understanding is to elucidate these enigmas. Another is the potential for increased utility through analogy, when applied as a tool to other disciplines within science and engineering. Without understanding why QM works it is difficult to make such analogies. A third reason is that many useful subtleties may emerge from a full understanding. In science, ignorance is not bliss, but rather what might be missing remains unknown without full awareness. Newtonian physics worked very well for more than two centuries; hardly a reason for Einstein to avoid attempts to improve it. Newton believed if one knew exactly the positions and velocities of all objects in a system one could predict its past and future with perfect accuracy. That might have been true if the system did not contain any smallest repeated constituent, but was continuous down to nothingness. However with a prevailing smallest unit, namely excess angle $\Delta\theta$ or the minimum energy-time/cycle = h = $\Delta\theta$ of this analysis, the system becomes described by a population of those units, rather than continuity. **That automatically injects indeterminism equaling the**

space between those smallest units into the process. Thus, Newton's oversight can equally attribute to his presumption of infinitesimal continuity via his own calculus and disregard of smallest populations, to his neglect of included indeterminism. Both breaches are equivalent, essentially being the same thing.

Photographs of figure 3 transparencies exhibited herein, plus various other feasible modes, result from the premise allowing orthogonal transition between axis integers. **That axiom is all it takes to validate the mathematical conclusions and all wave modes through time and space derivable from the various figures**. These modes supplant deterministic sine waves of exponential harmonics, a limit case of these unified waves when $\Delta\theta = 0$ and $n = \infty$. The analysis applies to non-divisible quanta or partially divisible entities, relating how elementary particles experience the unfamiliar dimensions brought forth herein. A surplus angle in space $\Delta\theta$ emerges consistent with that axiom. It coalesces the three motional conservation laws into a single law, which conserves traversal rate through $\Delta\theta$. Planck's constant is shown to specify an excess angle that applies for geometry of the physical world, as a possible fractional modifier to perimeter/diameter ratio π, for describing spacetime, for introducing indeterminism, and for waves of radiation and matter. Electromagnetic waves, photons, nuclear forces, and elementary particles are conjectured to sustain and propagate via the 8 new degrees of freedom described. Those "dimensions" are inferred as a possible basis for string theory. They constitute circulation around each axis to bypass all its irrational values in the process of variable advance. Unified waves alter time-reversal symmetry. They provide an arrow for time allowing unilateral increase in entropy at the microscopic level.

Quantum mechanics is a mathematical scheme whose calculations provide excellent correlation to physically measured parameters. It also introduces enigmas of interpretation. It is conjectured that those quandaries are the consequence of understandings from a vantage where the orthogonality axiom as introduced herein has been overlooked. Conceptual inclusion of that axiom should allegedly give further insight toward a logical foundation of atomic processes. Recognition of the axiom also provides various new predictions and a more rational picture of the physical world. It brings out new dimensions, times arrow, non-locality, and logical reasons for a probabilistic-distribution interpretation of phenomena. The axiom infers how motional conservation laws coalesce due to a fixed curvature of spacetime $\Delta\theta$. The inevitable indeterminism of a population bridges the classical/quantum divide underscoring how indeterminism must "conserve" in its various forms. Non-localization about the Euclidean axis indicates why no Newtonian point-by-point macrozone trajectory in X,Y,Z,t, occurs. An explanation emerges why irrational values cannot be part of physics though perfectly valid in the domain of mathematics. It infers how Schroedinger's equation might select out regular periodic modes from the chaos of random ones. It promises to clarify a few mysteries of objective reality.

Properties of natural numbers described here may have even greater relevance than examples of quantum mechanics or wave propagation. Waves are carriers of all physical force and provide a basis for characterizing elementary particles. All variables can subdivide into smallest constituents whose progression can be viewed as an increasing population. The mathematics can even portray populations of people. **In the sequence of integers n, perfect squares of (2n+1) are as basic as octaves are in music**. Cyclic repetition is intrinsic to population progression. Although the effects are not always observable, waves inherently embed within the advance of seemingly monotonic variables. The universe is immersed within a sea of waves. Throughout the cosmos, they provide the primary mechanism for information and phenomena transfer. The root of their existence stems from natural numbers, transformations of which also form the basis of mathematics, physics, stable system states, probabilistic descriptions, etc. Is it not plausible our historic failure to find intelligent signals from outer space stems from dissecting signals received via a different encoding criterion than the sender used? What method of information transfer would it likely employ other than the basis for waves intrinsically engrained within the sequence of numbers? Were this true, we may have been analyzing outer-space electromagnetic signals with "crystal sets", greatly reducing the probability of finding intelligence. That does not imply is a simple task to establish exactly what method or mode of information encoding an alien species might employ based on unified waves. However, a significant likelihood exists that a foundation feature of waves, mathematics, and physical laws, would partake in the information delivery process. In that context, SETI should delve into the possibilities. The payoff in emerging ideas and phenomena would likely justify the investigation. These conclusions apply so long as the mathematics presented herein is valid, independent of explanations given for phenomena. Even if those expositions were erroneous, or miss the mark, the consistent mathematics should provide a basis for extra-terrestrial encoding. That thesis is particularly relevant if that same mathematical encoding encompasses a description for DNA.

Hinging upon an **excess angle in spacetime**, unified wave theory embraces the concept of **space curvature** thereby implicating **association with a Cosmological constant** under general relativity. Discussions about figure 12 treated fixed ξ per mode without discussing effects of changing ξ, **which changes might entail acceleration**. The thesis treats **discrete monads** thus encompassing **quantum phenomena**. It is a theory of **symmetry conjoining particles and waves** having spins of up to at least ±2, in increments of ½. For that range of ½ integer spins it infers **supersymmetry between bosons and fermions**. Unified waves **originate at the impetus of the causal event** consistent with physical laws being governed by time intervals **between events** rather than absolute time devised from a "master clock" implied by sinusoids. Each discrete constituent "**frequency**" of exponential harmonics by contrast, **begins at minus infinity in time and extends to plus infinity**. Moreover, over the entire range $0 \leq \xi \leq 1$ (and greater), inherent symmetries place these new waves within the **same group** to **obey gauge transformations** and **exhibit a conserved quantity** under Noether's theorem. Accordingly, these waves appear to satisfy basic ingredients to foundation **a unified**

theory of physics underlying consolidation of all forces, particles and waves. Integrating **the roots of DNA** it goes one step further to include the substructure for **biology and living things**. Moreover, it ties all these disciplines and entities to **the bedrock substratum of natural numbers and mathematics**, whose intrinsic properties under the theory seem to constitute the **essence of physical laws**. It **eliminates perhaps six arbitrary constants** employed under other theories. It possesses beautiful **elegance, simplicity, cohesion, and logic**. Unified waves might potentially provide the underpinnings for a **Theory of Everything**.

Figures 3 and 50 indicate how every new cycle of the unified wave contributes one additional partition to each radian (each quadrant) of the four radians traversed per cycle. <u>For P prior cycles there will always be P partitions per radian</u>. Is there a limit to the number of partitions a radian can have, a limit to its resolution? The resolution of a meter is constrained to a Planck length. A radian's resolution is similarly constrained by Planck's constant. <u>A radian can have as many partitions as it has cycles P_{max}</u>, the reciprocal of Planck's constant since, $1/h = 1/\Delta\theta = P_{max}$, where $n_{max} = 2P_{max}^2 + 2P_{max} = 2/\Delta\theta^2$. Planck's constant stipulates the resolution limit of a radian as $1/P_{max} = h = \Delta\theta$, just as it limits the meter and second. Under this theory, Planck's constant sets a resolution threshold of a **<u>Planck-length</u>** for distance, a **<u>Planck-time</u>** for time, and a "**<u>Planck-radian</u>**" for rotation. Those are smallest subdivisions for three of the five fundamental physical variables, the other two being charge and mass.

This theory may be shown as invalid by proof that within a monadic population:

1)   The basic orthogonality axiom is false for advancement between integers.

2)   The geometric mean criterion for population change is invalid.

3)   Natural numbers cannot appropriately depict as concentric spherical surfaces.

4)   The Planck Angle representation of figure 18 in erroneous.

5)   Four quadrant harmonic symmetry does not sustain through all $0 \leq \xi \leq 1$ values.

6)   That angle θ_n from the origin does not alter with each population increase.

7)   The RMS criterion-of-growth to population n is invalid.

8)   Scalars rather than vectors adequately characterize population change.

9)   Assessment that the mathematics herein is incorrect.

10)   Demonstration that the excess-angle within spacetime is de facto zero.

11)   Invalidating equation 21*8 and the exact-periodicity conclusions of Table 4.

12)   The conjecture about homogeneous expansion may invalidate by verifying that the experimentally encountered value of perimeter/diameter $= \pi$ is not (mathematical-π)$(1 + \phi_{min})$, where $\phi_{min} = \Delta\theta/2$ is the actually determined excess-angle/cycle in appropriate dimensional units.

[Contrary proofs of the above items may aid cogency of the theory.]

24) SUMMARY

Under this theory, the existence of a smallest non-divisible unit of energy-time/cycle classically implies the following speculations:

1) Variables of this type define populations making orthonormal vector transitions obligatory between customary scalar-axis-integers.

2) Non-localized "waves" described via "ray" vectors from an origin and potentially occupying a multiplicity of locations, can represent conditions of time and space.

3) Unfamiliar tangential "dimensions" that may exhibit four-quadrant symmetry originate about each traditional axis. The added dimensions grant new degrees of freedom required by string theory to satisfy empirical observations.

4) A discontinuous gap associated with intangible tangential vectors manifests within the new dimensions. The smallest discernable embodiment of that quantized gap purportedly equals Planck's constant and provides one basis for the uncertainty principle.

5) A extraneous angle/cycle $\Delta\theta = h$ possibly affiliated with homogeneous expansion, infiltrates spacetime thereby creating a population of non-divisible energy-time units, injecting system indeterminism equaling the population's inter-integer interval.

6) The increment of uncertainty in energy-time issuing from the extraneous angle in spacetime appears preserved in different forms (but equivalent amounts) through a conservation law.

7) Three traditional motional conservation laws coalesce into one law.

8) An intrinsic periodicity inherent to the sequence of natural numbers to ∞ can furnish a generalized basis for harmonics in Science, Nature and Engineering.

9) A new type of wave and trigonometry result that might portray how all forces act and all particles interact.

10) Harmonic waves with undetermined amplitude can characterize quanta whose energy proportions to frequency rather than amplitude squared.

11) Cartesian coordinates and the imaginary operator $\sqrt{-1}$ may precisely address rectilinear Euclidean-space-mathematics, while being subtly insufficient toward meticulously characterizing spacetime with a surplus angle.

12) A distinction emerges between processes that can proceed to infinity in mathematics, and the finite limits of physical reality.

13) Following any origin event leading to a terminal event looms a maximum discernable progression of time and distance. Cognizance of time and distance does not continue unilaterally to infinity.

14) On the microscopic level time seems unidirectional. It bears an arrow though it need not advance uniformly for all observers.

15) An analytic basis for matter waves can materialize for all moving objects.

16) A classical foundation may surface for quantum mechanics.

17) Spacetime dimensions comprise feasibly occupied, linearly independent, orthogonal degrees of freedom. They allow a new form of non-local harmonic wave within unfamiliar dimensions. Any of the wave's characterizing variables contain all its available information. Exponential harmonics compose from an incomplete version of this more general unified wave.

18) Exponential harmonics though mathematically consistent and scientifically utilitarian are physically inappropriate via exponent division by 0.

19) Excess spatial angle $\Delta\theta = h$ for curved spacetime incorporates, and provides a more encompassing representation than, $\Delta\theta = 0$ of deterministic Cartesian-Coordinate spacetime.

20) Maintaining dimensional consistency is mandatory for the variable cycles, (or radians, degrees, or rotations) within modeling equations where the spacetime angle might differ from exactly 2π radians.

21) Populations can describe variables of reality that produce harmonic waves created by orthogonal transitions.

22) Response variables of analysis and all measurement quantities can be represented to always take on rational values. Indeterminate angle $\Delta\theta$ should further limit the maximum precision of physical variables to the number of digits that numerically depict $\Delta\theta$.

23) The degree to which an entity can be sub-divided into its smallest constituents may play an important role in the quantitative physical behavior of that entity. In spite of relativistic contraction for example, the fact that a sphere moving relativisticly maps into an observed sphere might provide one

example of that conclusion since a spherical object enjoys minimal sub-divisibility.

24)   Conventional harmonic-oscillator type-equations can be treated in terms of the accrual of discrete smallest entities and provide periodic solutions.

25)   Unified wave harmonics may furnish the double helix template for DNA and a basis for genetic encoding, self-organizing, and self-replicating systems.

26)   Two separate dimensional regions of spacetime classifiable as microzones and macrozones could plausibly provide the substructure for Bohr's complementarity interpretation of quantum systems.

27)   One added energy-time unit per cyclical quadrant might identify an origin for the non-commutation of matrix mechanics.

28)   Pure periodic fluctuations have quadrature axes that contain no prime integer population values of the enumerated constituents. All prime numbers partition into four symmetrically closed sets affiliated with the quadrants of waves. A prime number cannot associate with the peaks or zero crossing of a pure harmonic wave, which may be another way of describing periodicity!

29)   Units of energy-time/cycle h are seen to be the same as excess angle/cycle and angular-irresolution/cycle $\Delta\theta$, relating Planck's constant to geometric irresolution and analytic uncertainty.

30)   In a homogeneous expanding universe, the actual value of perimeter over diameter may differ from mathematical-π by an additive modifier ϕ_{min} [= h/2].

31)   The degree of homogeneous expansion or contraction might constrain dimensional units like energy-time/cycle to become non-arbitrary and scale in

relation to a smallest angle ϕ_{min}. Such dimensional units could only be arbitrary in Euclidean spacetime.

32) The standard unit-radian is more appropriately defined mathematically directly in terms of ***angular rotation*** through the orthogonality axiom rather than via π and the ratio of perimeter to diameter of a perfect circle on a perfect plane.

33) Irrational values seemingly do not partake in macrozone processes, being exclusively limited to the microzone.

34) It may be possible to characterize particles and force carriers in terms of waves generated by natural numbers interpreted under the orthogonality axiom.

35) The constant of proportionality between photon energy and frequency under the orthogonality axiom can be interpreted as an excess angle in spacetime.

36) As illustrated through Table 1, inherent periodicity within the sequence of natural numbers n provides an array of mathematical identities between integers to infinity affiliated with perfect squares of (2n+1) and with [2/(n+1)]. That periodicity alleges to be the basis of all waves and to harmonic phenomena comparable to the structural foundation of DNA.

37) Unified wave tangential vectors exhibiting periodic increases by 2 in s_n can cyclically represent time progression from an origin on a conic surface. Coherent with those cycles, summed chord-lengths formed between radial vectors on that cone that intersect a concentric unit-sphere surface will exhibit similar periodicity. This permits progression along unit-sphere chords

between the radial vectors to represent the time in analogue of how the exponential traverses the unit-circle.

38) A more appropriate analytic wave-generating expression than the harmonic exponential exp(iωt) exists. The expression furnishes a natural transformation from progression along an abscissa to advance along chord-lengths of a unit circle equaling a Planck time (or Planck length). The resultant waves come forth in an array of different modes intertwined with basic properties of ordinal numbers.

39) In physical reality the maximum number of sides on a polygon inscribed on a cone of apex angle $\sin^{-1}(2/\pi)$ at unit distance from its apex will not be infinite, but limited to (1+1/h). Here h is Planck's constant evaluated in dimensional units that accommodate an alleged excess angle in spacetime.

40) Waves depicting physical phenomena can be appropriately modeled through algebraic equations avoiding transcendental and imaginary constants like π, e, i, and ∞. Planck's constant Δθ can supplant those constants providing a simplistic solution for waves in the Occams razor sense of minimal arbitrary constants. That same excess-angle Δθ also defines two additional arbitrary constants of length and time, a Planck length and a Planck time, as minimal chord-lengths of the same angle/cycle Δθ about the conic unit circle.

41) Non-divisible quanta enumerate in a manner allowing the RMS sum of all prior transition magnitudes up to a given population n to equal that population.

42) It is plausible that propagating electromagnetic radiation associates with the process of "retardation delays" from the RMS accrual of smallest quanta comprising vacuum.

43) It is feasible to mathematically bridge Newtonian continuity for divisible things with Leibniz' monadality for quanta. Parameter ξ partitioning the space between integers of a population can quantitatively link the extremes, which variable also specifies the apex angle of a conic surface for allowed synchronous modes.

44) Every number of the natural number system has an inextricable bundle-of-possible-angles attached, related to the number n and the scatter parameter ξ.

45) For all ξ and n as defined herein, cumulative transitions can create synchronous repetitions describable on a conic surface of apex angle $\beta_s = \sin^{-1}[(2\sqrt{\xi})/\pi]$ and characterize $4\sqrt{\xi}$ radians of rotation per cycle.

46) The excess angle of spacetime overlooked by Euclid and Newton serves to limit actual variables of physical reality to finite multiple values of that angle.

47) Non-localized vectors depicted under the orthogonality axiom can enable analytic distinction between chaos and stable stationary state standing waves.

48) Under the orthogonality axiom the natural numbers characterize harmonic modes through the synchronous accrual of angles to virtually infinite accuracy. Any perturbation of the system may break that symmetry of those precarious angular sums thereby terminating interference and standing wave properties of the modes.

49) Any perimeter extent of 4 on the unit-sphere undergoes 4-radians of cumulated progression from the origin. Sums of broadened-s_n-line-segments in figure 14 signify time-increments equaling an exact extent of 2-per-cycle. Twice those vector-extents always fit and repeat precisely around any perimeter-of-4 on a unit-sphere providing perfect time-periodicity to unified waves. Uniformity bases on identical intervals between consecutive-odd-integers in s_n.

25) APPENDIX A. ADDITIONAL UNIFIED WAVE

CHARACTERISTICS

Table 1 is useful for visualizing a configuration of the DNA ladder. It is comprised of consecutive-rungs of four-serial-string-partitions (the 4 quadrants or two half-cycles) between a "pair of twisting rails" (the columns n_1 and n_5). Each triple-row of table-entries represents one ladder rung having 4-possible serial compartments or two half-cycle compartments. Perfect squares at n_1 and n_5 serve as the twisting rails. Each sequential quadrant constitutes an angular extent of essentially one mathematical radian. Employing a cone of apex angle equaling $\beta' = \arcsine (2/\pi+2\pi/10)$, relative position of "rails" n_1 and n_5 would "twist" ten rungs/helical cycle for 4 fixed rectilinear radians at $\beta = \arcsine (2/\pi)$. This is described by the circular polarization twist shown in figure 17. That is to say, when the spiral mode is plotted on a cone of apex angle 90° (a plane) as in figure 17, the rails signified by columns n_1 and n_5 spin cyclically in sync with the ladder rungs. Alternatively stated, overlapping perfect square locations at n_1 and n_5 would undergo "circular polarization" (twist) while maintaining 4 twisting quadrants (or two twisting half-cycles) between them during each consecutive 4-radian cycle. The cone is useful for visualization but unnecessary to express the four-serial-quadrant or two serial half-cycle ladder rungs. Reiterating, if the apex angle of the cone increased to $\pm \pi/2$ radians, (for the resulting spiral to be within a plane) polarization direction A and the quadrants simply "twist more rapidly", the twist being modified by apex angle increase.

The successive strings of four serial compartments constitute the four quadrants between n_1 and n_5 during each cycle (triple-row). In this context, ordinal numbers from $n = 0$ to ∞ incarnate a "helical composition" of sequenced 4-possible serial-compartments (or two half-cycle compartments) between a pair of "twisting rails". In short, they form a twisting DNA ladder accommodating two serial compartments of four identifiable possibilities per rung. That emerges as an innate property of the natural numbers being elucidated through this unified wave analysis. The paired perfect-squares (n_1 and n_5) become the ladder rails, additionally possessing an internal analytic marker [δn, or P_n], **which references position relative to the origin**. To amplify by example, if Table 1 were extended onto a very elongated paper sheet (like hinged computer printout paper) and were twisting continually like tinsel along its length, than n_1 and n_5 would constitute (in the form of numerical integers) the two DNA rails everywhere possessing two half-cycle compartment "ladder strings" of the four compartments Q1 to Q4 between them. The Table, (or the ordinal numbers) thus ostensibly coincide with the genetic encoding arrangement in all life forms, with added reference to the origin through n, s_n, δ_n, or P. Under the axiom of orthogonality, such a twisting ladder is as inherent to the natural numbers as 1, 2, 3.

Any two-bit 'modulation method' (e.g., as "controlled variability" in θ_n, s_n, n, triangle area, molecular structure, nucleotide base, etc.), yielding two distinguishable states of the four possible provides an information encryption process, [and a potentially efficient computational architecture]. Such encoding directly within integer numbers embodies all the previously discussed correlated-parameters of the unified wave. These include, tangential magnitude always registering synchronicity at the ½-integer markers in s_n, non-primes at the quadrant dividers, the RMS summation of prior transitions equaling n, polarization of the helix, angular directional scatter of ξ, the traversed cycles of P, etc. It suggests how DNA may have "originated from" (or be rooted within) inherent order in the natural numbers. Through input n, it suggests **absolute location-from-the-origin of each ladder rung likely contains encryption information, as well as the relative sequence in proximity**. All of the parameters within the functional structure of Table 1 intrinsically relate to the origin, as well as to each other, for example as given through equations 13*6. **The format provides a "periodic 4-quadrant scale" through the linear sequence of numbers that renders a manifestation of the DNA helix**. This characteristic innately embeds within the natural numbers as a most fundamental organizational theme for assimilating smallest entities. **It can be used to mathematically model the DNA sequence as a ladder or a wave**. A multidimensional interlaced scale intrinsically implants within the natural architecture of numbers and the format is essentially a DNA ladder. **That scale portrays how waves can diverge radially, tangentially, cyclically, helically, and in quadrature fashion from an origin event, *in the actual manner waves advance*.**

The DNA hierarchy simply portrays one interpretative form of that wave advance. In this context, ordinal numbers can configure atoms of quantum physics as well as the molecular information-base of life.

In these symmetrical cycles of 4 rectilinear quadrants, **tangential** magnitude s_n always progresses at twice the rate of **cyclic advance** $P_n+1 = \delta n = (s_n+1)/2$. **Integer values for magnitude s_n each symbolize a consecutive "rung position" of the DNA ladder**. The prior cycles plus unity furthermore equals the differential rate of **radial** progression δn. That occurs with angular directions unspecified for n or s_n so the represented vectors can diverge spherically, in a circular band, in a helix, in coherent or randomized locations, and while potentially exhibiting harmonic cadence.

Figures 2 and 3 depict a spiral sequence of discrete radial vectors-n that possess harmonic periodicity. Though portrayed on one plane the vectors are not constrained to that plane. A vector n emanating from the origin could be in any arbitrary radial direction of a hypothetical sphere surrounding that origin. The next (n+1) vector could also emerge from the origin at any arbitrary angle different from the direction of vector n by θ_n. Accordingly, the total number of possible angular positions surrounding the origin that n might materialize in equals 4π steridians divided by θ_{n-1}. Radial vector (n+1) might be in any one of $4\pi/\theta_n$ angular directions, and so forth. The angular direction between any n and its respective (n+1) must be θ_n, with $4\pi/\theta_{n-1}$ possible directions for vector n and $4\pi/\theta_n$ directions for vector (n+1).

When possibility for all vector n directions emanating from the origin are accounted for, the potential array consists of $4\pi/\theta_{n-1}$ radial vectors each having magnitude n separated from the others by angle θ_{n-1}. The situation might describe as $4\pi/\theta_{n-1}$ possible vector n node tips separated θ_{n-1} apart around a "spherical" surface. Vector (n+1) could portray $4\pi/\theta_n$ vector node tips spaced θ_n apart on a concentric spherical surface of radius (n+1). That "changing spherical surface" might therefore characterize all possibilities following an event at the origin.

Figure 16 indicates that D_s, the feasible number of "duplicate" radial n vectors within a sphere varies "linearly" with s_n, δ_n, and P_n. For any examined n-vector of the spiral, the gamut of its possible angles relative to the origin might fall in any spherical direction, so long as those directions are no closer than θ_{n-1} to alternative possible identical-magnitude n vectors. The feasible number of them D_S [their possible density per sphere] is thus $4\pi/\theta_{n-1} = 4\pi/\cos^{-1}[(n-1)/n]$. Figure 16 is plotted on what might be called "perfect square-linear" graph paper, indicating the quasi-linear relationship between density D_S and s_n, δ_n, and P_n. The graph statistically suggests how the spiral's successive radial vectors designated via abscissa n or s_n might occupy any of D_S feasible-angular-directions about the origin. The figure 2 spiral construction on a plane, for (n-1), n, (n+1), etc., illustrates successive sequences amongst that chain of vectors but it overlooked the potentially different angles those vectors might have about the origin.

Besides the possible density D_c portrayed below,
for all n around the conic-unit-circle,
$$\sqrt{(2n+2)} > 2/\cos^{-1}[n/(n+1)] > \sqrt{(2n+1)} = s_n \rightarrow (\text{"polygon" sides}_{@n}/2)$$

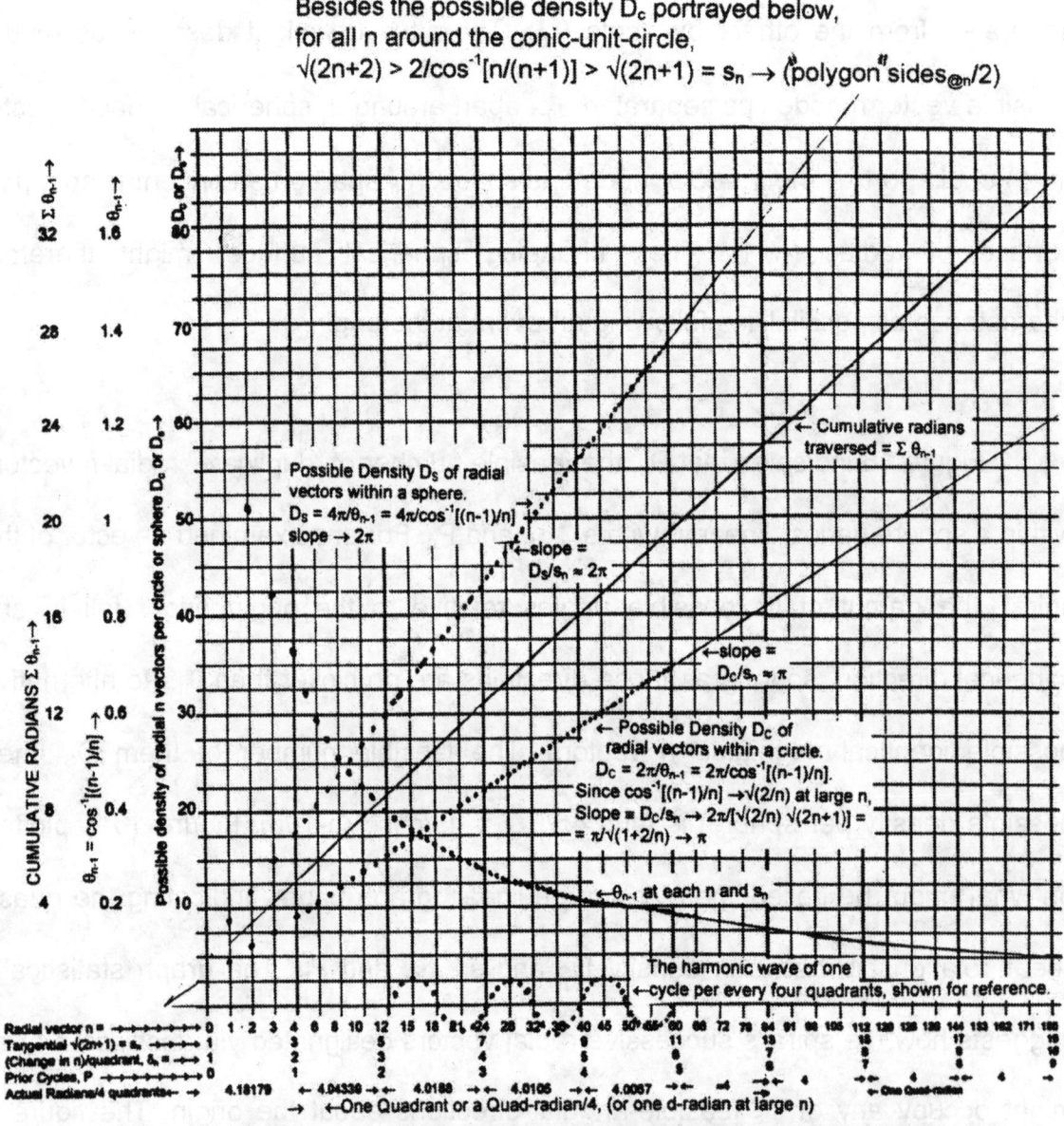

Figure 16 This plot shows the possible density of radial vectors within a sphere or circle for quadrants articulated whenever s_n becomes multiples of ½. The plot hides the slight elongation towards small n where actual radians/cycle > 4 although drawn here as 4. Thus, D_s and D_c appear perfectly linear with s_n although the few initial cycles actually elongate slightly as $n \rightarrow 1$. As the sphere or circle "expand" with increased n or s_n, a harmonic periodicity exists in the "possibility wave" about the origin.

For the moment, consider only values of n [called n_{PS}] where $(2n_{PS}+1)$ is a perfect square. Those abscissa conditions would articulate the slightly darkened vertical lines of figure 16. For any given n_{PS} a "fireworks type radial burst pattern" of D_S spherically emerging vectors from the origin **are feasible**, each of magnitude n_{PS}. Their precise locations would be unknown but in figure 16, D_S gives the number of possible radial spikes associated with any specific n_{PS}. A totality of $4\pi/\theta_{nps-1}$ such vectors are realizable; each statistically separated from the others by angle θ_{nps-1} around the 4π steridians of a sphere. Analogously for reference, if the density of conceivable n vectors were confined within a single plane instead of a sphere, their tips would form a circle or pinwheel comprised of $2\pi/\theta_{nps-1}$ radial vectors of magnitude n. Because the unified wave vectors are non-localized in angular direction [have non-specified or indeterminate compass bearing], the possible vector tips for each n [the attainable vector node-points in space] constitute a spherical pattern of dots surrounding the origin. At each subsequent perfect square value n_{PS} a new concentric ball of dots exists "further away" from the origin. Possible n vector locations thus designate a growing "globe of dots". While n influences θ_n and thus the number of dots, **it is δ_n or the ½ increment increases in s_n's magnitude that effectively establish the linear progression parameter of the wave (time)**. That tangential "control yardstick variable" differentiates quadrants wherever magnitude s_n increases by increments of ½. Each consecutive pair of

361

those ½ increments (each integer increase) also indicates "the next consecutive rung" in the DNA ladder.

If in a plane normal to the tip of a given n, a series of concentric circles spaced apart by ½ were drawn, each traversed circle in the magnitude of s_n marks off quadrants of the wave with the odd integers denoting complete cycles. Thus, the "expanding sphere" derives from magnitude s_n while the reciprocal of θ_n establishes the number of dots (sample points) on the sphere. From the slope of figure 16 the Density D_C of dots proportions to πs_n while the area of each concentric circle would be πs_n^2. Therefore the area associated with each dot is $\pi s_n^2 / \pi s_n = s_n$. Yet another meaning for s_n emerges as **the area per possible location on the surface of the expanding sphere** (timescale) surrounding the origin. Returning to figure 1 we note that s_n is the orthogonal traverse between integers. Then every ½ integer increase in s_n is effectively a measure of the tangential demarcation between cycles. Whether a value for s_n is less or more than those ½ integer increments sets the threshold of which category (the lower quadrant or the higher quadrant) applies. **Under the orthogonality axiom, the occurrence of ½-integer-magnitude transition values in the orthogonal direction, supplement the "half-way" points between Euclidean axis integers.** The geometric mean interpretation also grants a geometric "midpoint" connotation. Periodicity and linearity thus manifest in the tangential direction "rather than" along the radial axis direction. When characterizing something not sub-divisible, mid-integer markers in the magnitude of orthogonal transitions conceptually provide an effective scale of advance.

The unified wave can therefore entail a concentric expanding sphere representing a possible state of affairs. As figure 16 portrays, between each consecutive n_{PS} exist four quadrants of a harmonic wave. In the sphere's expansion, traversal of each subsequent n_{PS} denotes one more harmonic cycle [unity increase in P] quantifying an "equal time period". {The (actual radians)/(4-quadrants) abscissa scale shown at the bottom indicates the comparative straight-chord elongation required at small n.} In the mode illustrated, the possibility wave is therefore akin to an expanding spherical wave-front of a harmonic periodic wave, identifying with familiar electromagnetic waves at the far field of a radio antenna in space. As described for visualization, it is a wave characterizing possibility but can more generally provide the mathematical description of wave generation and propagation under the orthogonality axiom.

Any single drawn vector n of the spiral will legitimately have all other D_s virtual vectors noted in figure 1 as emanating possibilities around the 4π steridians of a growing sphere. This is so although in the expanding spiral depiction of figures 2 and 3 only one vector characterized the relative inter-comparison of increasing n vectors. That results because the orthogonality axiom allows triangle right-angles to be in any possible direction when progressing from one integer to the next. The spiral's vectors remain unconstrained in angular direction [other than relative to each other to degree θ_{n-1}]. In natural number terms, this portrays how waves generically emanate from an origin event. They can characterize a radio wave in spherical three-dimensional space, a (circular) surface wave on fluid, matter waves or the possibility waves of quantum electrodynamics. The orthogonality axiom that produces the harmonics has been the missing lynchpin unifying the number system of mathematics with the periodicity of waves. It can analytically apply to many more applications than waves. For example, being a linear function of s_n, δ_n, or P, the non-locatable expanding density of dots [possible vector node tips] passing through any given surface might be employed as a "controlled-variable random-distribution" in Monte-Carlo simulations.

Figure 16 suggests the wave might visualize as "cycling" about a perimeter of magnitude 2 (in s_n) instead of 2π, with possible virtual-node locations increasing at a rate π (= D_C/s_n) greater than perimeter traversed. Figure 17 repeats figure 2 with the wave's "circular polarization" $\equiv$ C, portrayed dotted between n = 40 and n = 60. It represents "rate of twist of the rectilinear format, (polarization)" between perfect squares **when plotted on a plane**. Since 4-radians prevail along the spiral path between those two perfect squares in (2n+1), the dotted line traverses (2π radians – 4 radians) of circular polarization twist occurring in the wave. Twist in the n_1 and n_5 "rails" of Table 1 enables formation of the DNA helix. If the figure 2 spiral's "coil" were pulled tighter as on a cone that proceeded from apex angle $\pm \pi/2 \equiv \alpha$ (as shown in figure 2) to $\pm \sin^{-1}(2/\pi) \equiv \beta$, (as shown in figure 4), circular polarization C ($\equiv$ radians "misalignment" between consecutive perfect square n vectors) would decrease from (2π radians - 4 radians) to zero. When the coil is pulled tighter as at apex angle β, all perfect square directions "line up" and zero circular polarization of the wave results.

Thus, circular polarization increases as apex angle χ differs from $\sin^{-1}(2/\pi) \equiv \beta$. Accordingly, angle χ indicates a degree of freedom wherein, at the expense of circular polarization, each of the D_S virtual n vectors in figure 1 might progress around any conic angle of the sphere surrounding the origin. Cones (or any radially emanating surfaces) of arbitrary net apex angle χ come out in all directions from the origin and all represent possible loci for the sequence of vectors. It can thus be visualized how the summed adding and canceling phases of all possible paths from an origin event can influence a remotely located outcome event. The net "throughput" in the QED phase canceling sense derives from the sum of all such possibilities. Since the represented units entail energy-time, solutions yielding specific energies only result where stationary standing waves can satisfy energy and time requirements imposed by the constraining potentials or fields in a given QM problem. The possibility waves merely characterize the numerical description of orthogonal (rather than Euclidean) progression between integers. The waves are inherently embedded within the periodicity of the number system and emerge when examined under the orthogonality axiom. Of academic interest is the gap G generically defined between n = 1 and n = 2 in figure 14 as the difference between s_n and where the hypotenuse of the (n+1) triangle intersects vector s_n. Gap G approaches zero rapidly as n increases making the "sawtooth" a viable approximation vehicle for increasing s_n at large n.

365

Ability to describe shapes and structures with a minimum number of arbitrary constants is an important earmark of mathematical and scientific models. The unified wave provides an ideal medium in this regard. For integers from n = 0 to ∞, the cone, plane, circle, polygon, spiral, sphere, helix, triangle, radian, randomness, interference, rectilinearity and the square, all are synthesizeable in one form or another. Combinations of these shapes like the cone and the plane, as conic sections, further enable the ellipse, parabola, and hyperbola. Many biological structures possess geometric forms following these mathematical descriptions. Monovalve mollusks have spiral-helix shells for example, easily mimicked by unified waves. These structures can emerge directly from the sequence of natural numbers under the orthogonality, axiom and that directness provides a likely correlation with the "origin" of such shapes in Nature. One argument of this thesis is that atomic and sub-atomic physical configurations, (as those of figures 6 to 11) occur in the form of these waves. Indeed, the quantum mechanical possibility of describing electron distribution clouds and the like, speculate as direct manifestations of the natural numbers under the orthogonality axiom. The universe is filled with examples of intricate geometry represented by these structures. Models necessary to simulate such geometric reality should abound in the Natural world. Thus, fruits of this theory can represent extreme simplicity and reductionism in mathematical representation. The generated wave packs highly ordered complexity into a simple logical and physical description.

26) APPENDIX B, CHORDS WITHIN THE UNIT CIRCLE ON A CONE

From figure 15, the number of straight-line chords or "polygon sides" that would fit around the conic unit circle can express as $4/\theta_n = 2/\phi_n = 4/\cos^{-1}[n/(n+1)]$. The segment of unit-circle perimeter must be constrained within the two limits of J-K and F-G shown

in the figure so $2/\sqrt{(2n+1)} > \theta_n = \cos^{-1}[n/(n+1)] > 2/\sqrt{(2n+2)}$ (B*1)

Now $2/\sqrt{(2n+1.5)}$ would be an even closer bound to θ_n than $2/\sqrt{(2n+1)}$.

Inverting and multiplying B*1by 4,

$\sqrt{(8n+6)} < 4/\cos^{-1}[n/(n+1)] < \sqrt{(8n+8)}$; applicable for all n   (B*2)

On a cone of apex angle $\sin^{-1}(2/\pi)$, as $n \rightarrow \infty$, exactly 4 radians comprise the circle that is everywhere unity from the conic origin. The number of chords (or sides) of a "**regular**" polygon on that circle, each intercepting angle θ_n at the cone apex would be $4/\theta_n \equiv 4/\cos^{-1}[n/(n+1)] \equiv X$. The number of chords that could fit into the unit circle is constrained by equation B*2 as

$\sqrt{(8n+6)} < 4/\theta_n < \sqrt{(8n+8)}$ (B*3)

This indicates the number of polygon sides around the 4 radian unit-circle, each of reiterated central angle θ_n, will always be non-integer unless $\sqrt{(8n+7)} \equiv X$ is an integer, {or $X^2 = 8n+7$}. That comes about because the lower bound of the inequality $\sqrt{(8n+6)}$ is always less than, and $\sqrt{(8n+8)}$ will invariably exceed, $4/\theta_n$. The only possible integer number of polygon sides between the two bounds could occur exclusively when $\sqrt{(8n+7)} \equiv X$ is integer. If integer values for both X and n could simultaneously satisfy $X^2 = 8n+7$, than a regular polygon of $4/\theta_n = \sqrt{(8n+7)}$ sides could fit within the conic unit circle. In fact, for all integers n **no pair of integers X and n exist** that provide a solution for $X^2 = 8n+7$. {Dr. Jim Farned, at <jfarned@rain.org>, can provide a rigorous proof.} Therefore, none of the chord-lengths formed from n = 0 to n $\rightarrow \infty$ could ever be an element of a **regular** polygon around that 4-radian circle on the cone. Accordingly, discrete synchronization would be impossible for straight-line chord-lengths of the spiral and the four absolute radians comprising the cycle, although perfect periodicity prevails for curved chord-lengths. Separate individual unity-change-in-n straight chord-length intervals s_n/n **never intermesh perfectly** with the overall 4-radian cadence. However collective sums of those normalized consecutive chord-length values maintain cycle-to-cycle rhythm. That cadence of 4-quadrants remains consistent with the peaks and zero-crossings of the wave since those quadrant locations never depict a prime number for n. **A non-prime number n means those defined points of the wave are always divisible by an earlier (smaller) n. That provides an alternate definition for periodicity**. Also, towards large n each chord about the conic unit circle approaches twice the respective widened-line segment of s_n in figure 14. Differentially added straight-line increments of time associated with unit advance in n are never an integer division of the 4 radian total harmonic period. This provides another indication for the one chord-length of uncertainty (the one Planck time or one Planck's constant action unit of indeterminism) inherent in the wave. The partitioning nature of the wave, exclusively at quadrant positions absent partitioning elsewhere, suggests divisibility plays a significant role in the underlying mechanism.

At first glance it might appear that for n = 0 and n = 1 (where $\theta_n = \pi/2 = 90°$ and $\pi/3 = 60°$ respectively), the ratio $4/\theta_n$ might be integer. That $4/\theta_n$ ratio represents radians/radians however with the tabulation of equation B*3 displaying its non-integer nature and convergence toward zero of $(\sqrt{(8n+7)} - 4/\theta_n)$ as $n \rightarrow \infty$.

The following summarizes the unified wave **exact periodic relationships** for cycles, radians, and position coordinates, on a unit-sphere with $\xi = 1$. [Example values for n =1000 are underlined].

Where: **n = any integer**; $s_n = \sqrt{(2n+1)}$; $P_n = (s_n-1)/2$; $\underline{P_n} \equiv$ **integer part of P_n**; $F_n \equiv$ **non-integer part of P_n.** One + total # of half-cycles traversed since the origin = $s_n = \sqrt{(2n+1)} = \sqrt{2001} = 44.73254..$half-cycles.

Total # of cycles traversed since the n = 0 origin = $P_n = (s_n-1)/2 = \sqrt{(2n+1)}-1]/2 =$ 21.86627..cycles. Total # of radians traversed since the n = 0 origin = $2\pi P_n =$ 137.38982..radians. [Also, $P_n \equiv \underline{P_n} + F_n$] Integer part of total # of cycles traversed from the origin $\equiv P_n$=21 cycles, [=$2\pi(21) = 131.946..$radians] Non-integer part of total cycles traversed $\equiv F_n = 0.86627...$cycles, [= $2\pi(0.86627) = 5.44293..$radians] Non-integer portion of the final cycle traversed in unit-circle degrees = $(360)F_n =$ 311.8572...°.

Cartesian-Coordinate components @ n: X= $\cos(2\pi F_n)=\cos(2\pi P_n)=0.66727.$; Y=$\sin(2\pi F_n) = -0.74481..$

Cumulative radians of error resulting from using straight-line chords in $\Sigma\,\theta_n = \Sigma\cos^{-1}[n/(n+1)]$ would be

Irwin Wunderman

n=n

$\Sigma \cos^{-1}[n/(n+1)] - [2(\sqrt{(2n+1)} -1]$ radians of error. This amounts to = 0.1066 for

n = 0

n = 1; 0.18179 for n = 4; 0.22519 for n = 12; 0.24406 for n = 24; 0.254 for n = 40,

0.261 for n = 60, etc., cumulating to about 0.3 radians of error toward n → ∞.

27) REFERENCES

1   I. Wunderman, (2000). *What is a Photon? A Unified Wave Theory Explained.* Wyndham Hall Press

2   P. *Coveney* & R. Highfield, (1990). *The Arrow of Time.* Ballantine Books

3   D. Freedman & P. Van Nieuwenhuizen, (1985). *The Hidden Dimensions of Spacetime.* Scientific American 252, P. 62

4   N. Bohr, (1963) *Atomic Physics and Human Knowledge.* New York, Wiley

5   J. S.Bell, (1966). *On the Problem of Hidden Variables in Quantum Mechanics.* Reviews of Modern Physics, 38, P. 447

6   W. Heisenberg, (1958). *Physics and Philosophy.* New York, Harper & Brothers

7   R. P. Feynman and A. R. Hibbs, (1965). *Quantum Mechanics and Path Integrals.* New York, Mcgraw Hill

8   R. P. Feynman, (1985). QED: *The Strange Theory of Light and Matter.* Princeton University Press

9   B. Greene, (2000). *The Elegant Universe.* Vintage Books

10   L. De broglie, (1923). Comptes Rendus de L'Academie des Sciences de Paris 11 A. Einstein, (1905). Annalen der Physik, 17, 891

12   A. Einstein, (1961). *Relativity,* New York, Crown

13   Godel, K. (1931) Uber Formal Unentscheidbare Satze der Principia Mathematica und Verwandter Systeme. I. Monatshaft fur Mathematik and Physik, 38, 173-98

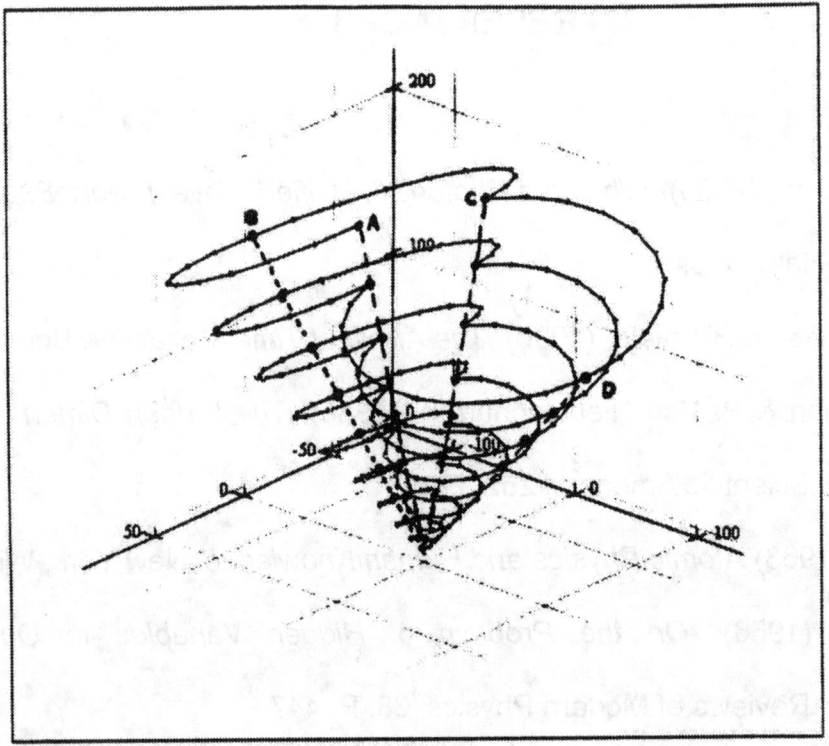

(x,y,z)

Figures 19 through 31 exemplify interactive computerized three-dimensional plots compiled by Allen Amaro, President of Wellspring Knowledge Corporation. These will be available on a CD-ROM as educational course material for electronic classroom teaching purposes. Ability to manipulate the figures in three-dimensional color on a screen provides a much more effective presentation than a static picture on a page. Important mathematical functions will also be available on the CD-Rom for use in various disciplines of physics, optics, electronics, engineering, cosmology, population studies, indeterminism analysis, prime number theory, and biology. The sample plots here are all based on straight-line chord-lengths.

Figure 19
Simply folding in directions A and C toward each other on the conic spiral will generate the pictures on this and the two following figures. Any degree of bringing together A and C or "squeezing" the lobes are possible and that flexibility is typical. This mode generates a pattern of two opposing lobe directions with peak lobe extensions at directions B and D. If directions B and D are also folded in toward the center four lobes result elaborating the quadrature nature of each cycle. These figures indicate but a very few of the enormous allowed-standing-wave mode patterns. One may note how resemblance to electron cloud patterns could emerge. Subsequent figures show views of the spiral in other modes.

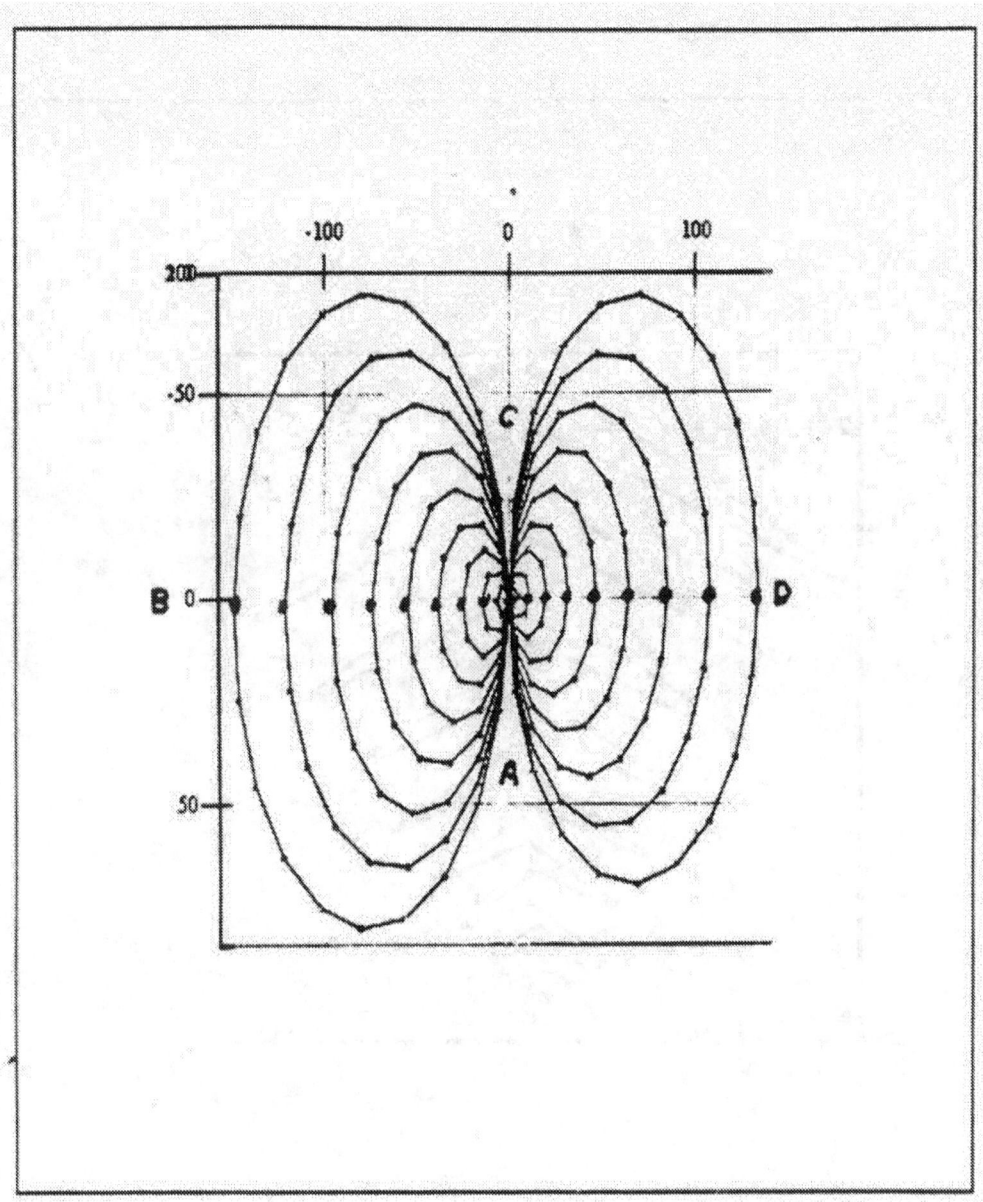

(x,y,z)

Figure 19a. This is a "top view" of figure 19 where directions A and C are brought together in contact. Each quadrant of a complete cycle is represented by the path extent between any two consecutive letters, i.e., A-to-B, B-to-C, C-to-D, D-to-A.

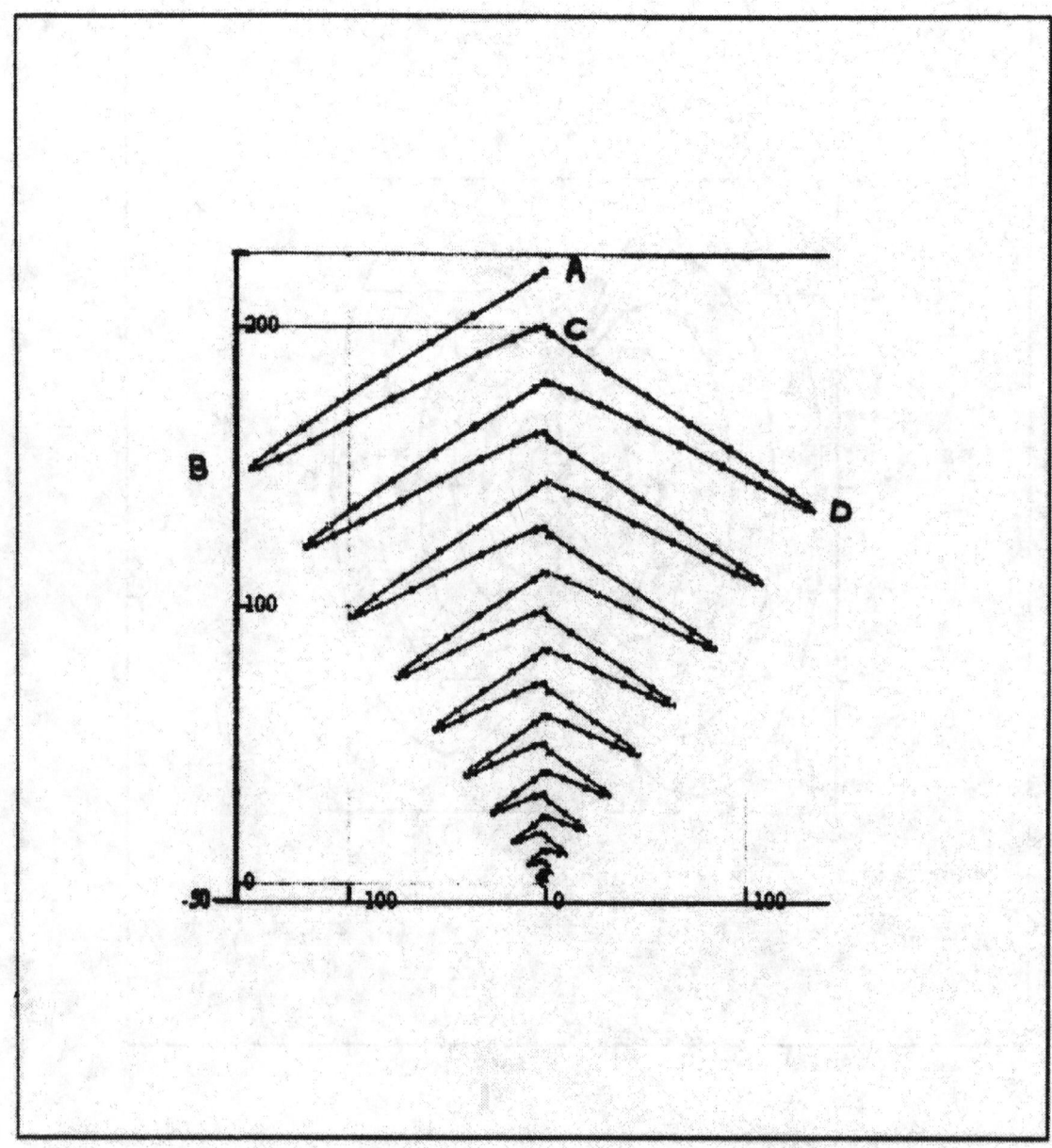

(x,y,z)

Figure 20. This is one "side view" of the mode of figure 19.

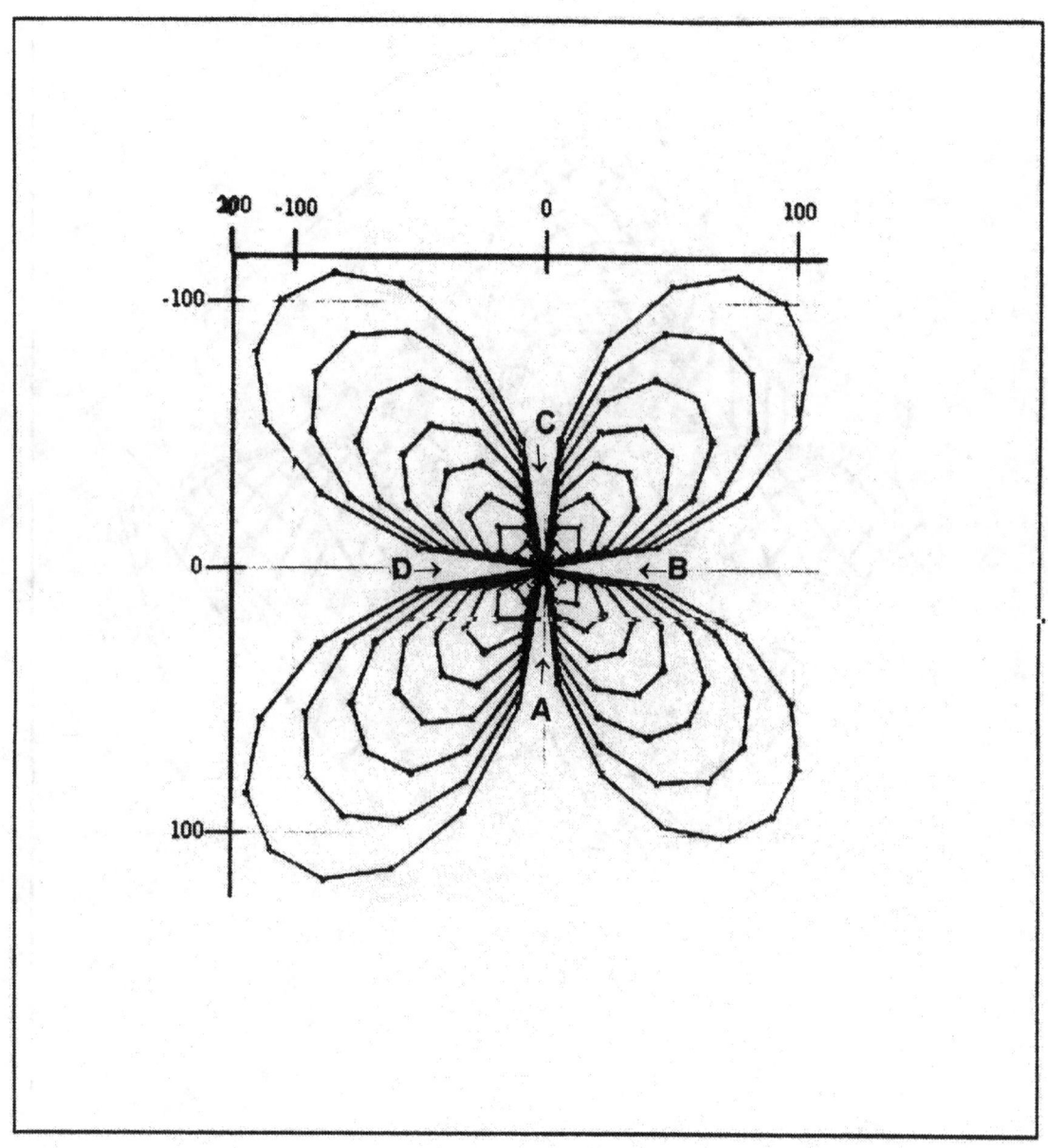

(x,y,z)

½ Spin 0-degrees

Figure 21. In this mode all four directions A, B, C, D, are brought into contact with this view "head on" along the axis.

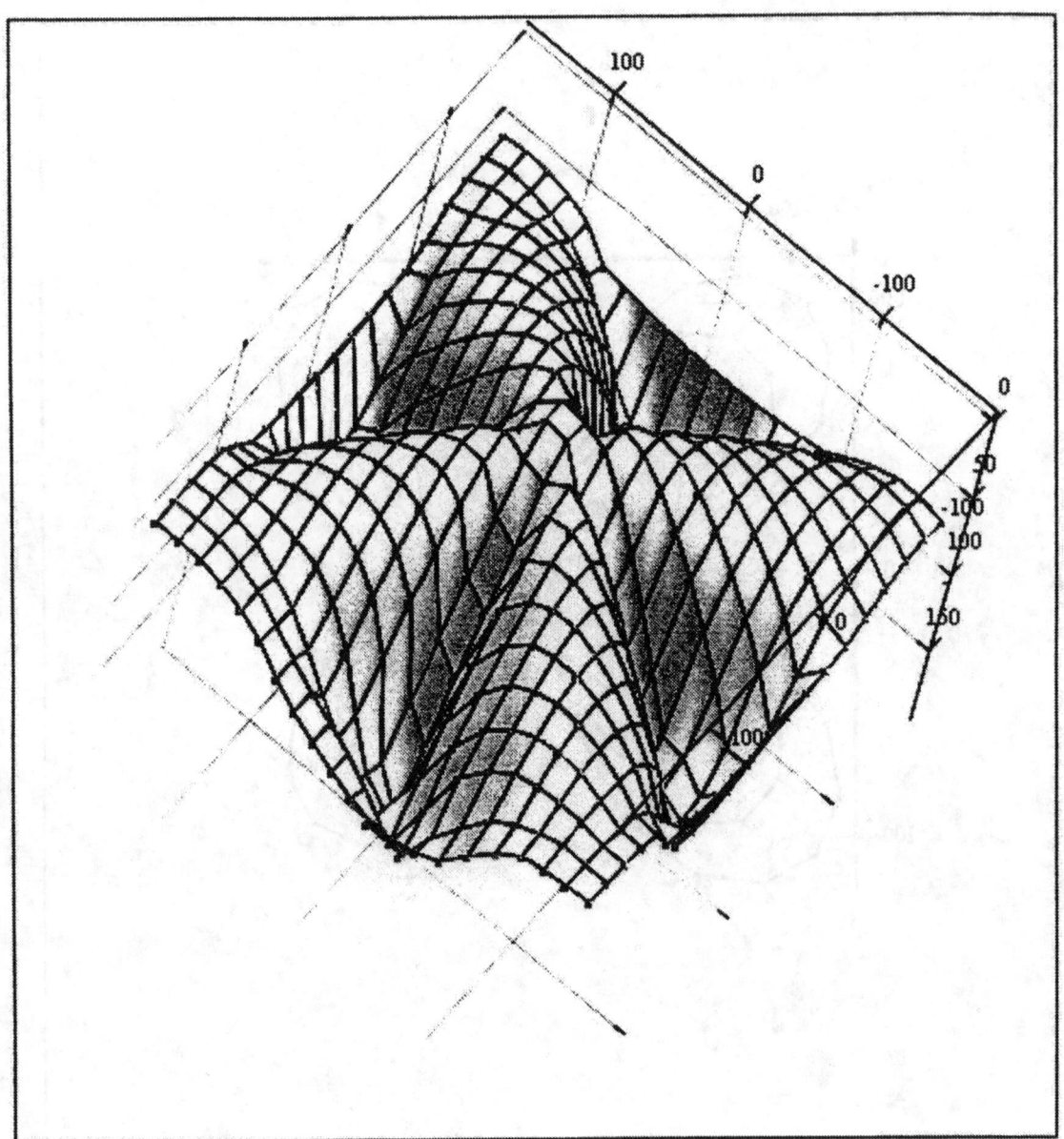

(x,y,z)

½-Spin Surface Plot (30-degrees)

Figure 22. This is the same mode configuration as figure 21 with the "surface" of created node points near the origin viewed from a 30° angle.

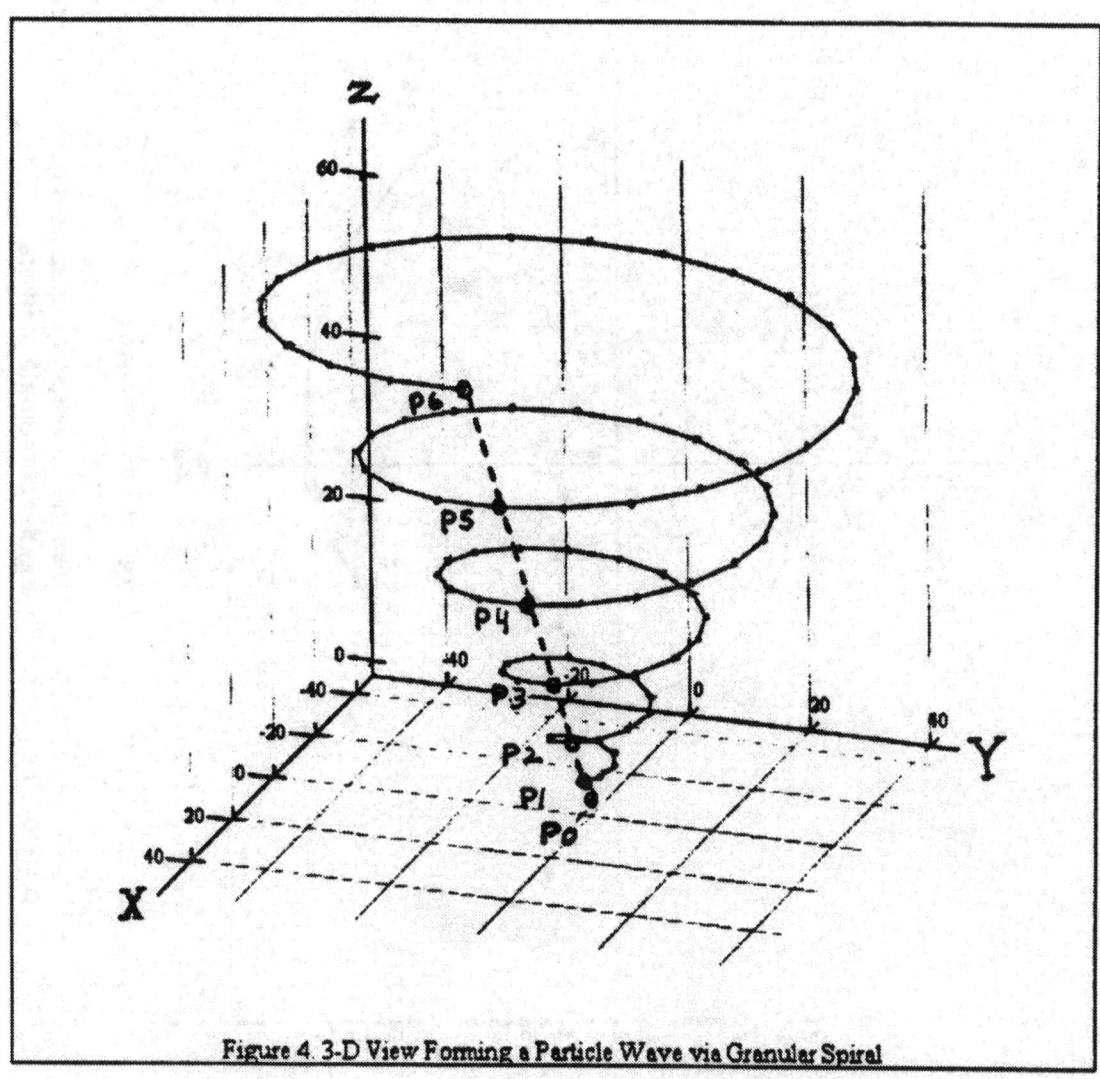

Figure 4. 3-D View Forming a Particle Wave via Granular Spiral

(x,y,z)

Figure 23. This mode shows alignment on the cone for each cycle up to P = 6

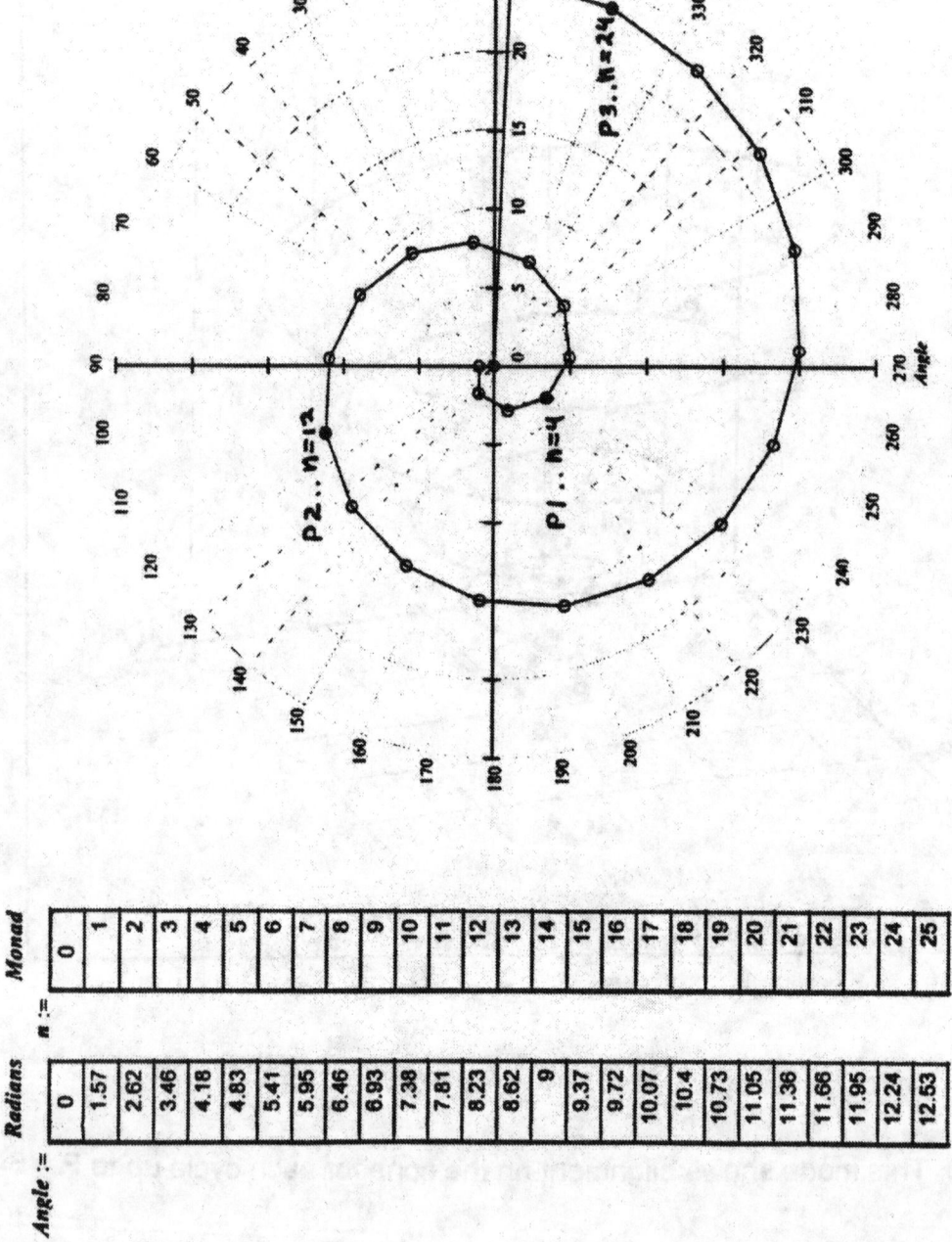

Plot of n vs. Cumulative Angle

Figure 24. This mode shows the 3 cycles of the spiral's cumulative angles plotted on a plane with angular coordinate markers.

Monad		Radians
n :=		Angle :=
0		0
1		1.57
2		2.62
3		3.46
4		4.18
5		4.83
6		5.41
7		5.95
8		6.46
9		6.93
10		7.38
11		7.81
12		8.23
13		8.62
14		9
15		9.37
16		9.72
17		10.07
18		10.4
19		10.73
20		11.05
21		11.36
22		11.66
23		11.95
24		12.24
25		12.53

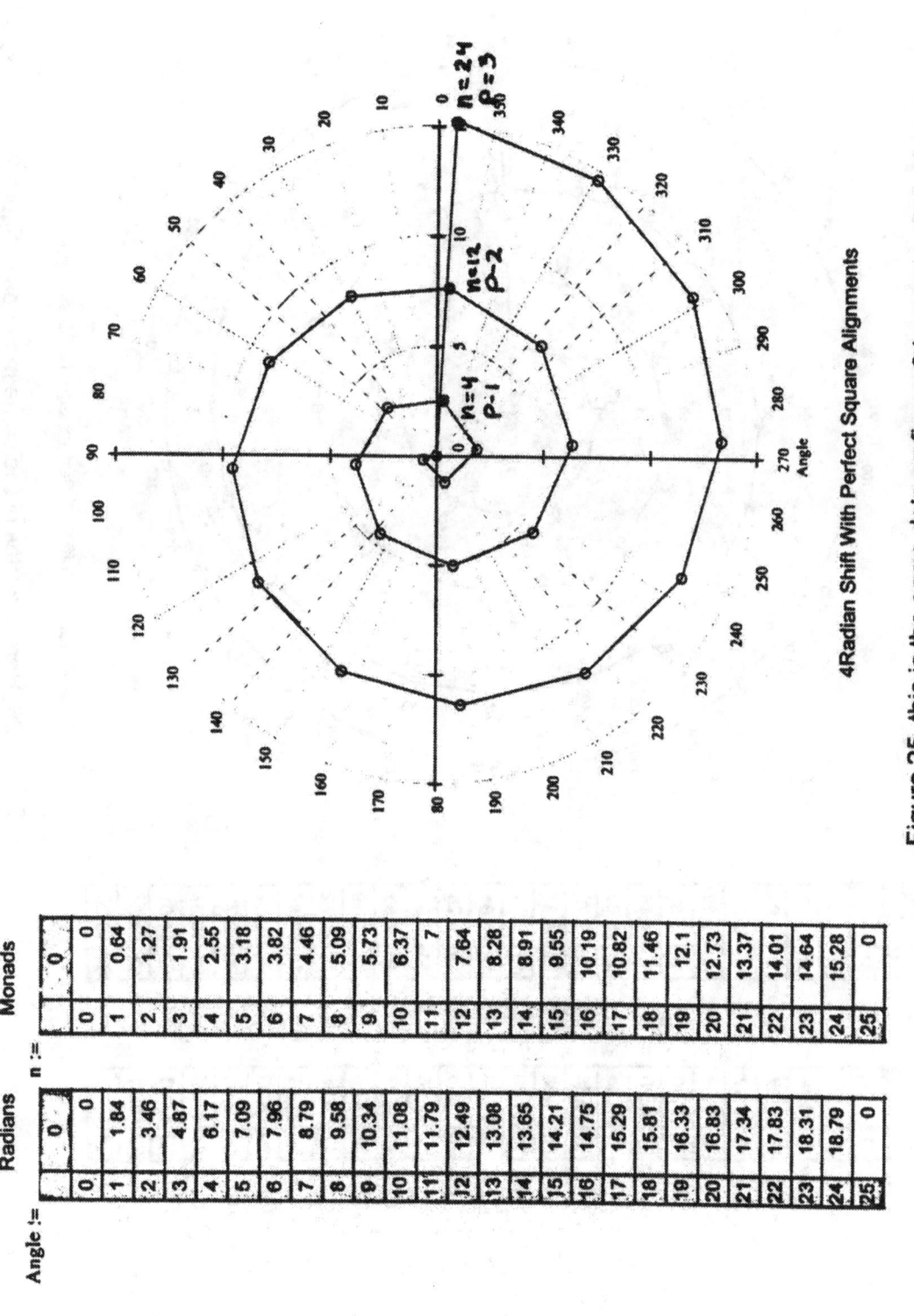

4Radian Shift With Perfect Square Alignments

Figure 25. this is the same data as figure 24 seen "end view' on a cone and showing alignment of the perfect squares of (2n+1) for 3 cycles.

Angle :=		Radians		n :=		Monads
0		0		0		0
0		0		0		0
1		1.84		1		0.64
2		3.46		2		1.27
3		4.87		3		1.91
4		6.17		4		2.55
5		7.09		5		3.18
6		7.96		6		3.82
7		8.79		7		4.46
8		9.58		8		5.09
9		10.34		9		5.73
10		11.08		10		6.37
11		11.79		11		7
12		12.49		12		7.64
13		13.08		13		8.28
14		13.65		14		8.91
15		14.21		15		9.55
16		14.75		16		10.19
17		15.29		17		10.82
18		15.81		18		11.46
19		16.33		19		12.1
20		16.83		20		12.73
21		17.34		21		13.37
22		17.83		22		14.01
23		18.31		23		14.64
24		18.79		24		15.28
25		0		25		0

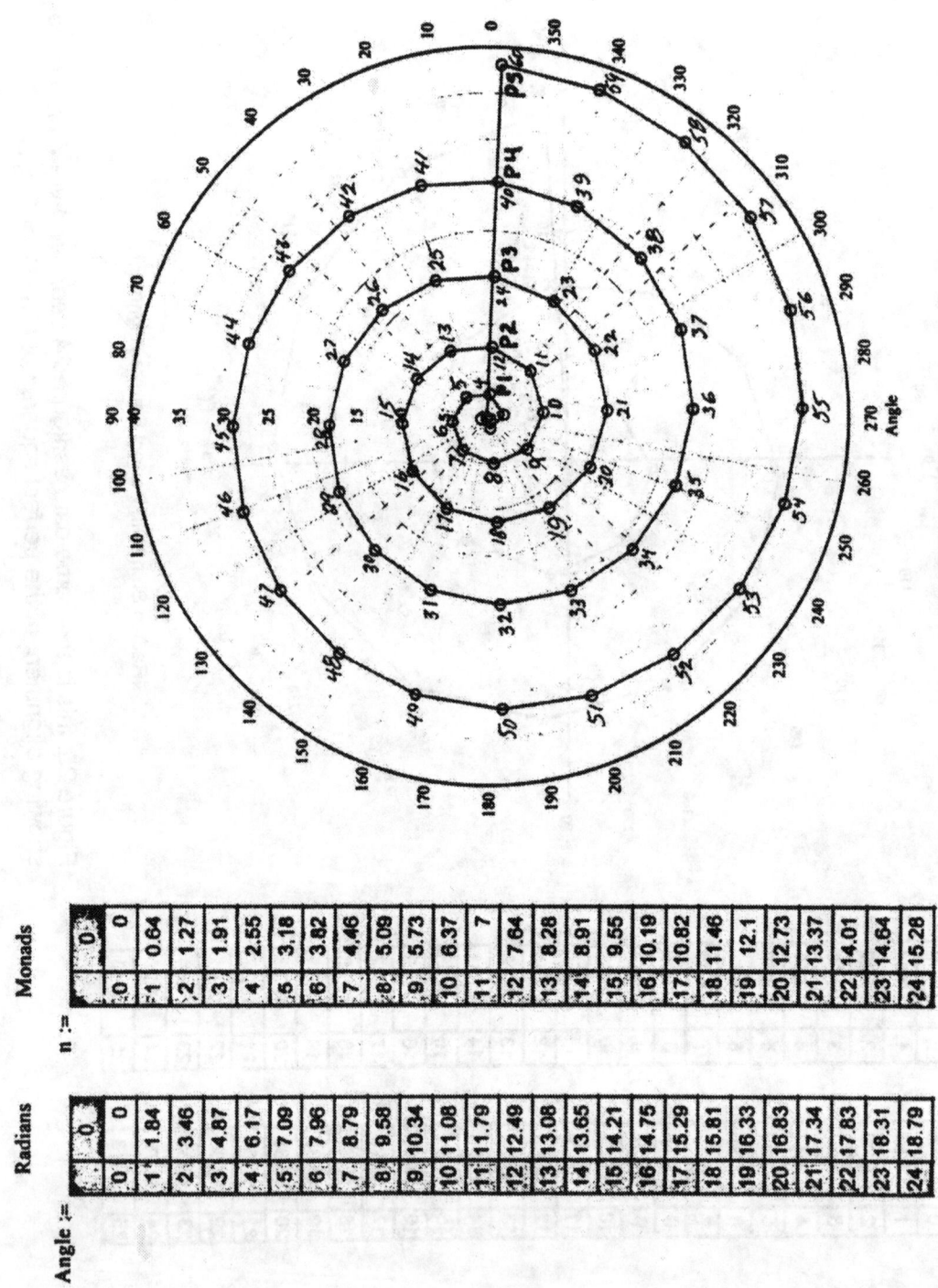

4 Radian Shift P=5 Perfect Square Alignments

Figure 26. Here 5 cycles of the same mode as in figure 25 are displayed.

Monads			
	0	:0	0
	0.64	=1	1
	1.27	.2	2
	1.91	3'	3
	2.55	4	4
	3.18	.5	5
	3.82	6	6
	4.46	7	7
	5.09	8'	8
	5.73	9,	9
	6.37	10	10
	7	11	11
	7.64	12	12
	8.28	13	13
	8.91	14	14
	9.55	15	15
	10.19	16	16
	10.82	17	17
	11.46	18	18
	12.1	19	19
	12.73	20	20
	13.37	21	21
	14.01	22	22
	14.64	23	23
	15.28	24	24

$n :=$

Radians			
	0	0	0
	1.84	1	1
	3.46	2	2
	4.87	3'	3
	6.17	4	4
	7.09	5'	5
	7.96	6.	6
	8.79	7	7
	9.58	8'	8
	10.34	9.	9
	11.08	10	10
	11.79	1'	11
	12.49	12	12
	13.08	13	13
	13.65	14	14
	14.21	15	15
	14.75	16	16
	15.29	17	17
	15.81	18	18
	16.33	19	19
	16.83	20	20
	17.34	21'	21
	17.83	22	22
	18.31	23	23
	18.79	24	24

Angle $:=$

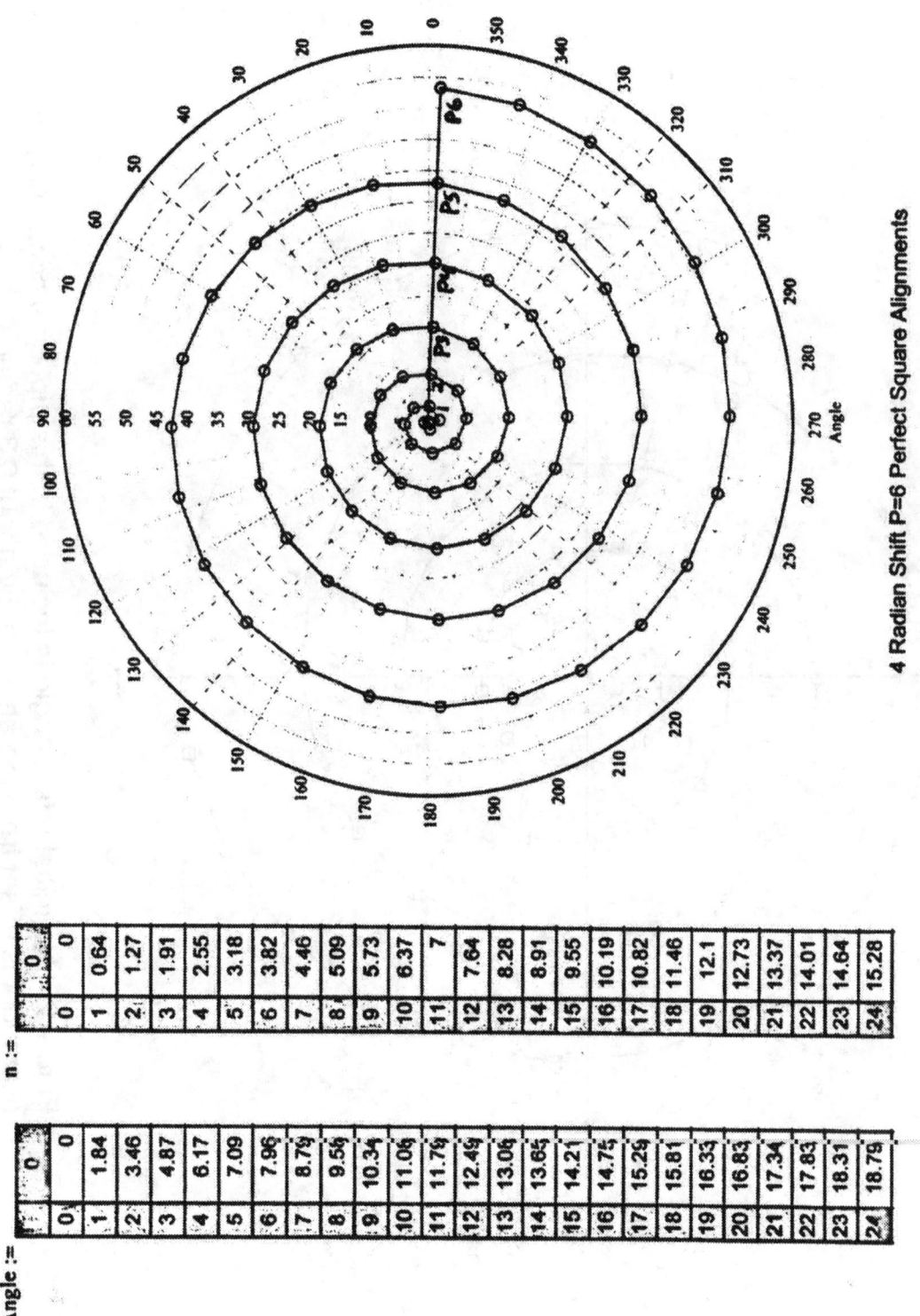

4 Radian Shift P=6 Perfect Square Alignments

Figure 27. This characterizes 6 cycles of the same mode as figures 25 and 26.

n :=

	0
0	0
1	0.64
2	1.27
3	1.91
4	2.55
5	3.18
6	3.82
7	4.46
8	5.09
9	5.73
10	6.37
11	7
12	7.64
13	8.28
14	8.91
15	9.55
16	10.19
17	10.82
18	11.46
19	12.1
20	12.73
21	13.37
22	14.01
23	14.64
24	15.28

Angle :=

	0
0	0
1	1.84
2	3.46
3	4.87
4	6.17
5	7.09
6	7.96
7	8.79
8	9.54
9	10.34
10	11.06
11	11.79
12	12.49
13	13.06
14	13.65
15	14.21
16	14.75
17	15.29
18	15.81
19	16.33
20	16.83
21	17.34
22	17.83
23	18.31
24	18.79

381

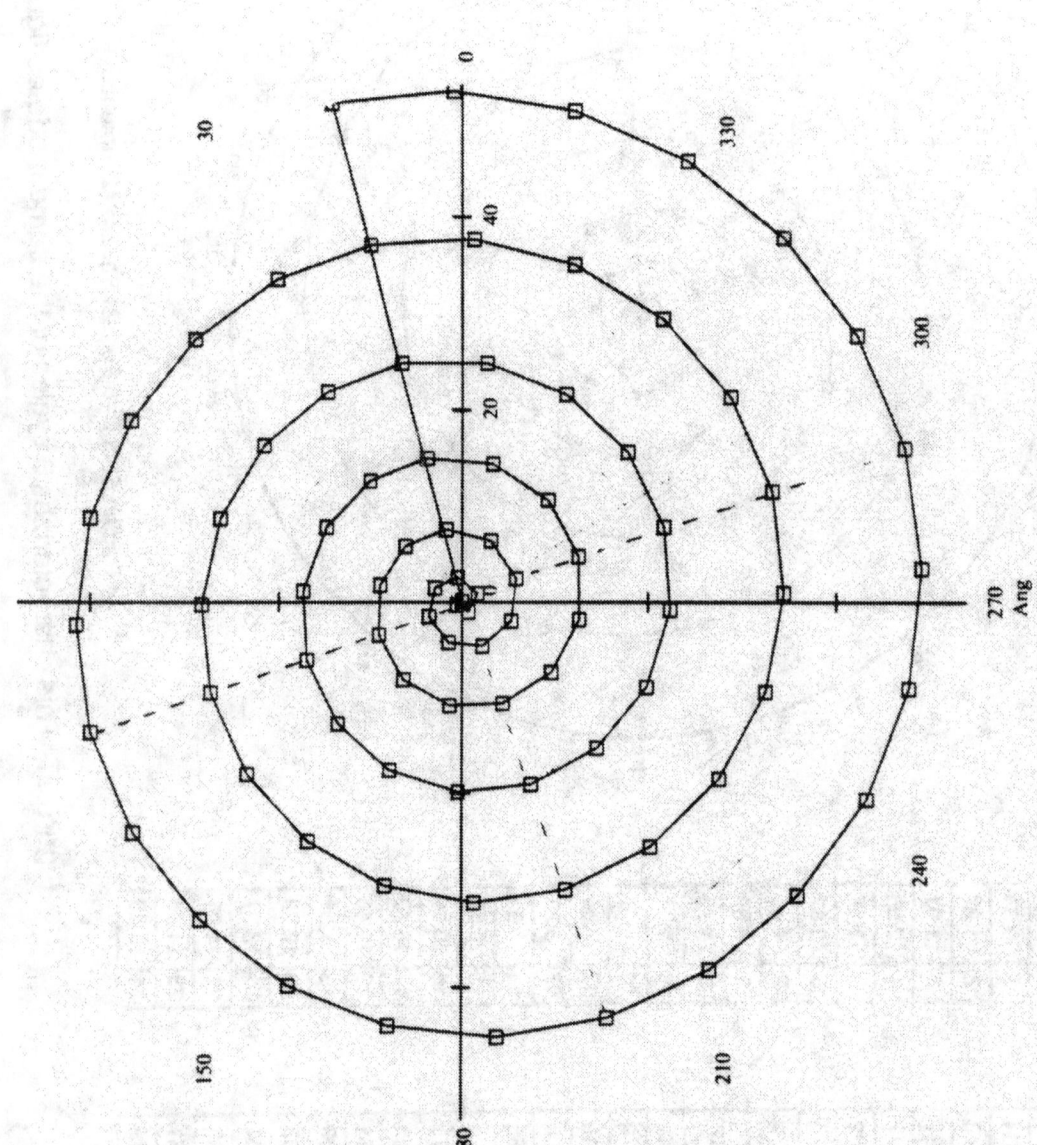

Figure 28. No initial offset angle is included in this plot on a cone such that alignment is offset from the abscissa by about 0.3 radians.

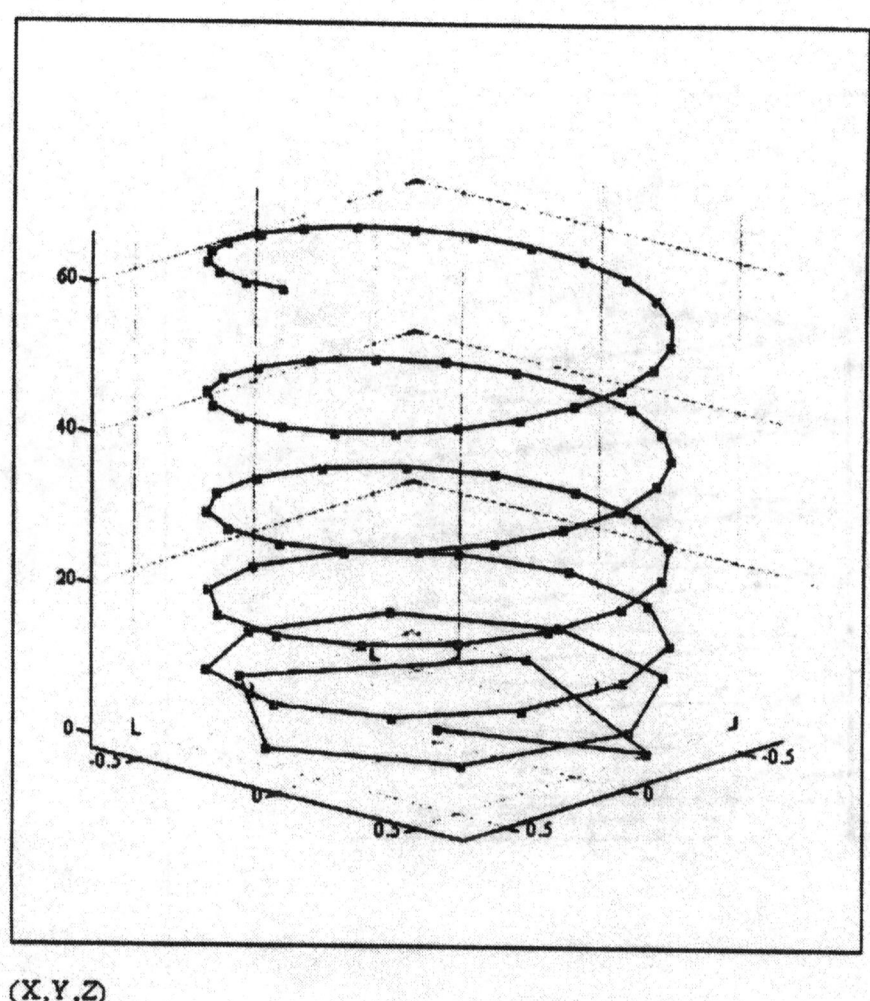

(X,Y,Z)

Figure 29. This is a view of the spiral not normalized in terms of cycles P and plotted to advance in a helical direction.

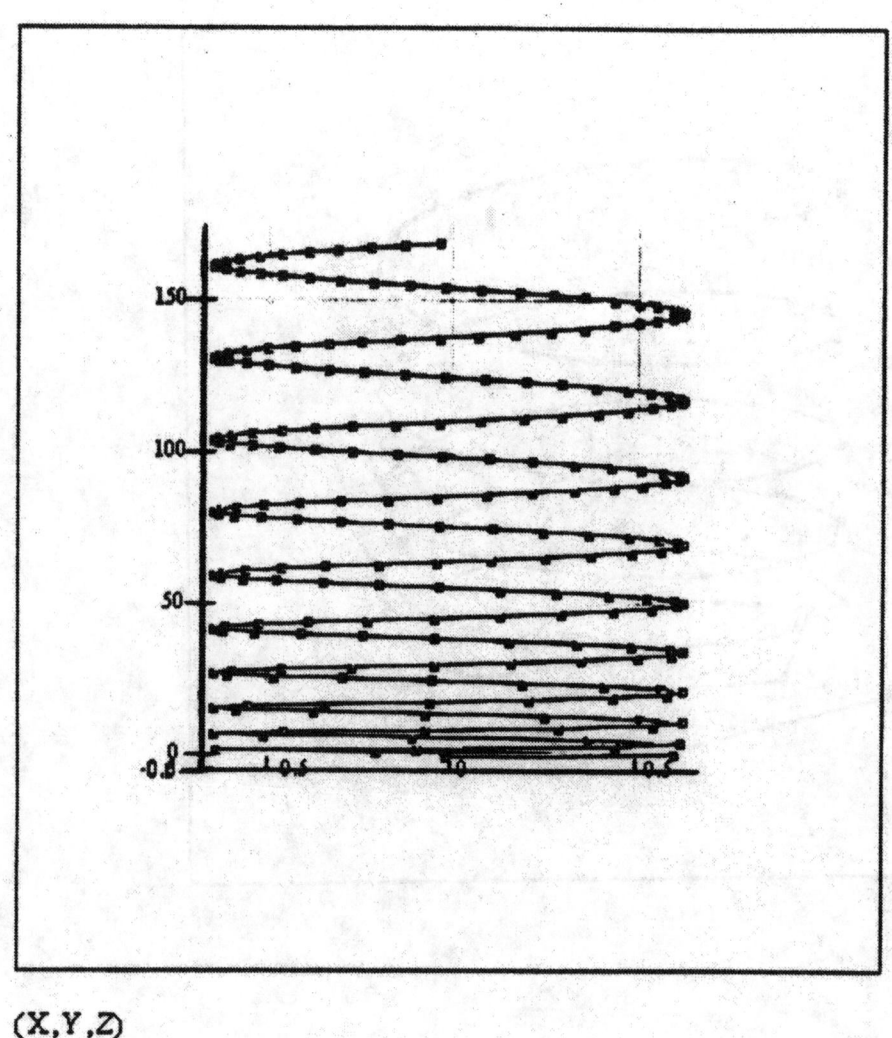

(X,Y,Z)

Figure 30. This is a side view of the helical direction plot of figure 29

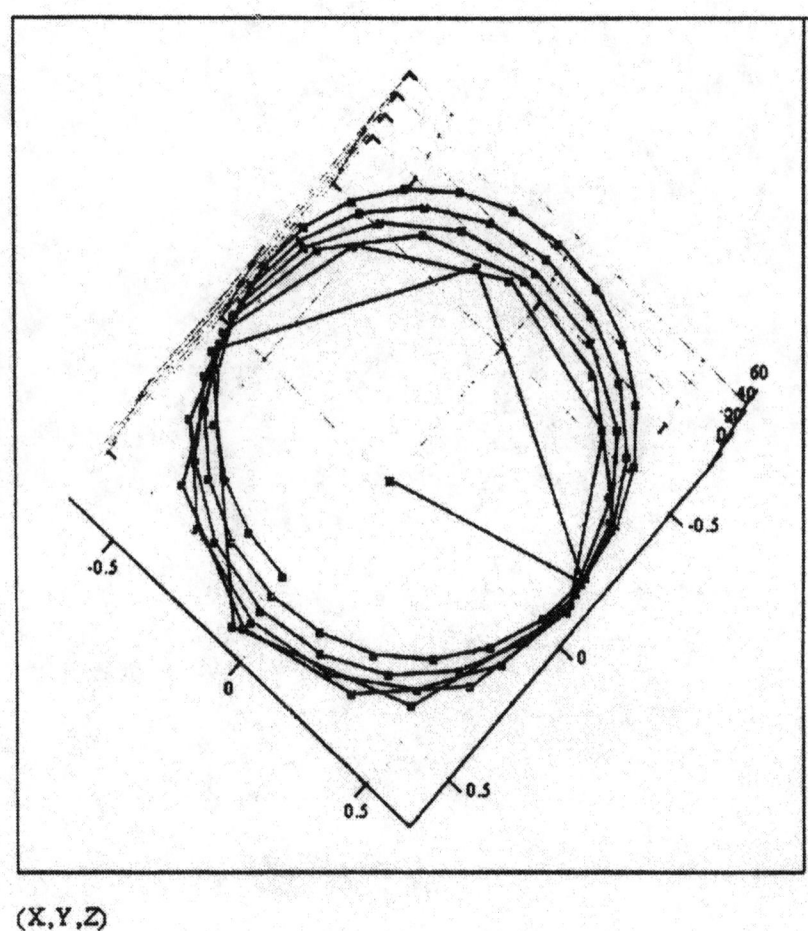

(X,Y,Z)

Figure 31. This is close to a "top view" of the spiral mode of figures 29 and 30.

28) PAGE INDEX FOR THE FIGURES

PROFILE OF IRWIN WUNDERMAN

Irwin Wunderman's technical pursuits focused on developing electronic instruments. He holds a Ph.D. in Electrical Engineering from Stanford University and Chaired the Northern California IEEE Professional Group on Instrumentation. His early professional career included responsibility for all solid-state research and development at Lockheed Aircraft Corp., and at Hewlett-Packard Co. He introduced transistorized circuitry at HP Corporate Labs and managed the transistorized digital instrument and optical instrument programs. As co-founder of Hewlett-Packard Associates he was Principal Investigator on U.S. Air Force contracts to find uses for junction luminescence and PH'd dissertation advisor to Air Force Academy Cadets. Through the early '60s his work included development of the first opto-couplers and fiber-optic communication links. Upon leaving HP in 1967 he founded Cintra Physics International as its President and CEO. Cintra developed a line of optical auto-ranging digital instruments and a compatible computer/calculator. He received the 1970 Industrial Research 100 Award for creating the first scientific computer/calculator to employ algebraic notation and having a data bus permitting real-time integration between digital instruments, keyboards, computers and network systems. Cintra was sold to Tektronix in 1971 who licensed the consumer-product version of the calculator to Texas Instruments. The Cintra computer-data-bus became prototype for the IEEE-488 bus standard and the calculator the basis of existing Texas Instruments scientific calculators. Dr. Wunderman received the 1968 commendation leadership letter from President Elect Richard Nixon. He holds 16 patents and authored 25 papers, some attaining international awards. This work culminates four decades of independent research attempting to establish how and why ordinary natural numbers and mathematical relationships could express fundamental physical laws through waves.

www.ingramcontent.com/pod-product-compliance
Lightning Source LLC
Chambersburg PA
CBHW081104170526
45165CB00008B/2316

* 9 7 8 1 4 1 0 7 5 7 9 1 3 *